OXFORD MATHEMATICAL MONOGRAPHS

Editors

OXFORD MATHEMATICAL MONOGRAPHS

EDDINGTON'S STATISTICAL THEORY
By C. W. KILMISTER and B. O. J. TUPPER. 1962

THE LAMINAR BOUNDARY LAYER EQUATIONS
By N. CURLE. 1962

THE THEORY OF LAMINAR BOUNDARY LAYERS IN COMPRESSIBLE FLUIDS
By K. STEWARTSON. 1964

MEROMORPHIC FUNCTIONS
By W. K. HAYMAN. 1964

HOMOGRAPHIES, QUATERNIONS AND ROTATIONS
By P. DU VAL. 1964

THE OPEN MAPPING AND CLOSED GRAPH THEOREMS IN TOPOLOGICAL VECTOR SPACES
By T. HUSAIN. 1965

CLASSICAL HARMONIC ANALYSIS AND LOCALLY COMPACT GROUPS

BY

HANS REITER

Mathematical Institute
University of Utrecht

OXFORD
AT THE CLARENDON PRESS
1968

Oxford University Press, Ely House, London W. 1
GLASGOW NEW YORK TORONTO MELBOURNE WELLINGTON
CAPE TOWN SALISBURY IBADAN NAIROBI LUSAKA ADDIS ABABA
BOMBAY CALCUTTA MADRAS KARACHI LAHORE DACCA
KUALA LUMPUR HONG KONG TOKYO

PRINTED IN GREAT BRITAIN
AT THE UNIVERSITY PRESS, OXFORD
BY VIVIAN RIDLER
PRINTER TO THE UNIVERSITY

PREFACE

THE pages that follow are an introduction to some of the methods, some of the ideas, contained in the books of N. Wiener (*The Fourier integral and certain of its applications*, Cambridge University Press, 1933), T. Carleman (*L'intégrale de Fourier et questions qui s'y rattachent*, Uppsala, 1944) and, above all, A. Weil (*L'intégration dans les groupes topologiques et ses applications*, Paris, 1940).

Weil's book contains the foundation for the development of analysis on locally compact groups. The ideas to be found there—hidden, perhaps, at times—are on the one hand very general; on the other hand they are also significant for classical analysis in euclidean space.

In the present volume topics exhibiting the development of analysis on groups are treated, with the aim of stimulating further research in this direction.†

Work on the manuscript was begun in 1962 at King's College (now The University), Newcastle upon Tyne, and continued since 1964 at the University of Utrecht. It has benefited much from remarks by students and colleagues alike: this is acknowledged more specifically in the text, but especial thanks are due to Professor J. Cigler (Groningen), Mr. G. van Dijk (Utrecht), Dr. J. E. Gilbert (Newcastle upon Tyne), Mr. J. D. Stegeman (Utrecht), and Professor H. de Vries (Nijmegen) who have greatly helped in developing the final version of the manuscript. The origin of the work, however, goes back to a temporary membership, during the academic year 1961–2, at the Institute for Advanced Study, Princeton. But now the following pages must speak for themselves.

H. R.

1967

† *Πρὸς δέ σε καὶ τόδ' ἄνωγα, τὰ μὴ πατέουσιν ἅμαξαι*
τὰ στείβειν, ἑτέρων ἴχνια μὴ καθ' ὁμά
δίφρον ἐλᾶν μηδ' οἶμον ἀνὰ πλατύν, ἀλλὰ κελεύθους
ἀτρίπτους, εἰ καὶ στεινοτέρην ἐλάσεις.

Callimachus, *Aetia*, i, Fr. 1, ll. 25–28
(ed. R. Pfeiffer, Clarendon Press, Oxford, 1949).

CONTENTS

READER'S GUIDE

THE monograph is designed to serve both as a general survey and as a detailed exposition of the topics it treats.

Proofs and more technical matters are set apart from the main text, wherever feasible, being printed in smaller type and in a quite different, clearly recognizable style. They can thus easily be omitted by a reader desiring general orientation about the principal trends of the theory before embarking on technicalities: the main text has been especially arranged for this purpose. The exposition proceeds from particular cases to general results, from classical theorems to recent ones; abstract concepts are illustrated by examples. In the formal exposition of proofs an attempt has been made to keep the underlying reasoning clearly visible beyond the technical details. A summary of contents is given below. The overlap with other books is slight; where it occurs, the reader may be interested to compare methods.

Terminology and notation are, generally, those of N. Bourbaki.† More specific information can be found by consulting the *Summary of notations* or the *Index.*

In the *Bibliography* the emphasis is not on completeness, but on those works through which the reader may trace the development, past and more recent, of the theory; to this end historical remarks are inserted in places. Some publications in which prerequisite matters or further developments can be found are likewise indicated. References to the bibliography are enclosed in square brackets.

Chapters are divided into sections and subsections; *cross-references* listing only section and subsection refer to the chapter in which they occur.

† Cf., e.g., his *Théorie des ensembles,* Fascicule des Résultats, 3rd edn, Hermann, Paris (1964).

SUMMARY OF CONTENTS

CHAPTER 1 discusses *Fourier transforms* of functions in $L^1(\mathbf{R}^\nu)$, the (complex-valued) Lebesgue integrable functions defined on ordinary ν-dimensional space $\mathbf{R}^\nu$. Then *Wiener's approximation theorem* in $L^1(\mathbf{R}^\nu)$ is proved, in as elementary and simple a way as possible. The methods of proof belong entirely to classical analysis, but the group-theoretic aspects are already emphasized here, in preparation for later developments. Various examples of *subalgebras of* $L^1(\mathbf{R}^\nu)$ are discussed; they prepare the ground for the more general approach in Chapters 3 and 6.

CHAPTER 2 introduces certain topological algebras, called *Wiener algebras*, of functions on a locally compact Hausdorff space. This is the abstract setting for *Wiener's theorem and its generalization.* Here the use of Banach algebras, so prevalent in the literature, has been entirely discarded; in this way some simplicity and generality are attained. Several examples of Wiener algebras, mainly from classical analysis, and their closed ideals are discussed and some open problems are mentioned.

CHAPTER 3 *is entirely independent of Chapters 1 and 2:* it concerns locally compact groups G, *Haar measure, the algebra* $L^1(G)$ *and the dual space* $L^\infty(G)$. The exposition is aimed at the applications, but the viewpoint is theoretical as well as practical; certain results are included that are required later, but seem not readily available in the literature. Quotient groups G/H, where H is a closed normal subgroup, are considered and the *relation between* $L^1(G)$ *and* $L^1(G/H)$ is studied. The restriction to the case of normal subgroups is made for the sake of simplicity, the consideration of arbitrary closed subgroups H and quotient spaces G/H being left for Chapter 8: this gradual approach will, it is hoped, make the exposition easier to follow. Finally certain subalgebras of $L^1(G)$ are studied, in continuation of the more classical discussion in Chapter 1 for $L^1(\mathbf{R}^\nu)$, but quite independently and from a more algebraic viewpoint.

CHAPTER 4 deals with *locally compact abelian groups.* A survey of their *duality theory* is given, entirely without proofs, but with detailed references to the literature. The analogy between locally compact abelian groups and Banach spaces is emphasized. Some examples are also treated, mainly projective and inductive limits. The *foundations of harmonic analysis* and the basic theorems are discussed in the form of a survey, with ample references. Then a general method of investigation, fundamental for the later work, is outlined, with examples: this is the *principle of relativization* for locally compact groups (abelian or not) which goes back to A. Weil's book.

CHAPTER 5 concerns *functions on locally compact abelian groups.* It treats mainly of properties used in the development of harmonic analysis. Some of these properties, however, extend to a larger class of locally compact groups, or to all such groups: this aspect is already emphasized here in order to gain the proper perspective and prepare the way for the developments in Chapter 8. A proof of *Poisson's formula* is also included.

CHAPTER 6 discusses *Wiener's theorem and its generalization* in the case of functions in $L^1(G)$, G being a locally compact abelian group. The main tools for the proof in this case have already been prepared in Chapters 2 and 5; the connexions with the classical case (Chap. 1) are pointed out. Then two classes of *subalgebras of $L^1(G)$* are studied; this is an extension of earlier work in Chapters 1 and 3 and includes some recent results.

CHAPTER 7 is devoted to the *spectrum* of a bounded measurable function and of more general functions. This concept—first investigated by Carleman in 1935 and now classical—has many applications in harmonic analysis. Recent developments in this field are discussed; here the methods of group theory and of classical analysis are combined, and used to the fullest extent. Some open problems are also mentioned.

CHAPTER 8 *depends only on Chapters 3 and 5;* it is concerned with *functions on general locally compact groups*. First the existence of a 'quasi-invariant' measure on a quotient space G/H is established (*theorem of Mackey–Bruhat*); this is applied to generalize several results in Chapter 3. Then *some properties of locally compact groups and their applications* are considered: more precisely, the extension of certain properties of locally compact abelian groups to a larger class of groups is investigated. This first arose as a problem on positive-definite functions in A. Weil's book in 1940; the problem was generalized by Godement in 1948 and is of considerable importance in the theory of unitary group representations. It turns out that it is connected, in fact equivalent, with a question about positive measures raised by Dieudonné in 1960, and also with the problem of the existence of a left invariant mean value for the bounded continuous functions on a locally compact group—a problem investigated in the case of discrete groups by von Neumann in 1929. Several recent results within this circle of ideas, and their applications, are treated in detail and new problems that arise are pointed out.

1

CLASSICAL HARMONIC ANALYSIS AND WIENER'S THEOREM

1. The algebra $L^1(\mathbf{R}^\nu)$; Fourier transforms

1.1. We write $\mathbf{R}$ for the real numbers, $\mathbf{R}^\nu$ for the ν-dimensional vector space over $\mathbf{R}$: the space $\mathbf{R}^\nu$ consists of all ν-tuples $x = (x_1,..., x_\nu)$ of real numbers, with the usual definitions of addition, subtraction, and multiplication by real scalars. If f is a complex-valued function on $\mathbf{R}^\nu$, integrable in the sense of Lebesgue over the whole of $\mathbf{R}^\nu$, we simply write

$$\int f(x)\,dx \tag{1}$$

for the ν-fold integral over $\mathbf{R}^\nu$. The Lebesgue integral (1) has the property

$$\int f(x-a)\,dx = \int f(x)\,dx \qquad a \in \mathbf{R}^\nu \tag{2}$$

and also satisfies

$$\int \rho^\nu f(\rho x)\,dx = \int f(x)\,dx \qquad \rho > 0. \tag{3}$$

The complex-valued functions on $\mathbf{R}^\nu$ which are Lebesgue integrable over $\mathbf{R}^\nu$ form a complete normed algebra, a so-called Banach algebra, over the complex numbers $\mathbf{C}$. This algebra is denoted by $L^1(\mathbf{R}^\nu)$; the norm is

$$\|f\|_1 = \int |f(x)|\,dx$$

and multiplication is defined by the convolution $f\star g$,

$$f\star g(x) = \int f(y)g(x-y)\,dy \qquad f, g \in L^1(\mathbf{R}^\nu),$$

which is commutative.†

$L^1(\mathbf{R}^\nu)$ contains the space $\mathscr{K}(\mathbf{R}^\nu)$ of all complex-valued continuous functions on $\mathbf{R}^\nu$ with compact supports (the *support* of a function f, $\operatorname{Supp} f$, is the closure of the set of all points x where $f(x) \neq 0$). Moreover, *$\mathscr{K}(\mathbf{R}^\nu)$ is dense in* $L^1(\mathbf{R}^\nu)$, in the metric of $L^1(\mathbf{R}^\nu)$; this fact is of fundamental significance.

† Two integrable functions that coincide almost everywhere (a.e.) on $\mathbf{R}^\nu$ are considered to be equivalent: they represent the same element of $L^1(\mathbf{R}^\nu)$.

In $L^1(\mathbf{R}^\nu)$ we have two important operators: the *translation operator* L_a

(4) $$L_a f(x) = f(x-a) \qquad a \in \mathbf{R}^\nu$$

and the *multiplication operator* M_ρ

(5) $$M_\rho f(x) = \rho^\nu f(\rho x) \qquad \rho > 0.$$

These operators are both isometric:

(6) $$\|L_a f\|_1 = \|f\|_1 \qquad a \in \mathbf{R}^\nu,$$

(7) $$\|M_\rho f\|_1 = \|f\|_1 \qquad \rho > 0.$$

1.2. The Fourier transform (F.t.) of a function $f \in L^1(\mathbf{R}^\nu)$ is

(8) $$\hat{f}(t) = \int f(x)\overline{\langle x, t\rangle}\, dx \qquad t \in \mathbf{R}^\nu,$$

where the bar indicates the complex conjugate and

$$\langle x, t\rangle = e^{2\pi i(x_1 t_1 + \ldots + x_\nu t_\nu)}.$$

In classical analysis the F.t. is usually defined with a slightly different normalization. The definition above is convenient from the viewpoint of group theory.

Fourier transforms have the following basic properties:

(9) $\hat{f}$ is continuous and $|\hat{f}(t)| \leqslant \|f\|_1$ for all $t \in \mathbf{R}^\nu$;

(10) $$(c_1 f_1 + c_2 f_2)\hat{} = c_1 \hat{f}_1 + c_2 \hat{f}_2 \qquad c_1, c_2 \in \mathbf{C};$$

(11) $$(f_1 \star f_2)\hat{} = \hat{f}_1 . \hat{f}_2;$$

(12) $$[L_a f]\hat{}(t) = \overline{\langle a, t\rangle} . \hat{f}(t) \qquad a \in \mathbf{R}^\nu;$$

(13) $$[\chi_{t_0} . f]\hat{} = L_{t_0} \hat{f} \qquad t_0 \in \mathbf{R}^\nu\ [\chi_{t_0}(x) = \langle x, t_0\rangle];$$

(14) $$[M_\rho f]\hat{}(t) = \hat{f}(t/\rho) \qquad \rho > 0.$$

Remark. We also note that $(f^*)\hat{} = \bar{\hat{f}}$, where $f^*(x) = \overline{f(-x)}$.

There are three fundamental theorems on Fourier transforms of functions in $L^1(\mathbf{R}^\nu)$:

(i) *Lemma of Riemann–Lebesgue.* The Fourier transform of a function in $L^1(\mathbf{R}^\nu)$ 'vanishes at infinity': $\hat{f}(t) \to 0$ when $|t| = \{t_1^2 + \ldots + t_\nu^2\}^{\frac{1}{2}} \to \infty$.

(ii) *Uniqueness theorem.* A function $f \in L^1(\mathbf{R}^\nu)$ is uniquely determined a.e. by its Fourier transform: if $\hat{f}(t) = 0$ for all $t \in \mathbf{R}^\nu$, then $f(x) = 0$ a.e.

(iii) *Inversion theorem.* Let $f \in L^1(\mathbf{R}^\nu)$ be continuous. If

$$\int |\hat{f}(t)|\, dt < \infty,$$

then

(15) $$f(x) = \int \langle x, t\rangle \hat{f}(t)\, dt \qquad x \in \mathbf{R}^\nu.$$

(15) differs only slightly from (8).

More refined theorems are, of course, proved in classical analysis, especially on the line. The results above are those most frequently used in practice; they are simple to state and to prove and carry over to groups (Chap. 4, § 4.1).

1.3. *Examples*

(i) For $A > 0$ let ϕ_A be the characteristic function of the interval $[-A, A]$: $\phi_A(x) = 1$ for $|x| \leqslant A$, $\phi_A(x) = 0$ for $|x| > A$ $(x \in \mathbf{R})$. The F.t. of ϕ_A is $\sin 2\pi At/\pi t$.

(ii) Let Δ_1 be the 'triangle' function defined for $x \in \mathbf{R}$ by $\Delta_1(x) = 1-|x|$ for $|x| \leqslant 1$, $\Delta_1(x) = 0$ for $|x| \geqslant 1$. This has the F.t. $(\sin \pi t/\pi t)^2$, as can be seen by direct computation or by observing that $\Delta_1 = \phi_{\frac{1}{2}} \star \phi_{\frac{1}{2}}$ (cf. (i)). By the inversion theorem

$$\Delta_1(x) = \int \left(\frac{\sin \pi t}{\pi t}\right)^2 e^{2\pi ixt}\, dt = \int \left(\frac{\sin \pi t}{\pi t}\right)^2 e^{-2\pi ixt}\, dt,$$

since Δ_1 is an even function. Now let us change the notation: define $\sigma_1 \in L^1(\mathbf{R})$ by

$$\sigma_1(x) = \left(\frac{\sin \pi x}{\pi x}\right)^2, \qquad x \neq 0, \qquad \sigma_1(0) = 1. \tag{16}$$

Then the F.t. of σ_1 is given by

$$\hat{\sigma}_1(t) = \Delta_1(t) = \begin{cases} 1-|t|, & |t| \leqslant 1 \\ 0, & |t| \geqslant 1. \end{cases} \tag{17}$$

(iii) Let T_1 be the 'trapezium' function defined for $x \in \mathbf{R}$ by $T_1(x) = 1$ for $|x| \leqslant 1$, $T_1(x) = 2-|x|$ for $1 \leqslant |x| \leqslant 2$, $T_1(x) = 0$ for $|x| \geqslant 2$. By direct computation—or by observing that $T_1 = \phi_{\frac{3}{2}} \star \phi_{\frac{1}{2}}$ (cf. (i))—we see that T_1 has the F.t. $(\sin 3\pi t \sin \pi t)/(\pi t)^2$. Applying the inversion theorem and changing the notation slightly, as in (ii), we obtain: let $\tau_1 \in L^1(\mathbf{R})$ be defined by

$$\tau_1(x) = \frac{\sin 3\pi x \sin \pi x}{(\pi x)^2}, \qquad x \neq 0, \qquad \tau_1(0) = 3; \tag{18}$$

then the F.t. $\hat{\tau}_1$ is given by

$$\hat{\tau}_1(t) = T_1(t) = \begin{cases} 1, & |t| \leqslant 1 \\ 2-|t|, & 1 \leqslant |t| \leqslant 2 \\ 0, & |t| \geqslant 2. \end{cases} \tag{19}$$

(iv) We obtain more general triangle or trapezium functions by putting $\Delta(t) = \Delta_1(t/\rho)$, $T(t) = T_1(t/\rho)$, for fixed $\rho > 0$. These are the Fourier transforms of $M_\rho \sigma_1$ respectively $M_\rho \tau_1$ (cf. (ii), (iii)). The functions Δ, T and their translates play an important role in harmonic analysis.

(v) A function on $\mathbf{R}$ with compact support and possessing a continuous nth derivative $(n \geqslant 2)$ is the F.t. of a continuous function $f \in L^1(\mathbf{R})$

such that $|f(x)| \leqslant A(1+|x|)^{-n}$, $x \in \mathbf{R}$, for some constant A. Actually, f is the restriction to the line of an entire analytic function and all derivatives of f likewise belong to $L^1(\mathbf{R})$.

(vi) For $x = (x_1,\ldots, x_\nu) \in \mathbf{R}^\nu$ put $f(x) = f_1(x_1).f_2(x_2).\ldots.f_\nu(x_\nu)$, where $f_1, f_2,\ldots, f_\nu \in L^1(\mathbf{R})$. Then $f \in L^1(\mathbf{R}^\nu)$ and $\hat{f}(t) = \hat{f}_1(t_1).\hat{f}_2(t_2).\ldots.\hat{f}_\nu(t_\nu)$. In this way we can use functions in $L^1(\mathbf{R})$ with certain properties to obtain functions in $L^1(\mathbf{R}^\nu)$ with analogous properties (especially concerning Fourier transforms).

1.4. The Fourier transforms of the functions in $L^1(\mathbf{R}^\nu)$ form an algebra of continuous, complex-valued functions, with the ordinary (pointwise) algebraic operations; this is the content of (9), (10), (11). Moreover, this algebra is isomorphic to $L^1(\mathbf{R}^\nu)$ by the uniqueness theorem 1.2 (ii). We denote it by $\mathscr{F}^1(\mathbf{R}^\nu)$. Thus we have another model for $L^1(\mathbf{R}^\nu)$—a simple model, in which the convolution is replaced by ordinary multiplication. We carry the L^1-norm over to $\mathscr{F}^1(\mathbf{R}^\nu)$ by putting

$$\|\hat{f}\| = \|f\|_1 \qquad \hat{f} \in \mathscr{F}^1(\mathbf{R}^\nu), \tag{20}$$

so that $\mathscr{F}^1(\mathbf{R}^\nu)$ becomes a complete normed algebra of functions, a Banach algebra (cf. § 1.1). In $\mathscr{F}^1(\mathbf{R}^\nu)$ we have again an isometric translation operator (cf. (13)):

$$\|L_{t_0}\hat{f}\| = \|\hat{f}\| \qquad t_0 \in \mathbf{R}^\nu. \tag{21}$$

1.5. Another familiar function algebra is formed by the Fourier series

$$\alpha(x) = \sum_{n\in\mathbf{Z}} a_n e^{inx} \qquad x \in \mathbf{R} \tag{22}$$

with complex coefficients such that

$$\|\alpha\| = \sum_{n\in\mathbf{Z}} |a_n| < \infty, \tag{23}$$

where $\mathbf{Z}$ denotes the integers. With the ordinary algebraic operations and the norm (23) the functions (22) form a commutative Banach algebra; moreover, they may be considered as functions on the (unit) circle $\mathbf{T}$. We denote this algebra by $\mathscr{F}^1(\mathbf{T})$. It is isomorphic to the algebra $L^1(\mathbf{Z})$ consisting of all complex sequences $a = (a_n)_{n\in\mathbf{Z}}$ with the norm $\|a\|_1 = \sum_{n\in\mathbf{Z}} |a_n| < \infty$; the multiplication of a, $b \in L^1(\mathbf{Z})$ is defined as convolution: $c = a \star b = (c_n)_{n\in\mathbf{Z}}$ with $c_n = \sum_{k\in\mathbf{Z}} a_k b_{n-k}$.

$L^1(\mathbf{Z})$ is the discrete analogue of $L^1(\mathbf{R})$. This analogy is most interesting from the viewpoint of the theory of groups, as will appear later. We also remark that the function algebras $\mathscr{F}^1(\mathbf{T})$ and $\mathscr{F}^1(\mathbf{R})$ are 'locally isomorphic': we shall discuss this in a more general context in Chapter 6, § 3.5.

1.6. In studying the classical literature on harmonic analysis, it is very interesting and highly instructive to observe how far the methods and results are based on special properties of **R** (or $\mathbf{R}^\nu$, **T**). In this connexion we note that the operator M_ρ (cf. (5))—which is of considerable importance in the applications—is rather special to $\mathbf{R}^\nu$: it is related to the (continuous) automorphisms of the additive group **R**, which are all of the form $x \to cx$ ($c \in \mathbf{R}$, $c \neq 0$).

References. [11], [12], [143, § 12, Lemma 6_{10}].

2. Functions in $L^1(\mathbf{R}^\nu)$

We consider here a few basic properties of the functions in $L^1(\mathbf{R}^\nu)$.

2.1. *Let f be in* $L^1(\mathbf{R}^\nu)$. *Then for any* $\epsilon > 0$ *there is a neighbourhood*† U_ϵ *of* 0 *in* $\mathbf{R}^\nu$ *such that*

(1) $$\|L_y f - f\|_1 < \epsilon \qquad \textit{for all } y \in U_\epsilon.$$

This result, due to Lebesgue, has many applications. The proof given by Hobson [76, § 431] uses a method of considerable interest, the '*reduction to* $\mathscr{K}(\mathbf{R}^\nu)$': (1) is almost obvious for functions in $\mathscr{K}(\mathbf{R}^\nu)$ and then follows for all $f \in L^1(\mathbf{R}^\nu)$ since $\mathscr{K}(\mathbf{R}^\nu)$ is dense in $L^1(\mathbf{R}^\nu)$ (§ 1.1). This method also works for groups (see Chap. 3, § 5.5).

It is easy to see that $\|L_y f - f\|_1$ cannot tend to zero too fast when $y \to 0$. In fact, *if* $f \in L^1(\mathbf{R}^\nu)$ *is not a.e. zero, then there is a constant* $C\ (= C_f) > 0$ *such that* $\|L_y f - f\|_1 \geqslant C|y|$ *for all sufficiently small* $y \in \mathbf{R}^\nu$, *where* $|y| = \{y_1^2 + \ldots + y_\nu^2\}^{\frac{1}{2}} > 0$.

Proof. There is a $t' \in \mathbf{R}^\nu$ such that $|\hat{f}(t')| > 0$, and hence there is a $\delta > 0$ such that $|\hat{f}(t)| \geqslant A > 0$ for all t with $|t - t'| \leqslant \delta$. Now, given *any* $y \in \mathbf{R}^\nu$, we can find $t_y \in \mathbf{R}^\nu$ such that $|t_y - t'| \leqslant \delta$ and $|(y \mid t_y)| \geqslant \frac{1}{2}\delta|y|$ (notation: $(a \mid b) = a_1 b_1 + \ldots + a_\nu b_\nu$ for $a, b \in \mathbf{R}^\nu$): indeed, if $|(y \mid t')| \geqslant \frac{1}{2}\delta|y|$, we may take $t_y = t'$, and if $|(y \mid t')| < \frac{1}{2}\delta|y|$, we may take $t_y = (\delta/|y|)y + t'$. Then

$$\|L_y f - f\|_1 \geqslant |[L_y f - f]^\wedge(t_y)|$$
$$\geqslant |\hat{f}(t_y)|\, 2|\sin\{\pi(y \mid t_y)\}| \geqslant 2A\,\delta|y|$$

for all $y \in \mathbf{R}^\nu$ with $|y| \leqslant \{2(|t'| + \delta)\}^{-1}$.

It is obvious that there are functions $f \in L^1(\mathbf{R}^\nu)$ such that

$$\|L_y f - f\|_1 = O(|y|) \qquad y \to 0$$

and $f(x)$ is not a.e. zero. We shall later construct an extensive class of such functions (Chap. 5, § 1.1, Remark).

† We shall often write 'nd.' for shortness.

2.2. Let f and h be functions in $L^1(\mathbf{R}^\nu)$. Then

$$\Big\|(M_\rho h)\star f-\Big[\int h(x)\,dx\Big]f\Big\|_1 \to 0 \qquad \rho\to\infty. \tag{2}$$

The value of the function $(M_\rho h)\star f-\big[\int h(x)\,dx\big]f$ at a point $y\in\mathbf{R}^\nu$ may be written in the form

$$\int M_\rho h(x)\{f(y-x)-f(y)\}\,dx.$$

Hence

$$\Big\|(M_\rho h)\star f-\Big[\int h(x)\,dx\Big]f\Big\|_1 \leqslant \int |M_\rho h(x)|\,.\,\|L_x f-f\|_1\,dx. \tag{*}$$

Put $\|L_x f-f\|_1 = \phi(x)$. Then $\int |M_\rho h(x)|\,\phi(x)\,dx = \int |h(x)|\,\phi(x/\rho)\,dx$ and it is an elementary exercise to show that this tends to 0 when $\rho\to\infty$, since ϕ is continuous, $\phi(0)=0$ and $0\leqslant\phi(x)\leqslant$ const. $(=2\|f\|_1)$ for all $x\in\mathbf{R}^\nu$.

2.3. As a first application of 2.2 we note: *for any* $f\in L^1(\mathbf{R}^\nu)$ *and any* $\epsilon>0$ *there is a* $v\in L^1(\mathbf{R}^\nu)$ *such that the F.t.* $\hat{v}$ *has compact support and*

$$\|v\star f-f\|_1<\epsilon. \tag{3}$$

Let $h\in L^1(\mathbf{R}^\nu)$ be such that $\hat{h}$ has compact support and $\hat{h}(0)=1$. Then we can put $v=M_\rho h$ for ρ large (cf. (2)).

In particular (3) says that *there are 'approximate units'* in $L^1(\mathbf{R}^\nu)$. There is, of course, no unit element in $L^1(\mathbf{R}^\nu)$, for this would necessarily have a F.t. equal to 1, which contradicts the Riemann–Lebesgue lemma.

2.4. Another application of 2.2 is the following. *Let f be a function in* $L^1(\mathbf{R}^\nu)$. *Then, given* $\epsilon>0$, *there is a function* $\tau\in L^1(\mathbf{R}^\nu)$ *such that*

(i) *the F.t.* $\hat{\tau}$ *has the value* 1 *near the origin*;

(ii) $\Big\|f\star\tau-\Big[\int f(x)\,dx\Big]\tau\Big\|_1<\epsilon.$

Let τ_1 be any function in $L^1(\mathbf{R}^\nu)$ satisfying (i). Then we obtain from (2) with a slight change of notation

$$\Big\|(M_\rho f)\star\tau_1-\Big[\int f(x)\,dx\Big]\tau_1\Big\|_1<\epsilon \qquad \text{for large } \rho. \tag{*}$$

Now

$$\Big\|(M_\rho f)\star\tau_1-\Big[\int f(x)\,dx\Big]\tau_1\Big\|_1 = \Big\|f\star M_{1/\rho}\tau_1-\Big[\int f(x)\,dx\Big]M_{1/\rho}\tau_1\Big\|_1, \tag{**}$$

as is readily verified by applying the operator $M_{1/\rho}$ to the left-hand side. Thus we can put $\tau=M_{1/\rho}\tau_1$, for large ρ; since $\hat{\tau}(t)=\hat{\tau}_1(\rho t)$, the condition (i) will still be fulfilled.

The significance of this fact will appear in § 3 and again later.

3. The theorem of Wiener–Lévy

3.1. The theorem of Wiener–Lévy expresses a fundamental fact concerning Fourier transforms. *Let* $\hat{f}$ *be in* $\mathscr{F}^1(\mathbf{R}^\nu)$. *Let* K *be a compact set in* $\mathbf{R}^\nu$ *and denote by* $\hat{f}(K)$ *the image of* K *under the map* $z=\hat{f}(t)$, $t\in K$. *Suppose* $z\to \mathrm{A}(z)$ *is a (single-valued) analytic function on an open* *neighbourhood of* $\hat{f}(K)$ *in* **C**. *Then there is a function* $\hat{g} \in \mathscr{F}^1(\mathbf{R}^\nu)$ *such that* $\hat{g}(t) = \mathrm{A}(\hat{f}(t))$ *for all* $t \in K$. The proof will be given in several steps.

3.2. First we prove the following important lemma. *Let* $\hat{f}$ *be in* $\mathscr{F}^1(\mathbf{R}^\nu)$. *Then, given any* $t_0 \in \mathbf{R}^\nu$, *there exists for every* $\epsilon > 0$ *a function* $\hat{\tau}_{t_0}$ $(= \hat{\tau}_{t_0,\epsilon})$ *in* $\mathscr{F}^1(\mathbf{R}^\nu)$ *such that*

(i) $\hat{\tau}_{t_0}(t) = 1$ *for* t *near* t_0;

(ii) $\|\{\hat{f}-\hat{f}(t_0)\}\hat{\tau}_{t_0}\| < \epsilon$.

In the special case $t_0 = 0$, this is merely 2.4, rewritten for $\mathscr{F}^1(\mathbf{R}^\nu)$. For general $t_0 \in \mathbf{R}^\nu$ we apply the special case to $L_{-t_0}\hat{f}$ $[L_{-t_0}\hat{f}(t) = \hat{f}(t+t_0)]$. Since for any $\hat{\tau} \in \mathscr{F}^1(\mathbf{R}^\nu)$

$$L_{t_0}[\{L_{-t_0}\hat{f}-L_{-t_0}\hat{f}(0)\}\hat{\tau}] = \{\hat{f}-\hat{f}(t_0)\}L_{t_0}\hat{\tau},$$

we can take $\hat{\tau}_{t_0} = L_{t_0}\hat{\tau}$, for appropriate $\hat{\tau}$ (cf. § 1.4 (21)).

3.3. Next we establish 3.1 'locally'. *Let* t_0 *be any point of* K. *Then there is a function* $\hat{g}_{t_0} \in \mathscr{F}^1(\mathbf{R}^\nu)$ *such that* $\hat{g}_{t_0}(t) = \mathrm{A}(\hat{f}(t))$ *for all* $t \in \mathbf{R}^\nu$ *near* t_0.

Put $z_0 = \hat{f}(t_0)$. Then for $|z-z_0| < \epsilon$, say, we have

$$\mathrm{A}(z) = \mathrm{A}(z_0)+\sum_{n\geqslant 1} c_n(z-z_0)^n.$$

Now choose $\hat{\tau}_{t_0}$ according to 3.2 and consider

$$\mathrm{A}(\hat{f}(t_0))\hat{\tau}_{t_0}+\sum_{n\geqslant 1} c_n[\{\hat{f}-\hat{f}(t_0)\}\hat{\tau}_{t_0}]^n. \tag{*}$$

The series converges in $\mathscr{F}^1(\mathbf{R}^\nu)$ by 3.2 (ii). Thus (*) is a function $\hat{g}_{t_0} \in \mathscr{F}^1(\mathbf{R}^\nu)$. Moreover, $\hat{g}_{t_0}(t) = \mathrm{A}(\hat{f}(t))$ for t near t_0 by 3.2 (i) (note that convergence in $\mathscr{F}^1(\mathbf{R}^\nu)$ implies pointwise convergence on $\mathbf{R}^\nu$).

3.4. Finally we extend the 'local' result 3.3 to the compact set K.

For each $t_0 \in K$ we have: (i) there is a $\hat{g}_{t_0} \in \mathscr{F}^1(\mathbf{R}^\nu)$ such that $\hat{g}_{t_0}(t) = \mathrm{A}(\hat{f}(t))$ for $t \in U_{t_0}$, a nd. of t_0; (ii) there is a function $\hat{\tau}'_{t_0} \in \mathscr{F}^1(\mathbf{R}^\nu)$ such that $\hat{\tau}'_{t_0}$ vanishes outside U_{t_0} and equals 1 in some small nd. U'_{t_0} (cf. § 1.3 (iii), (iv), (vi)). Now there are finitely many points $(t_n)_{1\leqslant n\leqslant N}$ in K such that the corresponding nds. $U'_n = U'_{t_n}$ cover K; we write also U_n, $\hat{g}_n$, $\hat{\tau}'_n$ for U_{t_n}, $\hat{g}_{t_n}$, $\hat{\tau}'_{t_n}$.

Put $\hat{e}_1 = \tau'_1$ and (if $N \geqslant 2$) $\hat{e}_2 = \hat{\tau}'_2.(1-\hat{\tau}'_1),\ldots,$ $\hat{e}_N = \hat{\tau}'_N.(1-\hat{\tau}'_1).\ldots.(1-\hat{\tau}'_{N-1})$. Then $\hat{e}_n$ is in $\mathscr{F}^1(\mathbf{R}^\nu)$, $1 \leqslant n \leqslant N$, and vanishes outside U_n. Moreover,

$$\sum_n \hat{e}_n = 1-(1-\hat{\tau}'_1).(1-\hat{\tau}'_2).\ldots.(1-\hat{\tau}'_N)$$

(as we see by induction) and hence $\sum_n \hat{e}_n(t) = 1$ for $t \in K$. Now consider the function

$$\hat{g} = \sum_n \hat{e}_n\hat{g}_n \in \mathscr{F}^1(\mathbf{R}^\nu).$$

For $t \in K$ we have $\hat{e}_n(t)\hat{g}_n(t) = \hat{e}_n(t)\mathrm{A}(\hat{f}(t))$, since $\hat{e}_n(t) = 0$ if $t \notin U_n$. Hence

$$\hat{g}(t) = \Big\{\sum_n \hat{e}_n(t)\Big\}\mathrm{A}(\hat{f}(t)) = \mathrm{A}(\hat{f}(t)) \qquad \text{for } t \in K.$$

Thus the proof of the theorem of Wiener–Lévy is complete.

3.5. The following *modification of the theorem of Wiener–Lévy* also holds. Let $\hat{f}$ be in $\mathscr{F}^1(\mathbf{R}^\nu)$ and let $\hat{f}(\mathbf{R}^\nu)$ be the image of $\mathbf{R}^\nu$ under the map $z = \hat{f}(t)$. If $z \to F(z)$ is (single-valued and) analytic on an open neighbourhood of $\hat{f}(\mathbf{R}^\nu) \cup \{0\}$ in $\mathbf{C}$ and if $F(0) = 0$, then there is a $\hat{g} \in \mathscr{F}^1(\mathbf{R}^\nu)$ such that $\hat{g}(t) = F(\hat{f}(t))$ for all $t \in \mathbf{R}^\nu$.

Since $F(0) = 0$, we have $F(z) = a_1 z + a_2 z^2 + \ldots$ for $|z| < \epsilon$, say. By 2.3 there is a $\hat{v} \in \mathscr{F}^1(\mathbf{R}^\nu)$ with compact support, say K, such that $\|\hat{v}\hat{f} - \hat{f}\| < \epsilon$; put $\hat{h} = \hat{f} - \hat{v}\hat{f}$. Then the function $\hat{g}_0 = \sum_{n \geqslant 1} a_n \hat{h}^n$ is in $\mathscr{F}^1(\mathbf{R}^\nu)$ and $\hat{g}_0(t) = F(\hat{f}(t))$ for $t \in \mathbf{R}^\nu - K$. Take a $\hat{\tau} \in \mathscr{F}^1(\mathbf{R}^\nu)$ which is 1 on K and has compact support, say U_K (cf. § 1.3 (iii), (iv), (vi)). By 3.1 there is a $\hat{g}_1 \in \mathscr{F}^1(\mathbf{R}^\nu)$ such that $\hat{g}_1(t) = F(\hat{f}(t))$ for $t \in U_K$. Then $\hat{g} = (1 - \hat{\tau})\hat{g}_0 + \hat{\tau}\hat{g}_1$ is in $\mathscr{F}^1(\mathbf{R}^\nu)$ and $\hat{g}(t) = F(\hat{f}(t))$, $t \in \mathbf{R}^\nu$.

3.6. The theorem of Wiener–Lévy (§ 3.1) was originally proved for the algebra $\mathscr{F}^1(\mathbf{T})$ (§ 1.5): first for $\mathrm{A}(z) = z^{-1}$ by Wiener [142, Lemma II e], then by P. Lévy [89, Theorem V] for analytic A. Theorem 3.1, for $\mathscr{F}^1(\mathbf{R})$, is contained implicitly in those papers (cf. [142, Lemma II f]). Carleman [20, p. 67] formulated Theorem 3.1 explicitly for the line and simplified its proof. This was also done, independently, by Ditkin [35, § 6, Theorem I], whose formulation is that in 3.5; see also [1].†

The essential part of the proof of Theorem 3.1 is 3.2 which depends on 2.4, and 2.4 was proved by means of a special property of $\mathbf{R}^\nu$, viz. the operator M_ρ. Once 3.2 is established, the proof of 3.1 offers no difficulty. Wiener himself worked with the circle $\mathbf{T}$ and proved the analogue of 3.2 for $\mathscr{F}^1(\mathbf{T})$, but in implicit form (cf. the proof of Lemma II d in [142]). In his book he formulated this analogue explicitly, giving a quite different proof [143, § 12, Lemma 6_{13}]. Still another proof appeared in 1935 in [144, 1st edn, § 6.51]. All these proofs simply reflect the problem of establishing an analogue of 2.4 for groups; this will be discussed in Chapter 6, § 1.1.

4. Wiener's theorem

4.1. In 1932 N. Wiener proved the following, now classical, theorem (for $\nu = 1$ [142, Theorem II]).

WIENER'S APPROXIMATION THEOREM. *Let f be in $L^1(\mathbf{R}^\nu)$. The linear combinations of translates of f,*

$$(1) \qquad \sum_n \lambda_n f(x - a_n) \qquad \lambda_n \in \mathbf{C},\ a_n \in \mathbf{R}^\nu,$$

are dense in $L^1(\mathbf{R}^\nu)$, that is, for every $g \in L^1(\mathbf{R}^\nu)$ and every $\epsilon > 0$ there is a

† The statement of the theorem in [1, § 101] should be modified; in the German translation pp. 232 and 233 are interchanged.

linear combination (1) *such that*

(2) $$\int \left| g(x) - \sum_n \lambda_n f(x-a_n) \right| dx < \epsilon,$$

if and only if the Fourier transform of f has no zeros:

(3) $$\hat{f}(t) \neq 0 \qquad \textit{for all } t \in \mathbf{R}^\nu.$$

The necessity of condition (3) is clear.

From (2) it follows (cf. § 1.2 (9) and (12)):

$$\left| \hat{g}(t) - \sum_n \lambda_n \overline{\langle a_n, t\rangle} \hat{f}(t) \right| < \epsilon \qquad \text{for all } t \in \mathbf{R}^\nu.$$

Now, if $\hat{f}(t_0) = 0$ for some $t_0 \in \mathbf{R}^\nu$, we can take a $g \in L^1(\mathbf{R}^\nu)$ such that $\hat{g}(t_0) \neq 0$ and then (2) cannot hold for $\epsilon \leqslant |\hat{g}(t_0)|$.

The sufficiency of (3) is proved in three steps.

(i) *Let g_0 be any function in $L^1(\mathbf{R}^\nu)$ whose F.t. $\hat{g}_0$ has compact support. Then there is an $h \in L^1(\mathbf{R}^\nu)$ such that $g_0 = h \star f$, if $\hat{f}$ satisfies* (3).

There is an $f_1 \in L^1(\mathbf{R}^\nu)$ whose F.t. $\hat{f}_1$ coincides with $1/\hat{f}$ on Supp $\hat{g}_0$ (§ 3.1, for $A(z) = z^{-1}$). We can put $h = g_0 \star f_1$ (observe $(g_0 \star f_1 \star f)\hat{} = \hat{g}_0 \cdot \hat{f}_1 \cdot \hat{f} = \hat{g}_0$).

(ii) *The functions $g_0 \in L^1(\mathbf{R}^\nu)$ whose Fourier transforms have compact support are dense in $L^1(\mathbf{R}^\nu)$.*

Given any $g \in L^1(\mathbf{R}^\nu)$ and $\epsilon > 0$, there is a $v \in L^1(\mathbf{R}^\nu)$ such that Supp $\hat{v}$ is compact and $\|g - v \star g\|_1 < \epsilon$ (§ 2.3). Then Supp$(v \star g)\hat{}$ is compact.

(iii) The proof is then completed by means of the proposition below.

4.2. *Let h, f be in $L^1(\mathbf{R}^\nu)$. The convolution $h \star f$ can be approximated in $L^1(\mathbf{R}^\nu)$, as closely as we please, by linear combinations* (1). That is, given $\epsilon > 0$, there is a finite sum $\sum_n \lambda_n L_{a_n} f$ such that $\left\| h \star f - \sum_n \lambda_n L_{a_n} f \right\|_1 < \epsilon$.

The proof is easily reduced to the case when h has compact support K. Given $\epsilon > 0$, choose $\delta > 0$ such that $\|L_y f - f\|_1 < \epsilon$ if $|y| < \delta$ (cf. § 2.1). There are finitely many points $a_n \in K$ such that the open balls U_n with centre a_n and radius δ cover K ($1 \leqslant n \leqslant N$, say). Let $A_1 = K \cap U_1$ and $A_n = \left(K - \bigcup_{r=1}^{n-1} A_r\right) \cap U_n$ for $2 \leqslant n \leqslant N$, if $N > 1$. Put $\lambda_n = \int_{A_n} h(x)\, dx$. Then

$$h \star f - \sum_n \lambda_n L_{a_n} f = \sum_n \int_{A_n} h(x)\{L_x f - L_{a_n} f\}\, dx,$$

hence

(*) $$\left\| h \star f - \sum_n \lambda_n L_{a_n} f \right\|_1 \leqslant \sum_n \int_{A_n} |h(x)| \cdot \|L_x f - L_{a_n} f\|_1\, dx < \|h\|_1 \cdot \epsilon,$$

since $\|L_x f - L_{a_n} f\|_1 < \epsilon$ for $x \in A_n$. This finishes the proof, since $\epsilon > 0$ was arbitrary.

Thus Wiener's approximation theorem is proved.

4.3. A (linear) subspace I of $L^1(\mathbf{R}^\nu)$ is said to be *invariant* under translations, or simply invariant, if $f \in I$ implies $L_a f \in I$ for all $a \in \mathbf{R}^\nu$.

Proposition 4.2 can now be stated as follows. Every closed invariant subspace of $L^1(\mathbf{R}^\nu)$ is an ideal. But the converse also holds: every closed ideal of $L^1(\mathbf{R}^\nu)$ is an invariant subspace.

This follows immediately from the existence of approximate units in $L^1(\mathbf{R}^\nu)$: given $f \in I$ and $\epsilon > 0$, there is a $u \in L^1(\mathbf{R}^\nu)$ such that $\|u \star f - f\|_1 < \epsilon$ (cf. § 2.3). We have $(L_a u) \star f = L_a(u \star f)$, $a \in \mathbf{R}^\nu$, so

$$\|(L_a u) \star f - L_a f\|_1 = \|u \star f - f\|_1 < \epsilon.$$

But $(L_a u) \star f \in I$ and I is closed, thus $L_a f \in I$.

Hence *the closed invariant subspaces of* $L^1(\mathbf{R}^\nu)$ *coincide with the closed ideals of* $L^1(\mathbf{R}^\nu)$.

4.4. There is a more general form of Wiener's theorem, already stated by Wiener [142, Theorem IV]. *Let $f, g,\ldots,$ be functions in $L^1(\mathbf{R}^\nu)$. If the Fourier transforms $\hat{f}, \hat{g},\ldots,$ have no common zero, then the closed invariant linear subspace spanned by these functions (i.e. the smallest such subspace containing them all) is $L^1(\mathbf{R}^\nu)$ itself, and conversely.*

Let I be this subspace. For every compact set $K \subset \mathbf{R}^\nu$ there are *finitely* many functions $f_n \in I$ such that at every point of K at least one of the Fourier transforms $\hat{f}_n$ is not zero. Put $f = \sum_n f_n \star f_n^*$, so that $\hat{f} = \sum_n |\hat{f}_n|^2$ (cf. § 1.2, Remark). Then $f \in I$, since I is an ideal (§ 4.3), and $\hat{f}(t) > 0$ for all $t \in K$. It follows that I contains all functions $g_0 \in L^1(\mathbf{R}^\nu)$ whose Fourier transforms vanish outside K (cf. the proof of § 4.1 (i)). By 4.1 (ii) this implies the assertion, since K was arbitrary and I is closed. The converse part is obvious.

Remark. In the proof above we have used the *involution* $f \to f^*$ which is a rather special property of the algebra L^1.

4.5. From 4.1 we obtain the following result.

Wiener's general Tauberian theorem [142, Theorem VIII], [143, § 10, Theorem 4]. *Let ϕ be a complex-valued, measurable, essentially bounded function on* $\mathbf{R}$. *Suppose there is an $f \in L^1(\mathbf{R})$ such that $\hat{f}(t) \neq 0$ for all $t \in \mathbf{R}$ and*

$$\int f(x-y)\phi(x)\,dx \to A \int f(x)\,dx \qquad y \to \infty.$$

Then

$$\int g(x-y)\phi(x)\,dx \to A \int g(x)\,dx \qquad y \to \infty$$

for every $g \in L^1(\mathbf{R})$.

Put $\psi = \phi - A$; we will prove the equivalent statement: if

$$\int f(x-y)\psi(x)\,dx \to 0 \qquad y \to \infty,$$

then

$$(*) \qquad \int g(x-y)\psi(x)\,dx \to 0 \qquad y \to \infty$$

for all $g \in L^1(\mathbf{R})$.

Clearly (*) holds if $g = \sum_n \lambda_n L_{a_n} f$. Let us show that, if (*) holds for g_n, $n = 1, 2, \ldots$, and $g_n \to g$ in $L^1(\mathbf{R})$ $(n \to \infty)$, then (*) also holds for g; by Wiener's approximation theorem (*) will then follow for all $g \in L^1(\mathbf{R})$. Now we can write for $n = 1, 2, \ldots$

$$\int L_y g(x)\psi(x)\,dx = \int \{L_y g(x) - L_y g_n(x)\}\psi(x)\,dx + \int L_y g_n(x)\psi(x)\,dx,$$

whence $$\left|\int L_y g(x)\psi(x)\,dx\right| \leqslant \|g-g_n\|_1 \cdot \|\psi\|_\infty + \left|\int L_y g_n(x)\psi(x)\,dx\right|.$$

Given $\epsilon > 0$ we can choose n so large that $\|g-g_n\|_1 \cdot \|\psi\|_\infty < \frac{1}{2}\epsilon$ and then C so large that, for this n, $\left|\int L_y g_n(x)\psi(x)\,dx\right| < \frac{1}{2}\epsilon$ for all $y > C$. This completes the proof.

The theorem also holds for $\mathbf{R}^\nu$, $\nu \geqslant 1$, if '$y \to \infty$' is replaced by '$|y| \to \infty$'; the proof is the same. An obvious generalization can be obtained from 4.4.

4.6. Wiener proved his celebrated approximation theorem in a rather long paper [142]; the proof itself only takes nine pages. In establishing his result, he was, perhaps, guided by an analogy with $L^2(\mathbf{R})$: there a similar theorem holds, but is much easier to prove ([142, Theorem I]; cf. also [12, Chap. IV, § 10]).

Wiener's theorem has been the starting point of many contemporary developments in harmonic analysis. The methods that Wiener used were entirely classical and he considered, of course, only the line. But the significance of these methods extends far beyond the original context in which Wiener invented them: they are capable of generalization—a generalization providing deeper insight and leading the way to new problems and new results (cf. Chaps. 2 and 6).

In 1935 Carleman gave a remarkable new proof of Wiener's theorem on the line, in a course of lectures at the Swedish Mittag–Leffler Institute that were published in 1944 [20]. Carleman used functional analysis and 'dualized' the whole theory by working with the space $L^\infty(\mathbf{R})$. In this way he also proved some important results extending those of Wiener. The proofs of these results, as given by Carleman, are not simple, but the methods which he introduced are very powerful and, like those of Wiener, admit generalization. Carleman's ideas have exerted a profound influence on later work connected with Wiener's theorem; more details will be given in Chapter 6, § 1.7 and Chapter 7, § 1.1.

Additional references. [122], [104].

5. Some subalgebras of $L^1(\mathbf{R}^\nu)$

We shall consider here certain subalgebras of $L^1(\mathbf{R}^\nu)$ that are themselves Banach algebras with respect to some norm. Multiplication is always the usual convolution in $L^1(\mathbf{R}^\nu)$.

(i) The continuous functions $f \in L^1(\mathbf{R}^\nu)$ that vanish at infinity form a Banach algebra under the norm $\|f\| = \|f\|_1 + \|f\|_\infty$.

This example can also be described as the algebra of all *uniformly* continuous $f \in L^1(\mathbf{R}^\nu)$.

(ii) Let p, $1 < p < \infty$, be given. The (equivalence classes of) functions belonging to both $L^1(\mathbf{R}^\nu)$ and $L^p(\mathbf{R}^\nu)$ form a Banach algebra under the norm $\|f\| = \|f\|_1 + \|f\|_p$.

(iii) Let K be the 'box' $\{x \mid x = (x_1,\ldots, x_\nu),\ 0 \leqslant x_j \leqslant 1,\ j = 1,\ldots, \nu\}$ in $\mathbf{R}^\nu$. Let $\mathbf{Z}^\nu$ be the subgroup of all points in $\mathbf{R}^\nu$ with coordinates in $\mathbf{Z}$. The complex-valued continuous functions f on $\mathbf{R}^\nu$ such that

$$\|f\| = \sup_{y \in \mathbf{R}^\nu} \sum_{n \in \mathbf{Z}^\nu} \max_{x \in K} |f(y+x+n)| < \infty \tag{1}$$

form a Banach algebra with $\|.\|$ as norm, contained in $L^1(\mathbf{R}^\nu)$.

We note that

$$\sum_{n \in \mathbf{Z}^\nu} \max_{x \in K} |f(x+n)| \leqslant \|f\| \leqslant 2^\nu \sum_{n \in \mathbf{Z}^\nu} \max_{x \in K} |f(x+n)|. \tag{*}$$

If $\|f\| < \infty$, then the series in (1) is a continuous function of y, and since it is obviously periodic with respect to $\mathbf{Z}^\nu$, we may replace in (1) '$\sup_{y \in \mathbf{R}^\nu}$' by '$\max_{y \in K}$'.

We remark that the *right-hand* side of (*) may also be used as a norm for this algebra, but this norm is not translation invariant. The algebra was introduced by Wiener ([142, § 3], [143, § 10]) on the line; he used (essentially) the norm just mentioned.

(iv) The continuous functions $f \in L^1(\mathbf{R}^\nu)$ such that the Fourier transforms $\hat{f}$ also belong to $L^1(\mathbf{R}^\nu)$ form a Banach algebra under the norm $\|f\| = \|f\|_1 + \|\hat{f}\|_1$. Similarly we have for fixed p, $1 < p < \infty$, the algebra of (equivalence classes of) functions $f \in L^1(\mathbf{R}^\nu)$ such that $\hat{f} \in L^p(\mathbf{R}^\nu)$, with the norm $\|f\| = \|f\|_1 + \|\hat{f}\|_p$.

(v) The functions $f \in L^1(\mathbf{R})$ which are absolutely continuous and such that f' is integrable over $\mathbf{R}$ form a Banach algebra under the norm $\|f\| = \|f\|_1 + \|f'\|_1$.

The methods used in §§ 1–3 for $L^1(\mathbf{R}^\nu)$ are also applicable to these examples.

For each subalgebra the operators L_a and M_ρ still apply. L_a is also isometric in the new norm, but M_ρ is not: instead of 1.1 (7) we note that

$$\|M_{1/\rho} f\| \leqslant \|f\| \qquad \text{for } \rho \geqslant 1.$$

In example (iii) we let ρ be an integer.

The propositions 2.1–2.4 continue to hold for the algebras above and the corresponding norms. The analogue of 2.2 (*) can be verified directly in each case. In 2.4 (**) the equality sign is replaced by '$\geqslant$'.

The analogues of the results in § 3 hold. Their proofs are the same, with an obvious interpretation of the norm $\|\hat{f}\|$. The function τ_1 of 1.3 (18) belongs to each of the subalgebras above if $\nu = 1$; likewise for $\nu > 1$ (cf. § 1.3 (vi)). It is to be observed that in the case of example (v) the relation 1.4 (21) is replaced by $\|L_{t_0}\hat{f}\| \leqslant c_{t_0}\|\hat{f}\|$, where c_{t_0} is independent of $\hat{f}$; this entails a slight change in the proof of 3.2 for example (v).

The results in 4.1–4.4 *hold unchanged, with the same proofs, for the algebras* (i)–(v) *above.*

The norm $\|.\|_1$ is simply replaced by $\|.\|$. The estimate 4.2 (*) can again be verified in each case.

As to the analogue of 4.5, this is of interest mainly for example (iii). For an explicit formulation see [142, Theorem IX] or [143, § 10, Theorem 5]; the proof is as in 4.5.

The examples above appear, at first sight, to be rather different from one another. Actually they have the same 'abstract' structure; this will be discussed in Chapter 6, § 2.

6. The algebras $L^1_\alpha(\mathbf{R}^\nu)$; Beurling algebras

6.1. Let α be a positive number (or zero) and consider the (equivalence classes of) functions $f \in L^1(\mathbf{R}^\nu)$ such that

(1) $$\|f\|_{1,\alpha} = \int |f(x)|.(1+|x|)^\alpha \, dx < \infty,$$

where $$|x| = \{x_1^2+\ldots+x_\nu^2\}^{\frac{1}{2}} \qquad x \in \mathbf{R}^\nu.$$

We shall show that these functions form a Banach algebra under convolution, with (1) as norm; we denote this algebra by $L^1_\alpha(\mathbf{R}^\nu)$.

It will be convenient to consider the matter from a more general viewpoint. Put

(2) $$w(x) = (1+|x|)^\alpha \qquad x \in \mathbf{R}^\nu.$$

Then w has the following properties:

(i) $1 \leqslant w(x) < \infty$ for all $x \in \mathbf{R}^\nu$;

(ii) $w(x+y) \leqslant w(x)w(y)$ for all $x, y \in \mathbf{R}^\nu$;

(iii) w is measurable.

Any function w satisfying (i)–(iii) is said to be a *weight function.*

A weight function w *defines a subalgebra* $L^1_w(\mathbf{R}^\nu)$ *of* $L^1(\mathbf{R}^\nu)$ *which is a Banach algebra under the norm*

(3) $$\|f\|_{1,w} = \int |f(x)|\, w(x)\, dx.$$

$L^1_w(\mathbf{R}^\nu)$ consists, by definition, of all $f \in L^1(\mathbf{R}^\nu)$ such that (3) is finite.

$L^1_w(\mathbf{R}^\nu)$ is clearly a Banach space. Also, if $f, g \in L^1_w(\mathbf{R}^\nu)$, then by (ii)

$$\int \left| \int f(y)g(x-y)\, dy \right| w(x)\, dx \leqslant \int \left\{ \int |g(x-y)|\, w(x-y)\, dx \right\} |f(y)|\, w(y)\, dy,$$

hence $f \star g \in L^1_w(\mathbf{R}^\nu)$ and $\|f \star g\|_{1,w} \leqslant \|f\|_{1,w} \cdot \|g\|_{1,w}$.

These algebras are termed *Beurling algebras*: they were introduced by Beurling in [7, § 2].

We now state two further conditions for w which obviously hold for (2).

(iv) $w(x/\rho) \leqslant w(x)$ for all $\rho \geqslant 1$, $x \in \mathbf{R}^\nu$.

REMARK 1. From (i), (ii), (iv) it follows readily that *w is bounded on compact sets.*†

(v) There are constants $C \geqslant 1$ and $\alpha > 0$ such that $w(\rho x) \leqslant C\rho^\alpha w(x)$ for all $\rho \geqslant 1$, $x \in \mathbf{R}^\nu$.

REMARK 2. The condition (v) implies: $w(x) \leqslant C_1 |x|^\alpha$ for $|x| \geqslant 1$, where $C_1 = C \sup_{|x|=1} w(x)$.

We observe that $L^1_w(\mathbf{R}^\nu)$ contains $\mathscr{K}(\mathbf{R}^\nu)$, since w is bounded on compact sets. Moreover, *$\mathscr{K}(\mathbf{R}^\nu)$ is dense in $L^1_w(\mathbf{R}^\nu)$* (cf. also § 1.1).

Proof. Given $f \in L^1_w(\mathbf{R}^\nu)$ and $\epsilon > 0$, take $k_1 \in \mathscr{K}(\mathbf{R}^\nu)$ such that $\int |fw - k_1| < \frac{1}{2}\epsilon$. Let K_1 be a compact set containing $\operatorname{Supp} k_1$ in its interior. Then $w(x) \leqslant A$, say, if $x \in K_1$. There is a $k \in \mathscr{K}(\mathbf{R}^\nu)$ such that $\operatorname{Supp} k \subset K_1$ and $\int |(k_1/w) - k| < \epsilon/(2A)$. Then $\int |f - k| w < \epsilon$. For *continuous* w the proof is of course quicker.

We consider now some other examples that satisfy the conditions stated.

Example 1. Let δ be any measurable function on $\mathbf{R}^\nu$ such that $\delta(x) \geqslant 0$, $\delta(x+y) \leqslant \delta(x) + \delta(y)$, $\delta(x/\rho) \leqslant \delta(x)$ $(x, y \in \mathbf{R}^\nu, \rho \geqslant 1)$. Then we can put $w(x) = \{1 + \delta(x)\}^\alpha$ for any $\alpha \geqslant 0$ ((i)–(iv) obviously hold; also, if n_ρ is the smallest integer $\geqslant \rho$, then

$$\delta(\rho x) \leqslant \delta(n_\rho x) \leqslant n_\rho \delta(x) \leqslant 2\rho\, \delta(x),$$

hence (v) holds). For instance, we can take $\delta(x) = f_1(|x|)$, where $u \to f_1(u)$, $0 \leqslant u < \infty$, is any function such that $f_1(u) \geqslant 0$, $f_1(u+v) \leqslant f_1(u) + f_1(v)$ for all $u, v \geqslant 0$, and $f_1(u_1) \leqslant f_1(u_2)$ if $0 \leqslant u_1 \leqslant u_2$. Examples of such functions f_1 can be obtained as follows. Let A be an additive semi-group of positive real numbers (i.e. $\lambda_1, \lambda_2 \in A$ implies $\lambda_1 + \lambda_2 \in A$); let ϕ be any real, positive function on A such that $\phi(\lambda_1 + \lambda_2) \leqslant \phi(\lambda_1) + \phi(\lambda_2)$ for all $\lambda_1, \lambda_2 \in A$ and put for $u \geqslant 0$ $f_1(u) = \inf_{\lambda \in A, \lambda > u} \phi(\lambda)$. Then f_1 clearly has the required properties; we note that f_1 is upper semi-continuous and hence so is the corresponding weight function w. In particular, let (ξ_ι) be a

† Actually this is already a consequence of the conditions (i)–(iii): see [75, §§ 7.6.5 and 7.13.1].

finite or infinite set of positive real numbers linearly independent over the rationals, and let A consist of all linear combinations $\lambda = \sum_\iota n_\iota \xi_\iota$, where the coefficients n_ι are integers, $n_\iota \geqslant 0$ for all ι, and $n_\iota > 0$ only for finitely many ι; then we can put $\phi(\lambda) = \sum_\iota n_\iota$.

Example 2. Let δ_1 be any function on $\mathbf{R}$ such that $\delta_1(u) \geqslant 0$, $\delta_1(u+v) \leqslant \delta_1(u)+\delta_1(v)$, $\delta_1(u/\rho) \leqslant \delta_1(u)$ $(u, v \in \mathbf{R}, \rho \geqslant 1)$. For instance, take $\delta_1(u) = \frac{1}{2}(u+|u|)$ or let, more generally, $\delta_1(u) = 0$ for $u < 0$, $\delta_1(u) = f_1(u)$ for $u \geqslant 0$, where f_1 is as in Example 1. We can put for any $\alpha \geqslant 0$

$$w(x) = \{1+\delta_1(x)\}^\alpha \qquad x \in \mathbf{R}.$$

This extends readily to $\mathbf{R}^\nu$: if $w_1,..., w_\nu$ are weight functions on $\mathbf{R}$, then

$$w(x) = \prod_{1\leqslant j\leqslant \nu} w_j(x_j) \qquad x = (x_1,..., x_\nu) \in \mathbf{R}^\nu$$

is a weight function on $\mathbf{R}^\nu$.

A weight function w may be bounded, even if it is not constant, e.g. $w(x) = 1+|x|/(1+|x|)$; then, of course, $L^1_w(\mathbf{R}^\nu)$ is essentially the same as $L^1(\mathbf{R}^\nu)$.

6.2. We are now going to show that *the methods and results in* §§ 1–4 *can be extended to the algebras* $L^1_w(\mathbf{R}^\nu)$ with only minor modifications. Here w is any weight function with the properties 6.1 (i–v). The algebras $L^1_\alpha(\mathbf{R}^\nu)$ corresponding to the weight functions (2) will be of particular interest later, but it will also be instructive to consider other examples.

6.3. The *translation operator* L_a is still defined in $L^1_w(\mathbf{R}^\nu)$, but is not isometric (unless w is constant): we only have, by 6.1 (ii),

$$(4) \qquad \|L_a f\|_{1,w} \leqslant w(a)\|f\|_{1,w} \qquad a \in \mathbf{R}^\nu.$$

The analogue of 2.1 continues to hold: given $f \in L^1_w(\mathbf{R}^\nu)$, we have

$$(5) \qquad \|L_y f - f\|_{1,w} < \epsilon \qquad \text{for } y \in U_\epsilon.$$

The proof is by reduction to $\mathscr{K}(\mathbf{R}^\nu)$ (cf. § 2.1). Let V be a compact nd. of 0 in $\mathbf{R}^\nu$: then $w(x) \leqslant A$, say, for $x \in V$ (§ 6.1, Remark 1). Given $f \in L^1_w(\mathbf{R}^\nu)$ and $\epsilon > 0$, take $k \in \mathscr{K}(\mathbf{R}^\nu)$ such that $\|f-k\|_{1,w} < \epsilon/\{2(A+1)\}$. There is a nd. U_ϵ of 0 contained in V such that $\|L_y k - k\|_{1,w} < \frac{1}{2}\epsilon$ for $y \in U_\epsilon$. Then

$$\begin{aligned}\|L_y f - f\|_{1,w} &\leqslant \|L_y f - L_y k\|_{1,w} + \|L_y k - k\|_{1,w} + \|k-f\|_{1,w} \\ &< w(y)\epsilon/\{2(A+1)\} + \tfrac{1}{2}\epsilon + \epsilon/\{2(A+1)\} \\ &\leqslant \epsilon \qquad \text{for } y \in U_\epsilon.\end{aligned}$$

6.4. The multiplication operator M_ρ (§ 1.1 (5)) is also defined in $L^1_w(\mathbf{R}^\nu)$, for all $\rho > 0$: we have by 6.1 (iv)

$$(6) \qquad \|M_\rho f\|_{1,w} \leqslant \|f\|_{1,w} \qquad \rho \geqslant 1$$

and by 6.1 (v)

$$(7) \qquad \|M_{1/\rho}f\|_{1,w} \leqslant C_\rho \|f\|_{1,w} \qquad \rho \geqslant 1,$$

where C_ρ is independent of $f \in L^1_w(\mathbf{R}^\nu)$.

REMARK. In the case of the weight function (2) we have:

$$\|M_{1/\rho}f\|_{1,w} \to \infty \qquad \rho \to \infty,$$

unless $f = 0$. This happens also for more general weight functions.

The limit relation (2) in 2.2 still holds in $L^1_w(\mathbf{R}^\nu)$: for $h, f \in L^1_w(\mathbf{R}^\nu)$

$$(8) \qquad \left\|(M_\rho h) \star f - \left[\int h(x)\, dx\right] f\right\|_{1,w} \to 0 \qquad \rho \to \infty.$$

The proof in 2.2 remains valid if $\|\cdot\|_1$ is replaced by $\|\cdot\|_{1,w}$. The function $\phi(x) = \|L_x f - f\|_{1,w}$ is still continuous and $0 \leqslant \phi(x) \leqslant Cw(x)$ (cf. (4), (5)). The limit for $\rho \to \infty$ is the same as in 2.2, since $\phi(x/\rho) \leqslant Cw(x/\rho) \leqslant Cw(x)$ for $\rho \geqslant 1$ and $\int |h(x)|\, w(x)\, dx < \infty$.

From (8) it follows at once that Proposition 2.3 holds also for $L^1_w(\mathbf{R}^\nu)$; in particular, $L^1_w(\mathbf{R}^\nu)$ possesses approximate units.

Observe that $L^1_w(\mathbf{R}^\nu)$ does contain functions v such that $\hat{v}$ has compact support: this follows from 6.1, Remark 2, and 1.3 (v) and (vi).

The proof in 2.4, however, breaks down in the case of $L^1_w(\mathbf{R}^\nu)$: compare, in this respect, the remark above.

6.5. Let $\mathscr{F}^1_w(\mathbf{R}^\nu)$ be the algebra of Fourier transforms of the functions in $L^1_w(\mathbf{R}^\nu)$; in the case of $L^1_\alpha(\mathbf{R}^\nu)$ (§ 6.1) we simply write $\mathscr{F}^1_\alpha(\mathbf{R}^\nu)$. We define a norm in $\mathscr{F}^1_w(\mathbf{R}^\nu)$ by $\|\hat{f}\| = \|f\|_{1,w}$, $\hat{f}$ being the F.t. of $f \in L^1_w(\mathbf{R}^\nu)$; in this way $\mathscr{F}^1_w(\mathbf{R}^\nu)$ becomes a complete, normed function algebra isomorphic to $L^1_w(\mathbf{R}^\nu)$. This is quite analogous to 1.4.

We now prove that *the analogue of the theorem of Wiener–Lévy* (§ 3.1) *holds for* $\mathscr{F}^1_w(\mathbf{R}^\nu)$.

First we establish the analogue of 3.3. To do this, we have to proceed in a slightly different way: the proof in 3.2 does not carry over to $\mathscr{F}^1_w(\mathbf{R}^\nu)$, since we have no analogue of 2.4 at our disposal (cf. § 6.4).

The new proof is based on (i) the change of variable $t \to \lambda t + t_0$ ($\lambda > 0$, $t_0 \in \mathbf{R}^\nu$) in $\mathbf{R}^\nu$, which transforms functions in $\mathscr{F}^1_w(\mathbf{R}^\nu)$ into functions in $\mathscr{F}^1_w(\mathbf{R}^\nu)$; (ii) the behaviour of the norm in $\mathscr{F}^1_w(\mathbf{R}^\nu)$ under this transformation.

Let $\hat{\tau}$ be a function in $\mathscr{F}^1_w(\mathbf{R}^\nu)$ which is 1 near the origin (cf. § 6.1, Remark 2, and § 1.3 (v) and (vi)).

We shall use the same notation as in 3.3. Consider $\hat{f} \in \mathscr{F}^1_w(\mathbf{R}^\nu)$ and put $z_0 = \hat{f}(t_0)$, where $t_0 \in K$. We have

$$\mathrm{A}(z) = \mathrm{A}(z_0) + \sum_{n \geqslant 1} c_n (z - z_0)^n \quad \text{for } |z - z_0| < \epsilon.$$

Put, for $\rho > 0$, $\qquad \hat{a}_\rho(t) = \{\hat{f}(t/\rho + t_0) - \hat{f}(t_0)\}\hat{\tau}(t).$

Then $\hat{a}_\rho \in \mathscr{F}^1_w(\mathbf{R}^\nu)$ and, if ρ is large, $\|\hat{a}_\rho\| < \epsilon$ (cf. (8)). Thus for large ρ the function

$$\hat{b}_\rho = \mathrm{A}(z_0)\hat{\tau} + \sum_{n\geqslant 1} c_n \hat{a}_\rho{}^n$$

is in $\mathscr{F}^1_w(\mathbf{R}^\nu)$, since the series on the right converges. Now put $\hat{g}_{t_0}(t) = \hat{b}_\rho(\rho(t-t_0))$: then $\hat{g}_{t_0} \in \mathscr{F}^1_w(\mathbf{R}^\nu)$ and $\hat{g}_{t_0}(t) = \mathrm{A}(\hat{f}(t))$ for t near t_0.

Secondly we observe that the proofs in 3.4 and 3.5 carry over without change. The required auxiliary functions are available in $\mathscr{F}^1_w(\mathbf{R}^\nu)$.

Thus the theorem of Wiener–Lévy has been proved also for $\mathscr{F}^1_w(\mathbf{R}^\nu)$, by rather elementary methods.

6.6. We can now show that the analogue of *Wiener's approximation theorem* (§ 4.1) *holds for* $L^1_w(\mathbf{R}^\nu)$, *if* w *satisfies the conditions* 6.1 (i–v). The method of proof is exactly the same as in 4.1. We note that the analogue of 4.2 is true in $L^1_w(\mathbf{R}^\nu)$.

Let $f \in L^1_w(\mathbf{R}^\nu)$ be fixed. Given any $A > 0$ and $\epsilon > 0$, there is a $\delta = \delta_{A,\epsilon} > 0$ such that $\|L_x f - L_y f\|_{1,w} < \epsilon$, whenever $|x| \leqslant A$ and $|x-y| < \delta$: this follows from (4) and (5). The proof in 4.2 then simply carries over.

We also observe that *the closed invariant subspaces of* $L^1_w(\mathbf{R}^\nu)$ *coincide with the closed ideals of* $L^1_w(\mathbf{R}^\nu)$: the proof in 4.3 applies to $L^1_w(\mathbf{R}^\nu)$ almost unchanged.

The more general form of Wiener's theorem in 4.4 *likewise holds for* $L^1_w(\mathbf{R}^\nu)$.

This can be proved in exactly the same way as in 4.4 if the weight function w is symmetric $[w(-x) = w(x)]$, as in (2). But if w is not symmetric (cf., e.g., § 6.1, Example 2), we cannot use the involution $f \to f^*$. The required modification of the proof is contained in the theory expounded in Chapter 2, §§ 1, 2.

6.7. Let w be a function on $\mathbf{R}$ satisfying the conditions 6.1 (i–v). Then we can prove the following analogue of Wiener's general Tauberian theorem (§ 4.5). Let ϕ be a complex-valued, measurable function on $\mathbf{R}$ such that $|\phi(x)| \leqslant Cw(x)$ a.e., for some constant C. Suppose there is an $f \in L^1_w(\mathbf{R})$ such that $\hat{f}(t) \neq 0$ for all $t \in \mathbf{R}$ and

$$\frac{1}{w(y)} \int f(x-y)\phi(x)\,dx \to 0 \qquad y \to \infty.$$

Then for all $g \in L^1_w(\mathbf{R})$

$$\frac{1}{w(y)} \int g(x-y)\phi(x)\,dx \to 0 \qquad y \to \infty.$$

The proof is as in 4.5. A similar result holds for $L^1_w(\mathbf{R}^\nu)$, $\nu \geqslant 1$.

More general results than those in 6.6 and 6.7 were proved by Beurling [7, Theorems V B and VI]. We shall consider other generalizations in Chapter 3, § 7 and Chapter 6, § 3.

2

FUNCTION ALGEBRAS AND THE GENERALIZATION OF WIENER'S THEOREM

1. Function algebras

1.1. HERE we shall discuss, in considerable generality, algebras of functions and their ideals. The concrete basis for our study is the algebra $\mathscr{F}^1(\mathbf{R}^\nu)$ (Chap. 1, § 1.4): we want to develop methods for extending Wiener's theorem to groups and for generalizing it. The abstract approach will make the proofs simpler and more lucid.

Let X be a locally compact (l.c.) Hausdorff space.† Let $\mathscr{A}(X)$ be an algebra of complex-valued continuous functions on X with the ordinary pointwise algebraic operations.‡

$\mathscr{A}(X)$ need not contain any constant functions besides 0 and thus may not have a unit element. This happens precisely in the case of $\mathscr{F}^1(\mathbf{R}^\nu)$ (cf. Chap. 1, § 1.2 (i)).

We introduce now a concept that forms the basis of the further developments.

DEFINITION. $\mathscr{A}(X)$ is a *standard function algebra* (standard algebra for short) if it has the following 'standard properties':

(i) If $f \in \mathscr{A}(X)$ and $f(a) \neq 0$ at a point $a \in X$, then there is a $g \in \mathscr{A}(X)$ such that $g(x) = 1/f(x)$ for all x in some neighbourhood of a.

(ii) For any closed set $E \subset X$ and any point $a \in X-E$ there is an $f \in \mathscr{A}(X)$ vanishing on E and such that $f(a) \neq 0$.

The first condition is of an algebraic nature: $\mathscr{A}(X)$ comes as near as possible to being a field. The second condition is topological: there are enough functions in $\mathscr{A}(X)$ to separate the points and closed sets of X.

$\mathscr{F}^1(\mathbf{R}^\nu)$ is a standard algebra: (i) is, essentially, Wiener's part of the theorem of Wiener–Lévy (cf. Chap. 1, §§ 3.1 and 3.6) and (ii) is clear by Chapter 1, § 1.3 (ii), (iv), (vi).

The properties (i) and (ii) imply at once:

(iii) For any $a \in X$ and any neighbourhood U_a of a there is a function

† We shall write 'l.c. space', for shortness. In the Russian literature the term 'locally bicompact' is used.

‡ We denote functions in $\mathscr{A}(X)$ by italic or Greek letters. In the case of $\mathscr{F}^1(\mathbf{R}^\nu)$ a circumflex should of course be added.

$\tau_a \in \mathscr{A}(X)$ which is 1 near a, but 0 outside U_a (or, more strongly: $\operatorname{Supp} \tau_a \subset U_a$).

1.2. Let I be an ideal in a function algebra $\mathscr{A}(X)$ and F *any* (complex-valued) function on X. We say that F *belongs locally to I at the point* $a \in X$ if there is a function $f_a \in I$ coinciding with F near a. If X is not compact, we say that F *belongs locally to I at infinity* if there is an $f \in I$ coinciding with F outside some compact set.

1.3. Let $\mathscr{A}(X)$ be a standard algebra or, more generally, a function algebra with the property 1.1 (iii). Let I be an ideal of $\mathscr{A}(X)$. Suppose F is a function on X belonging locally to I at every point of X and also at infinity, if X is not compact. Then F is contained in I.

This is called the *localization lemma*; it is very useful for the study of ideals in standard function algebras.

Proof. There is an $f_0 \in I$ coinciding with F outside some compact set $K \subset X$ (if X is compact, let $f_0 = 0$ and $K = X$). For every $a \in K$ there is an $f_a \in I$ coinciding with F in a nd. U_a. There is also a $\tau_a \in \mathscr{A}(X)$ vanishing outside U_a and equal to 1 near a, say in an open nd. V_a. Since K is compact, there are finitely many points, say $(a_n)_{1 \leqslant n \leqslant N}$, of K such that the corresponding nds. $V_n = V_{a_n}$ cover K. Let f_n, τ_n stand for f_{a_n}, τ_{a_n}, respectively. We define now $e_1 = \tau_1$ and (if $N \geqslant 2$) $e_2 = \tau_2 . (1-\tau_1), \ldots, e_N = \tau_N . (1-\tau_1) . \ldots . (1-\tau_{N-1})$. Then $e_n \in \mathscr{A}(X)$, $1 \leqslant n \leqslant N$; also (by induction) $\sum_n e_n = 1 - \prod_n (1-\tau_n)$, so that $\sum_n e_n$ is 1 on K. Consider the function $h = \left(1 - \sum_n e_n\right) f_0 + \sum_n e_n f_n$ which belongs to I. Here $e_n f_n = e_n F$, since $e_n(x) \neq 0$ implies $f_n(x) = F(x)$; likewise $\left(1 - \sum_n e_n\right) f_0 = \left(1 - \sum_n e_n\right) F$. Hence $h = F$ and F is in I. This proof is, of course, just a combination and general formulation of those in Chapter 1, §§ 3.4 and 3.5.

Remark. The method of proof above can also be used to show that a standard algebra $\mathscr{A}(X)$—or any function algebra satisfying 1.1 (iii)—contains for any given compact set $K \subset X$ and any open nd. U_K of K a function that is 1 on K and has its support in U_K. More generally, if $(U_n)_{1 \leqslant n \leqslant N}$ is any finite open covering of K, there exist N functions $e_n \in \mathscr{A}(X)$ such that $\sum_n e_n$ is 1 on K and $\operatorname{Supp} e_n \subset U_n$, $1 \leqslant n \leqslant N$.

Corollary. If $\mathscr{A}(X)$ is a standard function algebra and $f \in \mathscr{A}(X)$ does not take the value zero on the compact set $K \subset X$, then there is a $g \in \mathscr{A}(X)$ such that $g(x) = 1/f(x)$ for all $x \in K$.

We have $f(x) \neq 0$ for $x \in K'$, where K' is some compact set containing K in its interior. There is an $f_1 \in \mathscr{A}(X)$ which is 1 on K and 0 outside K' (cf. the remark above). Define g by $g(x) = f_1(x)/f(x)$ if $f(x) \neq 0$, $g(x) = 0$ if $f(x) = 0$. Then g belongs locally to $\mathscr{A}(X)$ at every point of X and at infinity, and we can apply the localization lemma with $I = \mathscr{A}(X)$.

1.4. Let I be an ideal in a standard algebra $\mathscr{A}(X)$. The set of points of X where all functions in I vanish will be called the *cospectrum* of I and denoted $\operatorname{cosp} I$. We also write $\operatorname{cosp} f$ for the set of zeros of a function $f \in \mathscr{A}(X)$. The cospectrum is a closed set, possibly empty. It is of considerable importance in the theory of ideals; the name stems from harmonic analysis (cf. Chap. 6, § 1.3, footnote).

(i) If I is an ideal in a standard algebra $\mathscr{A}(X)$, then a function $f \in \mathscr{A}(X)$ belongs locally to I at every point of X outside $\operatorname{cosp} I$.

If $a \in X - \operatorname{cosp} I$, then there is an $h \in I$ with $h(a) \neq 0$. By 1.1 (i) there is an $h_1 \in \mathscr{A}(X)$ such that $h_1(x) = 1/h(x)$ in some nd. U_a. Then $hh_1 f$ is in I and equals f on U_a.

(ii) An ideal I in a standard algebra $\mathscr{A}(X)$ contains every function in $\mathscr{A}(X)$ that has compact support disjoint from $\operatorname{cosp} I$. In particular, if $\operatorname{cosp} I$ is empty, I contains all functions in $\mathscr{A}(X)$ with compact support.

This follows at once from (i) and the localization lemma (§ 1.3), since such a function obviously belongs locally to I at every point of $\operatorname{cosp} I$ and at infinity.

Remark. (i) and (ii) contain, in particular, the following fact: an ideal I in a standard algebra $\mathscr{A}(X)$ is itself a standard algebra if the functions in I are restricted to $X - \operatorname{cosp} I$ (and $\operatorname{cosp} I \neq X$).

The proposition (ii) above is basic in the ideal theory of standard algebras. It can be formulated also as follows.

(iii) Let $\mathscr{A}(X)$ be a standard algebra and let E be a closed set in X. The ideal of all functions in $\mathscr{A}(X)$ with compact supports disjoint from E is the smallest ideal of $\mathscr{A}(X)$ with cospectrum E (i.e. contained in every other ideal with cospectrum E).

1.5. The definition in 1.2 has its roots in P. Lévy's paper [89, § 3] and was given explicitly, together with the localization lemma 1.3, by Ditkin [35, § 4] and Shilov [125, Chap. 1, § 2, Theorem 4] who considered only compact spaces X; the extension to locally compact spaces was carried out by Mackey [95, Theorem 36 and p. 113]. For 1.4 (i, ii, iii) see [125, Chap. 1, § 2, Theorem 5] and [95, Theorem 37].

2. Topological function algebras and Wiener's theorem

2.1. An algebra $\mathscr{A}(X)$ as defined in 1.1 is said to be a *topological function algebra* if it has the following properties:

(i) $\mathscr{A}(X)$ is a topological vector space.

(ii) The product fg is continuous in $f \in \mathscr{A}(X)$ for fixed $g \in \mathscr{A}(X)$.

(iii) For each $a \in X$ the linear functional $f \to f(a)$ on $\mathscr{A}(X)$ is continuous.

From (ii) it follows that the closure of an ideal of $\mathscr{A}(X)$ is again an ideal; (iii) implies: for any subset E of X the set of all $f \in \mathscr{A}(X)$ vanishing on E is a *closed* ideal of $\mathscr{A}(X)$.

For the theory of topological vector spaces and topological algebras see [13 *b*] and [101, § 8].

2.2. $\mathscr{A}(X)$ is said to be a *normed function algebra* if the following holds:

(i) $\mathscr{A}(X)$ is a normed vector space, with norm $\|.\|$.

(ii) $\|fg\| \leqslant \|f\|.\|g\|$ for all $f, g \in \mathscr{A}(X)$.

(iii) For each $a \in X$ the linear functional $f \to f(a)$ is continuous.

The condition (iii) means, in the present case, that for each $a \in X$ there is a constant C_a such that $|f(a)| \leqslant C_a\|f\|$ for all $f \in \mathscr{A}(X)$. This actually implies

$$(1) \qquad \sup_{x \in X}|f(x)| \leqslant \|f\| \qquad \text{for all } f \in \mathscr{A}(X).$$

Suppose $|f(a)| > \|f\|$ for some $f \in \mathscr{A}(X)$ and some $a \in X$. Put $g = f/f(a)$: then $\|g\| < 1$ and $1 = |g(a)^n| \leqslant C_a\|g^n\| \leqslant C_a\|g\|^n \to 0$ $(n \to \infty)$, a contradiction.

Any normed function algebra can be extended to a complete normed function algebra by the usual process of completion (note that a Cauchy sequence in $\mathscr{A}(X)$ relative to the norm $\|.\|$ converges *uniformly* on X, by (1), so that the completion yields continuous functions).

If we assume to begin with in (i) above that $\mathscr{A}(X)$ is complete with respect to the given norm, then (i) and (ii) already imply (1) and in particular (iii).

Suppose again $|f(a)| > \|f\|$ for some $f \in \mathscr{A}(X)$ and some $a \in X$. Let $g = f/f(a)$: then $h = \sum_{n \geqslant 1} g^n$ is in $\mathscr{A}(X)$ and satisfies $g+gh = h$, whence the contradiction $1+h(a) = h(a)$.

2.3. It is clear what is to be understood by a *topological standard algebra* and a *normed standard algebra* (cf. § 1.1). They will form the object of our study.

In topological standard algebras it is natural to investigate the *closed* ideals. We obtain at once from 1.4 (iii):

(i) Let $\mathscr{A}(X)$ be a topological standard algebra. Let E be a closed set in X and let J_E be the closure of the ideal of all functions with compact supports disjoint from E. Then J_E is the smallest closed ideal of $\mathscr{A}(X)$ with cospectrum E, that is, J_E is contained in every closed ideal $I \subset \mathscr{A}(X)$ such that $\operatorname{cosp} I = E$.

In particular we have for $E = \emptyset$:

(ii) In a topological standard algebra $\mathscr{A}(X)$ the closure of the ideal of all functions in $\mathscr{A}(X)$ with compact supports is the smallest closed ideal with empty cospectrum: it is contained in every closed ideal of $\mathscr{A}(X)$ with empty cospectrum.

2.4. The last result is of particular interest if $\mathscr{A}(X)$ has the following '*density property*': the functions with compact supports are dense in $\mathscr{A}(X)$. A topological standard algebra with this property will be called a *Wiener algebra*. $\mathscr{F}^1(\mathbf{R}^\nu)$ is a Wiener algebra (cf. Chap. 1, § 4.1 (ii)).

From 2.3 (ii) we now obtain that *in a Wiener algebra $\mathscr{A}(X)$ a closed ideal with empty cospectrum coincides with $\mathscr{A}(X)$.* This is the abstract form of Wiener's theorem.

References. [95, Theorem 38], [101, § 15, No. 5]; the treatment there is based on the theory of Banach algebras.

For topological standard algebras which are not Wiener algebras Wiener's theorem obviously does not hold.

2.5. Now the following problems arise. How far is a closed ideal in a Wiener algebra—and in particular in $\mathscr{F}^1(\mathbf{R}^\nu)$—determined by its cospectrum if the cospectrum is not empty? What closed ideals are there in a given Wiener algebra? These problems have been the object of much research in recent years. They are far from being solved, especially in the case of $\mathscr{F}^1(\mathbf{R}^\nu)$. An account of some of the results obtained will be given in this chapter and in Chapters 6 and 7.

3. Examples of topological function algebras

3.1. As a first illustration of the preceding theory we consider the following examples of complete normed Wiener algebras.

(i) The algebra $\mathscr{C}^0(X)$ of all complex-valued continuous functions f on a locally compact space X that '*vanish at infinity*',† with the norm $\|f\|_\infty = \max_{x\in X}|f(x)|$. For non-compact X, $\mathscr{C}^0(X)$ is the completion of $\mathscr{K}(X)$, the algebra of continuous functions on X with compact supports; if X is compact, $\mathscr{C}^0(X) = \mathscr{K}(X)$.

The algebra $\mathscr{K}(X)$ satisfies the condition 1.1 (iii), even in the following somewhat stronger form: given any point $a \in X$ and a nd. U_a, there is an $f \in \mathscr{K}(X)$ such that f is real, $0 \leqslant f(x) \leqslant 1$ for all $x \in X$, f is 1 near a and $\operatorname{Supp} f \subset U_a$. Clearly $\mathscr{K}(X)$ is even a standard algebra and so is $\mathscr{C}^0(X)$.

REMARK. The method in 1.3, combined with the special property above, can be used to obtain the familiar result: if K is any compact set in X and $(U_n)_{1\leqslant n\leqslant N}$

† A continuous function f on X is said to vanish at infinity if for every $\epsilon > 0$ the set $\{x \mid x \in X, |f(x)| \geqslant \epsilon\}$ is compact.

a finite open covering of K, then there are N functions $e_n \in \mathscr{K}(X)$ such that e_n is real, $0 \leqslant e_n(x) \leqslant 1$ for all $x \in X$, $\operatorname{Supp} e_n \subset U_n$ $(1 \leqslant n \leqslant N)$ and $\sum_n e_n$ is 1 on K.

(ii) The algebra of all functions $f \in \mathscr{C}^0(X)$ such that $f \in L^p(X, \mu)$ (p fixed, $1 \leqslant p < \infty$, μ a given positive measure on X), with the norm $\|f\| = \|f\|_\infty + \|f\|_p$ (e.g. $X = \mathbf{R}^\nu$ and $d\mu(x) = dx$).

(iii) The algebra formed by the functions $f \in \mathscr{C}^0(\mathbf{R})$ which are absolutely continuous† and such that f' is pth-power integrable over $\mathbf{R}$, with the norm $\|f\| = \|f\|_\infty + \|f'\|_p$ (p fixed, $1 \leqslant p < \infty$).

(iv) The algebra of all functions $f \in \mathscr{C}^0(\mathbf{R})$ which are absolutely continuous and such that $f \in L^1(\mathbf{R})$ and the derivative f' is pth-power integrable over $\mathbf{R}$ (p fixed, $1 \leqslant p < \infty$), with the norm

$$\|f\| = \|f\|_\infty + \|f\|_1 + \|f'\|_p.$$

In each of the algebras (i)–(iv) a closed ideal I is completely characterized by its cospectrum: it consists of *all* functions in the algebra vanishing on $\operatorname{cosp} I$.

(i) Suppose $f \in \mathscr{C}^0(X)$ vanishes on $\operatorname{cosp} I$. Let $T_\epsilon(t)$ be 1 for $0 \leqslant t \leqslant \frac{1}{2}\epsilon$, 0 for $t \geqslant \epsilon$, linear for $\frac{1}{2}\epsilon \leqslant t \leqslant \epsilon$ $(\epsilon > 0)$, and put $u_\epsilon(x) = 1 - T_\epsilon(|f(x)|)$, $x \in X$. Then u_ϵ is in $\mathscr{K}(X)$ and vanishes near $\operatorname{cosp} I$, hence (§ 1.4 (ii)) $u_\epsilon \in I$. Since $\|f - fu_\epsilon\|_\infty \leqslant \epsilon$ and the ideal I is closed, $f \in I$.

(ii) The proof is as in (i): since $|f(x)|\{1 - u_\epsilon(x)\} \leqslant |f(x)|$ and

$$|f(x)|\{1 - u_\epsilon(x)\} \to 0 \quad (\epsilon \downarrow 0)$$

for each x (even uniformly), it follows that $\|f - fu_\epsilon\| \to 0$ $(\epsilon \downarrow 0)$.

(iii) Suppose a function f in this algebra vanishes on $\operatorname{cosp} I$. Let $f = f_1 + if_2$ with f_j real $(j = 1, 2)$ and define $f_{j,n}(x) = f_j(x)$ if $|f_j(x)| \leqslant 1/n$, $f_{j,n}(x) = 1/n$ if $f_j(x) \geqslant 1/n$, $f_{j,n}(x) = -1/n$ if $f_j(x) \leqslant -1/n$ $(n = 1, 2, \ldots)$. Put $f_n = f_{1,n} + if_{2,n}$; then f_n is in the algebra and $f - f_n$ has compact support disjoint from $\operatorname{cosp} I$. Hence (§ 1.4 (ii)) $f - f_n \in I$. Moreover, $\|f_n\| \to 0$ $(n \to \infty)$ (note that $\|f'_n\|_p \to 0$ $(n \to \infty)$, since (as is readily seen) $f'_n(x) \to 0$ a.e. and $|f'_n(x)| \leqslant |f'(x)|$ a.e.).

(iv) The proof is analogous to that of (iii).

References. (i) [129, Theorem 11], [125, Chap. I, § 1, Theorem 1]; (iii) [125, Chap. I, § 2, Theorem 7].

3.2. Let X be a locally compact, but not compact, space. Let $\mathscr{C}(X)$ be the algebra of all complex-valued continuous functions on X, with the topology of uniform convergence on compact sets (i.e. the topology determined by the family of semi-norms $f \to \max_{x \in K} |f(x)|$, with K ranging over all compact sets in X). In this algebra a closed ideal is again competely determined by its cospectrum.

† More precisely: absolutely continuous in every bounded interval. This implies that f' exists a.e. Compare also [133, §§ 11.7 and 11.71].

For any compact set $K \subset X$ there is a $\tau_K \in \mathscr{C}(X)$ which is 1 on K and has compact support. If f vanishes on cosp I, then $f\tau_K \in I$: this is shown as for the algebra 3.1 (i). Hence $f \in I$, since $f\tau_K$ is arbitrarily near to f.

3.3. We discuss now some Wiener algebras that are subalgebras of $\mathscr{F}^1(\mathbf{R}^\nu)$.

(i) For $1 \leqslant p \leqslant \infty$ let $\mathscr{F}_{(p)}$ consist of all $\hat{f} \in \mathscr{F}^1(\mathbf{R})$ such that $\hat{f}$ is absolutely continuous† and the derivative $\hat{f}'$ is pth-power integrable respectively essentially bounded if $p = \infty$, in every *finite* interval. The topology of $\mathscr{F}_{(p)}$ is defined by the sequence of norms

$$\|f\|_1 + \left\{\int_{-n}^{n} |\hat{f}'(t)|^p\,dt\right\}^{1/p} \text{ respectively } \|f\|_1 + \operatorname*{ess.sup}_{|t| \leqslant n} |\hat{f}'(t)|,\ n = 1, 2,...,$$

$\hat{f}$ being the F.t. of $f \in L^1(\mathbf{R})$.

(ii) The algebra formed by the Fourier transforms of all $f \in L^1(\mathbf{R}^\nu)$ such that $f \in L^p(\mathbf{R}^\nu)$ for *all* p, $1 < p < \infty$, with the topology induced by the family of norms $\|f\|_1 + \|f\|_p$, $1 < p < \infty$.

(iii) The algebra of Fourier transforms of all $f \in L^1(\mathbf{R})$ such that f is indefinitely differentiable and $f^{(n)} \in L^1(\mathbf{R})$ for all $n \geqslant 1$, with the topology defined by the norms $\|f\|_1 + \|f^{(n)}\|_1$.

These three examples are intersections of normed algebras (cf. also Chap. 1, § 5 (ii) and (v)). (i) and (iii) are countably normed spaces in the sense of I. M. Gelfand and G. E. Shilov [52]. It can be shown that in (ii) it also suffices to take $p = 2, 3,...$.

The algebras (ii) and (iii) are actually ideals of $\mathscr{F}^1(\mathbf{R}^\nu)$ respectively $\mathscr{F}^1(\mathbf{R})$, but (i) is not.‡

The examples (i), (ii), (iii) *are Wiener algebras.*

(i) $\mathscr{F}_{(p)}$ contains all absolutely continuous $\hat{f} \in \mathscr{F}^1(\mathbf{R})$ with compact supports and such that $\hat{f}' \in L^p(\mathbf{R})$. It is an exercise in analysis to show that the algebras (ii) and (iii) contain all $\hat{f}$ in $\mathscr{F}^1(\mathbf{R}^\nu)$ respectively $\mathscr{F}^1(\mathbf{R})$ with compact supports. It follows that (i), (ii), and (iii) are topological standard algebras. To prove that they are Wiener algebras we proceed essentially as in Chapter 1, §§ 2.2 and 2.3. Let τ_1 be defined as in Chapter 1, § 1.3 (iii) and put $\tau_\rho = M_\rho \tau_1$ for $\rho > 0$, so that $\hat{\tau}_\rho(t) = \hat{\tau}_1(t/\rho)$, $t \in \mathbf{R}$. Consider any $\hat{f} \in \mathscr{F}_{(p)}$. We have $\|f - f \star \tau_\rho\|_1 \to 0$ $(\rho \to \infty)$ by Chapter 1, § 2.2. Also, for each $n = 1, 2,...$, $\hat{\tau}_\rho(t) = 1$ for $|t| \leqslant n$ if $\rho \geqslant n$, and the desired property follows at once for $\mathscr{F}_{(p)}$. The proof for (ii) and (iii) is analogous (cf. also the examples in Chap. 1, § 5 (ii) and (v)).

† Cf. n. †, p. 23.

‡ For the last part we observe that one can easily construct an $\hat{f} \in \mathscr{F}^1(\mathbf{R})$ coinciding in [0, 1] with Weierstrass's non-differentiable function [133, § 11.22] and $\mathscr{F}_{(p)}$ contains, of course, functions that are 1 on [0, 1].

3.4. Finally we consider some subalgebras of $\mathscr{F}^1(\mathbf{R})$ which are themselves Banach algebras.

(i) The algebra $\mathscr{F}_p$, $1 \leqslant p \leqslant \infty$, formed by all $\hat{f} \in \mathscr{F}^1(\mathbf{R})$ which are absolutely continuous and such that $\hat{f}'$ is pth-power integrable (essentially bounded if $p = \infty$) over $\mathbf{R}$, with the norm $\|\hat{f}\| = \|f\|_1 + \|\hat{f}'\|_p$, $\hat{f}$ being the F.t. of $f \in L^1(\mathbf{R})$.

(ii) The algebra $\mathscr{F}_{1,p}$, $1 \leqslant p \leqslant \infty$, consisting of all absolutely continuous $\hat{f} \in \mathscr{F}^1(\mathbf{R})$ such that $\hat{f}$ is integrable and $\hat{f}'$ is pth-power integrable (essentially bounded, if $p = \infty$) over $\mathbf{R}$, with the norm

$$\|\hat{f}\| = \|f\|_1 + \|\hat{f}\|_1 + \|\hat{f}'\|_p.$$

$\mathscr{F}_p$ and $\mathscr{F}_{1,p}$ are topological standard algebras for all p. They are Wiener algebras for $p < \infty$, but not for $p = \infty$.

$\mathscr{F}_p$ and $\mathscr{F}_{1,p}$ contain all absolutely continuous $\hat{f} \in \mathscr{F}^1(\mathbf{R})$ with compact supports and such that $\hat{f}'$ is pth-power integrable (respectively essentially bounded, for $p = \infty$) over $\mathbf{R}$; hence they are topological standard algebras. Moreover, the functions with compact supports are dense if $1 \leqslant p < \infty$. This is proved essentially as in 3.3: let $\hat{\tau}_\rho$ be as stated there, then for $p < \infty$

$$\|\hat{f}' - (\hat{f}\hat{\tau}_\rho)'\|_p \leqslant \|\hat{f}'.(1-\hat{\tau}_\rho)\|_p + 2 \max_{\rho \leqslant |t| \leqslant 2\rho} |\hat{f}(t)| \to 0 \quad (\rho \to \infty)$$

and the assertion follows readily for both $\mathscr{F}_p$ and $\mathscr{F}_{1,p}$. Now let

$$\hat{f}(t) = \sum_{n \geqslant 1} n^{-2}\Delta_1(n^2[t-n^2]),$$

with Δ_1 as in Chapter 1, § 1.3 (ii). Then $\hat{f}$ is in $\mathscr{F}_\infty$ and in $\mathscr{F}_{1,\infty}$: thus the functions with compact supports are *not* dense in these algebras.

Let L^1_p and $L^1_{1,p}$ be the subalgebras of $L^1(\mathbf{R})$ isomorphic to $\mathscr{F}_p$ and $\mathscr{F}_{1,p}$ via the Fourier transformation. These algebras are rather different from those considered in Chapter 1, §§ 5 and 6. They are not ideals† of $L^1(\mathbf{R})$. The translation operator is not applicable in L^1_p; in $L^1_{1,p}$ it is applicable, but the norm is not invariant.

Let $c > 0$ and define for $a > 0$: $f_a(x) = x^{a-1}e^{-cx}$ $(x > 0)$, $f_a(x) = 0$ $(x \leqslant 0)$. Then $\hat{f}_a(t) = \Gamma(a)/(c+2\pi i t)^a$, the determination being such that $\hat{f}_a(0) > 0$. Thus f_a is in L^1_p for every $a > 0$, but if $a \leqslant 1/p$, then $L_y f_a \notin L^1_p$ for $y \neq 0$.

3.5. We note here a lemma useful for the construction of examples.

LEMMA. If $\mathscr{A}$ and $\mathscr{B}$ are complete normed function algebras on the same space X, with norms $\|.\|_{\mathscr{A}}$ and $\|.\|_{\mathscr{B}}$, then $\mathscr{A} \cap \mathscr{B}$ is a complete normed function algebra under the norm $\|f\| = \|f\|_{\mathscr{A}} + \|f\|_{\mathscr{B}}$, $f \in \mathscr{A} \cap \mathscr{B}$.

Clearly $\mathscr{A} \cap \mathscr{B}$ is a normed algebra under $\|.\|$. Let $(f_n)_{n \geqslant 1}$ be a Cauchy sequence in $\mathscr{A} \cap \mathscr{B}$: then it is also a Cauchy sequence in each algebra $\mathscr{A}$, $\mathscr{B}$ and hence has a limit in each, say $a \in \mathscr{A}$ and $b \in \mathscr{B}$. But then also $f_n(x) \to a(x)$ and $f_n(x) \to b(x)$ for every $x \in X$ $(n \to \infty)$ (cf. § 2.2 (1)) and hence $a = b \in \mathscr{A} \cap \mathscr{B}$.

† Cf. n. ‡, p. 24.

In practice the norm of $\mathscr{A} \cap \mathscr{B}$ can often be replaced by another equivalent norm.

Examples. 3.1 (iv) is obtained from 3.1 (ii) and (iii); the norms in 3.3 (i) and 3.4 (i) are a combination of those in $\mathscr{F}^1(\mathbf{R})$ and in 3.1 (iii), and similarly for 3.4 (ii).

3.6. Corollary. Let $\mathscr{A}$ be a complete normed function algebra with norm $\|.\|_{\mathscr{A}}$. Suppose $\mathscr{B}$ is a subalgebra of $\mathscr{A}$ and is itself a complete normed algebra under some norm $\|.\|_{\mathscr{B}}$. Then there is a constant c such that $\|f\|_{\mathscr{A}} \leqslant c\|f\|_{\mathscr{B}}$ for all† $f \in \mathscr{B}$. In particular, if $\|.\|_*$ is any other norm on $\mathscr{A}$ under which $\mathscr{A}$ is complete, then for some c we have

$$c^{-1}\|f\|_{\mathscr{A}} \leqslant \|f\|_* \leqslant c\|f\|_{\mathscr{A}}$$

for all $f \in \mathscr{A}$.

By 3.5 $\mathscr{B}$ is also a complete normed algebra under the norm $\|f\| = \|f\|_{\mathscr{A}} + \|f\|_{\mathscr{B}}$. Since $\|f\| \geqslant \|f\|_{\mathscr{B}}$ for all $f \in \mathscr{B}$, we can apply a familiar theorem on Banach spaces: there is a c such that $\|f\| \leqslant c\|f\|_{\mathscr{B}}$ for all $f \in \mathscr{B}$, and the result follows.

4. The generalization of Wiener's theorem

4.1. The algebra $\mathscr{F}^1(\mathbf{R}^\nu)$ that serves us as a model has a certain property which we formulate here abstractly. We say that a topological function algebra $\mathscr{A}(X)$ *possesses approximate units* if for every $f \in \mathscr{A}(X)$ and every neighbourhood $\mathscr{U}_0$ of 0 in‡ $\mathscr{A}(X)$ there is a $u \in \mathscr{A}(X)$ such that $uf \in f+\mathscr{U}_0$; here u may depend on f as well as on $\mathscr{U}_0$. For the case of $\mathscr{F}^1(\mathbf{R}^\nu)$ see Chapter 1, § 2.3.

If the functions with compact support are dense in $\mathscr{A}(X)$—in particular, if $\mathscr{A}(X)$ is a Wiener algebra—and if $\mathscr{A}(X)$ possesses approximate units, then $\mathscr{A}(X)$ also has approximate units with compact support: given any $f \in \mathscr{A}(X)$ and any $\mathscr{U}_0$, there is a $v \in \mathscr{A}(X)$ with compact support such that $vf \in f+\mathscr{U}_0$.

Let $\mathscr{U}_0'$ be such that $\mathscr{U}_0'+\mathscr{U}_0' \subset \mathscr{U}_0$ and choose $u \in \mathscr{A}(X)$ so that $uf \in f+\mathscr{U}_0'$. There is a $\mathscr{V}_0$ such that $\mathscr{V}_0.f \subset \mathscr{U}_0'$. Choose $v \in \mathscr{A}(X)$ with compact support so that $v \in u+\mathscr{V}_0$. Then

$$vf = uf+(v-u)f \in f+\mathscr{U}_0'+\mathscr{V}_0.f \subset f+\mathscr{U}_0'+\mathscr{U}_0' \subset f+\mathscr{U}_0.$$

The existence of approximate units in $\mathscr{A}(X)$ means that every $f \in \mathscr{A}(X)$ is contained in the closed ideal generated by f. Clearly, if a Wiener algebra does not have approximate units, then it contains distinct closed ideals having the same cospectrum.§

† Cf. [51, § 9, Theorem 1], where, however, an additional assumption is made.

‡ The letters $\mathscr{U}_0$ or $\mathscr{V}_0$ will always denote such nds. from now on.

§ H. Mirkil [99] has constructed a Wiener algebra without approximate units, actually a complete normed algebra of functions on $\mathbf{Z}$.

4.2. Let $\mathscr{A}(X)$ be a Wiener algebra with approximate units. If I is a closed ideal of $\mathscr{A}(X)$ and $f \in \mathscr{A}(X)$ belongs locally to I at all points of cosp I, then $f \in I$. In particular I contains all functions in $\mathscr{A}(X)$ that vanish near cosp I.

If $v \in \mathscr{A}(X)$ has compact support, then $vf \in I$, by 1.4 (i) and the localization lemma 1.3, and by 4.1 we can take v also such that vf is arbitrarily near to f. Hence $f \in I$, since I is closed.

This means that, under the conditions stated, the point at infinity need not be considered. The particular case shows that for any closed set $E \subset X$ the smallest closed ideal J_E of $\mathscr{A}(X)$ with cospectrum E is the closure of the ideal of all $f \in \mathscr{A}(X)$ vanishing near E. This is a stronger statement than 2.3 (i), useful in the applications.

4.3. We now introduce another property of function algebras that holds for our model $\mathscr{F}^1(\mathbf{R}^\nu)$ and plays an important part in the generalization of Wiener's theorem.

A topological standard algebra $\mathscr{A}(X)$ is said to satisfy the *condition of Wiener–Ditkin* (condition (WD)) if for every point $a \in X$ the following holds: for any function $f \in \mathscr{A}(X)$ vanishing at a and any neighbourhood $\mathscr{U}_0$ there is a $\tau \in \mathscr{A}(X)$ such that (i) τ is 1 near a, (ii) $f\tau \in \mathscr{U}_0$.

We note that, if $\mathscr{U}_0$ is chosen small, then τ itself cannot lie in $\mathscr{U}_0$, by 2.1 (iii); for normed standard algebras this can be stated more precisely, by 2.2 (1).

Remark. If $\mathscr{A}(X)$ satisfies the condition (WD), then we can even choose a $\tau \in \mathscr{A}(X)$ satisfying (i) and (ii) and, in addition, vanishing outside any pre-assigned neighbourhood of† a.

$\mathscr{F}^1(\mathbf{R}^\nu)$ satisfies this rather remarkable condition (Chap. 1, § 3.2).

The condition (WD) was established for $\mathscr{F}^1(\mathbf{T})$ by Wiener (cf. Chap. 1, § 3.6). Its general significance appeared first in a paper of Ditkin [35] in the case of $\mathscr{F}^1(\mathbf{R})$ and of some other function algebras. Later it was formulated in a much more abstract setting by Shilov [125, pp. 52 and 110]; the formulation there differs slightly from that given here. See also [95, p. 117].

4.4. We can now prove the following important proposition due, essentially, to Shilov [125, Chap. I, § 4, Theorem 13], [95, Theorem 49]. Let I be a closed ideal in a topological standard algebra $\mathscr{A}(X)$ satisfying the condition of Wiener–Ditkin. Suppose $f \in \mathscr{A}(X)$ vanishes on cosp I and let $P(f, I)$ be the set of all points $a \in X$ such that f does *not* belong

† By 1.1 (iii) there is for any U_a a $\tau_a \in \mathscr{A}(X)$ vanishing outside U_a and equal to 1 near a. Then, given f and $\mathscr{U}_0$ in $\mathscr{A}(X)$, there is a $\tau' \in \mathscr{A}(X)$ such that τ' is 1 near a and $(f\tau_a)\tau' \in \mathscr{U}_0$. Thus we can take $\tau = \tau_a \tau'$.

locally to I at a (cf. § 1.2). Then $P(f, I)$ is a perfect set contained in† $\operatorname{Bdr} \operatorname{cosp} f \cap \operatorname{Bdr} \operatorname{cosp} I$.

Clearly f belongs locally to I at every interior point of $\operatorname{cosp} f$ and also at every point of $X - \operatorname{cosp} I$ (§ 1.4 (i)). Hence (since $\operatorname{cosp} f \supset \operatorname{cosp} I$) $P(f, I)$ lies in

$$\operatorname{Bdr} \operatorname{cosp} f \cap \operatorname{Bdr} \operatorname{cosp} I.$$

Also $P(f, I)$ is closed, for $\complement P(f, I)$ is clearly open. Let us show that $P(f, I)$ has no isolated points. Suppose a is in $\operatorname{Bdr} \operatorname{cosp} f \cap \operatorname{Bdr} \operatorname{cosp} I$ and there is some nd. U_a such that f belongs locally to I at every point of $U_a - \{a\}$. Let $\tau_a \in \mathscr{A}(X)$ be such that τ_a is 1 near a and has compact support contained in U_a. By the condition (WD) there is a $\tau \in \mathscr{A}(X)$ which is 1 near a and such that $(f\tau_a)\tau$ is close to $0 \in \mathscr{A}(X)$. The difference $f\tau_a - (f\tau_a)\tau$ belongs locally to I at *every* point of X and has compact support; hence it is contained in I (§ 1.3). Moreover, $f\tau_a - (f\tau_a)\tau$ is arbitrarily close to $f\tau_a$ (by proper choice of τ). Hence $f\tau_a$ is in I, since I is closed, or f belongs locally to I at a, that is, $a \notin P(f, I)$.

4.5. Let $\mathscr{A}(X)$ be a topological standard algebra satisfying the condition of Wiener–Ditkin and let I be a closed ideal of $\mathscr{A}(X)$. Let f be a function in $\mathscr{A}(X)$ vanishing on $\operatorname{cosp} I$ and with compact support. If

$$\operatorname{Bdr} \operatorname{cosp} f \cap \operatorname{Bdr} \operatorname{cosp} I$$

contains no (non-empty) perfect set, then f is in I. This follows from 4.4 and the localization lemma 1.3.

A set that contains no (non-empty) perfect subset is said to be *scattered*.

In a locally compact (Hausdorff) space every countable set is scattered: a non-empty perfect set in such a space is not countable. In a Hausdorff space (not necessarily locally compact) with a countable basis for the open sets every closed scattered set is countable.

4.6. *Suppose now that $\mathscr{A}(X)$ is a Wiener algebra with approximate units and satisfies the condition of Wiener–Ditkin. Then a closed ideal I of $\mathscr{A}(X)$ contains all functions $f \in \mathscr{A}(X)$ vanishing on* $\operatorname{cosp} I$ *and such that* $\operatorname{Bdr} \operatorname{cosp} f \cap \operatorname{Bdr} \operatorname{cosp} I$ *is a scattered set* (generalization of Wiener's theorem). This follows at once from 4.4 and 4.2.

In particular we now have a generalization of Wiener's theorem for $\mathscr{F}^1(\mathbf{R}^\nu)$ (Chap. 1, § 4); we shall see later that this extends to groups (Chap. 6, § 1).

4.7. Let $\mathscr{A}(X)$ be a topological standard algebra and E a closed set in X. We say that E is a *Wiener set* for $\mathscr{A}(X)$, or W-set for short, if there is only one closed ideal of $\mathscr{A}(X)$ having E as cospectrum. There

† The boundary of a closed set $E \subset X$, $\operatorname{Bdr} E$, consists of those points of E which are not interior points of E. For $\operatorname{cosp} f$ see 1.4.

is clearly at least one such ideal, namely

$$I_E = \{f \mid f \in \mathscr{A}(X), f \text{ vanishes on } E\},$$

the largest (closed) ideal with cospectrum E. There is also a smallest such ideal, J_E (cf. § 4.2).

Remark. If $E \neq X$ and if we restrict the functions in I_E to $X-E$, then I_E is a topological standard algebra (cf. § 1.4, Remark). E is a Wiener set if and only if I_E, considered as a topological standard algebra on $X-E$, is a Wiener algebra.

We can now state a particular case of Theorem 4.6 as follows. *If $\mathscr{A}(X)$ satisfies the hypothesis of Theorem* 4.6, *then every closed set in X with a scattered boundary, in particular every closed scattered or countable set, is a Wiener set for $\mathscr{A}(X)$.*

In $\mathbf{R}^\nu$, $\nu \geqslant 2$, a closed proper subset with a scattered boundary is itself countable.

4.8. The result in 4.7 was proved for $\mathscr{F}^1(\mathbf{R})$ and some other function algebras by Ditkin† [35, § 7, Theorem A]. It was extended to complete normed function algebras by Shilov and Mackey [125, Chap. I, § 4, Theorem 13], [95, Theorem 50]. Their methods also yield Theorem 4.6, but the first explicit formulation of 4.6 is due—for $\mathscr{F}^1(\mathbf{R})$ and independently of the authors cited—to Mandelbrojt and Agmon [98, Theorem II].

In 4.6 and 4.7 the conditions on $\mathscr{A}(X)$ are very general. Now the following question arises: given a closed set E with a non-scattered boundary in a locally compact space X, does there exist a Wiener algebra $\mathscr{A}(X)$ satisfying these conditions and for which E is not a Wiener set? A study of this problem may yield some new insight into the significance of the result in 4.7 and the generalization of Wiener's theorem (§ 4.6).

Remark. The result in 4.7 appears here as a special case of Theorem 4.6. We shall see later that it is also contained in another result of a rather different nature (cf. § 5).

4.9. As examples we consider the Wiener algebras $\mathscr{F}_{(p)}$ (§ 3.3 (i)) which have approximate units. $\mathscr{F}_{(p)}$ satisfies the condition of Wiener–Ditkin if $1 \leqslant p < \infty$: for $p = 1$ this is easy to prove, but the general proof, given below, is more delicate and is due to J. D. Stegeman. For $p = \infty$, however, the condition of Wiener–Ditkin does not hold.

† A somewhat weaker result than Ditkin's was given (independently of Ditkin's work) by Segal [123, Theorem 2.7]. It is instructive to look at Segal's proof of the condition (WD) for $\mathscr{F}^1(\mathbf{R})$ [123, Lemma 2.7.1].

Suppose $\hat{f} \in \mathscr{F}_{(p)}$ vanishes for $a \in \mathbf{R}$. Put for $t \in \mathbf{R}$ $\hat{\tau}_{a,\lambda}(t) = \hat{\tau}_1([t-a]/\lambda)$, $\lambda > 0$ (Chap. 1, § 1.3, (iii) and (iv)). In $L^1(\mathbf{R})$ we have: $\|f \star \tau_{a,\lambda}\|_1 \to 0$ $(\lambda \downarrow 0)$, since $\hat{f}(a) = 0$ (Chap. 1, §§ 2.4 and 3.2). Also for each $n = 1, 2,...$

$$\left[\int_{-n}^{n} \left|\frac{d}{dt}\{\hat{f}(t)\hat{\tau}_{a,\lambda}(t)\}\right|^p dt\right]^{1/p} \leqslant A_p(\lambda)+B_p(\lambda),$$

where

$$A_p(\lambda) = \left[\int_{a-2\lambda}^{a+2\lambda} |\hat{f}'(t)|^p\, dt\right]^{1/p}, \qquad B_p(\lambda) = \frac{1}{\lambda}\left[\left\{\int_{a-2\lambda}^{a-\lambda} + \int_{a+\lambda}^{a+2\lambda}\right\}|\hat{f}(t)|^p\, dt\right]^{1/p}.$$

$A_p(\lambda) \to 0$ $(\lambda \downarrow 0)$, since $\hat{f}'$ is locally pth-power integrable (essentially bounded if $p = \infty$). Clearly $B_p(\lambda) \to 0$ $(\lambda \downarrow 0)$ if $p = 1$, since $\hat{f}(a) = 0$. For $1 < p < \infty$ observe that for any $t \in \mathbf{R}$

$$|\hat{f}(t)| = \left|\int_a^t \hat{f}'(u)\, du\right| \leqslant \left|\int_a^t |\hat{f}'(u)|^p\, du\right|^{1/p} . |t-a|^{1/p'},$$

where $1/p+1/p' = 1$, and hence

$$|\hat{f}(t)| \leqslant A_p(\lambda) . |t-a|^{1/p'} \quad \text{for } a-2\lambda \leqslant t \leqslant a+2\lambda.$$

Therefore

$$B_p(\lambda) \leqslant \frac{1}{\lambda} A_p(\lambda)\left[\int_{a-2\lambda}^{a+2\lambda} |t-a|^{p/p'}\, dt\right]^{1/p} = \frac{1}{\lambda} A_p(\lambda) C_p \lambda,$$

where C_p is a constant. Thus also $B_p(\lambda) \to 0$ $(\lambda \downarrow 0)$ *if* $1 \leqslant p < \infty$. It follows that $\mathscr{F}_{(p)}$ satisfies the condition (WD) if $1 \leqslant p < \infty$. For $p = \infty$ see below, § 6.2, Example 1.

The algebras $\mathscr{F}_p$ and $\mathscr{F}_{1,p}$ (§ 3.4) also satisfy the condition of Wiener–Ditkin $(1 \leqslant p < \infty)$. The proof is the same.

5. Wiener–Ditkin sets for function algebras

5.1. Let $\mathscr{A}(X)$ be a Wiener algebra. If $\mathscr{A}(X)$ has approximate units, then the condition of Wiener–Ditkin can be stated in a somewhat different but equivalent form as follows. Given any $a \in X$ and $f \in \mathscr{A}(X)$ such that $f(a) = 0$, there is for every nd. $\mathscr{U}_0$ a $v \in \mathscr{A}(X)$ *with support in* $X-\{a\}$ such that $vf \in f+\mathscr{U}_0$.

Suppose $\mathscr{A}(X)$ satisfies (WD). Let a and f be as above. Given a nd. $\mathscr{U}_0$—which we may assume symmetric—there is a $v_0 \in \mathscr{A}(X)$ such that $v_0 f \in f+\frac{1}{2}\mathscr{U}_0$. By (WD) there is a $\tau \in \mathscr{A}(X)$ such that τ is 1 near a and $(v_0 f)\tau \in \frac{1}{2}\mathscr{U}_0$. Put $v = v_0 - v_0\tau$; then $\operatorname{Supp} v \subset X-\{a\}$ and $vf \in f+\mathscr{U}_0$.

Conversely, assume the stated condition holds. Given $a \in X$, choose any $\tau_a \in \mathscr{A}(X)$ such that τ_a is 1 near a. Then, given $f \in \mathscr{A}(X)$ with $f(a) = 0$ and a symmetric nd. $\mathscr{U}_0$, there is a $v \in \mathscr{A}(X)$ vanishing near a such that $vf\tau_a \in f\tau_a+\mathscr{U}_0$. Put $\tau = \tau_a - \tau_a v$; then τ is 1 near a and $f\tau \in \mathscr{U}_0$, so $\mathscr{A}(X)$ satisfies (WD).

5.2. (i) The foregoing leads to the following definition. Let $\mathscr{A}(X)$ be a Wiener algebra with approximate units. We call a closed set $E \subset X$ a *Wiener–Ditkin set* for $\mathscr{A}(X)$, or WD-set for short, if, given any $f \in \mathscr{A}(X)$

vanishing on E and any $\mathscr{U}_0$ in $\mathscr{A}(X)$, there is a $v \in \mathscr{A}(X)$ with support in $X-E$ such that $vf \in f+\mathscr{U}_0$.

(ii) In this terminology the condition of Wiener–Ditkin simply means that single points are Wiener–Ditkin sets (cf. § 5.1).

The method of proof in 5.1 shows, more generally, that a *compact* set $K \subset X$ is a WD-set if and only if, given any $f \in \mathscr{A}(X)$ vanishing on K, there is for every nd. $\mathscr{U}_0$ a $\tau \in \mathscr{A}(X)$ which is 1 near K and satisfies $f\tau \in \mathscr{U}_0$.

Wiener–Ditkin sets are, in particular, Wiener sets; this follows at once from 4.2.

REMARK. A closed set $E \subset X$ is thus a Wiener–Ditkin set for $\mathscr{A}(X)$ if and only if the ideal I_E, considered as a function algebra on $X-E$, is a Wiener algebra with approximate units (cf. § 4.7, Remark). The existence of approximate units in closed ideals of $\mathscr{A}(X)$—considered as function algebras in their own right—is a problem that still awaits investigation.

(iii) It is not known—especially in the case of $\mathscr{F}^1(\mathbf{R}^\nu)$—whether, conversely, Wiener sets are also Wiener–Ditkin sets. But *if* ALL *closed subsets of X are Wiener sets for $\mathscr{A}(X)$, then they are also Wiener–Ditkin sets*. More generally, suppose $\mathscr{A}(X)$ contains a closed ideal I which, considered as a topological function algebra on its own (cf. § 1.4, Remark), does not contain approximate units. Then cosp I is contained in a closed set that is not a Wiener set for $\mathscr{A}(X)$ (cf. [125, p. 52], [95, Theorem 48]).

Proof. There is an $f \in I$ not contained in the closure I_f of $\{fh \mid h \in I\}$. But I_f is an ideal even in $\mathscr{A}(X)$ and cosp I_f = cosp $f \supset$ cosp I. A special case has already been considered in 4.1.

5.3. The main results concerning Wiener–Ditkin sets are as follows. Let $\mathscr{A}(X)$ be a Wiener algebra possessing approximate units.

(i) If $\mathscr{A}(X)$ satisfies the condition of Wiener–Ditkin, then a closed scattered set in X is a Wiener–Ditkin set for $\mathscr{A}(X)$.

(ii) A closed set in X whose boundary is a Wiener–Ditkin set for $\mathscr{A}(X)$ is itself a Wiener–Ditkin set for $\mathscr{A}(X)$.

(iii) A closed set in X which is the union of countably many Wiener–Ditkin sets for $\mathscr{A}(X)$ is a Wiener–Ditkin set for $\mathscr{A}(X)$.

If we combine (i) and (ii) we obtain a result which contains that in 4.7 (cf. § 5.2 (ii) and also § 4.8, Remark).

The proofs of (i), (ii), (iii) follow.

(i) First we introduce two definitions. Let E be a closed set in X. Then we call a closed subset F of E *distinguished* (relative to E) if for every $f \in \mathscr{A}(X)$

vanishing on E and for every nd. $\mathscr{U}_0$ there is a $v \in \mathscr{A}(X)$ vanishing near F and such that $vf \in f+\mathscr{U}_0$. This definition 'relativizes' that of a WD-set (which corresponds to the case $F = E$). Next we call a point of E *regular* if it has a closed relative nd. in E which is distinguished relative to E. We shall prove:

(*) *Every closed subset of E containing only regular points is distinguished.*

(**) *The set of non-regular points of E is perfect.*

Clearly (*) and (**) imply (i).

Proof of (*). Suppose each point of the closed set $F \subset E$ has a closed relative nd. in E which is distinguished. Let $f \in \mathscr{A}(X)$ vanish on E and let $\mathscr{U}_0$ be given. There is a $v_0 \in \mathscr{A}(X)$ with compact support such that $v_0 f \in f+\frac{1}{2}\mathscr{U}_0$. Put

$$C = F \cap \operatorname{Supp} v_0.$$

Since C is compact, there are finitely many distinguished closed relative nds., say $(U_n)_{1 \leqslant n \leqslant N}$, covering C. Take $v_1 \in \mathscr{A}(X)$ vanishing near U_1 and such that $v_1(v_0 f) \in v_0 f+(2N)^{-1}\mathscr{U}_0$ and then take, inductively, $v_n \in \mathscr{A}(X)$ vanishing near U_n and such that

$$v_n(v_0 v_1 \dots v_{n-1} f) \in v_0 v_1 \dots v_{n-1} f+(2N)^{-1}\mathscr{U}_0, \qquad 1 \leqslant n \leqslant N,$$

if $N > 1$. Put $v = v_0 v_1 \dots v_N$. Then v vanishes near F and $vf \in f+\mathscr{U}_0$ which proves (*).

Proof of (**). The set of non-regular points of E is certainly closed. To show that it has no isolated point, let a be a point of E with a closed relative nd. U_a such that $U_a - \{a\}$ consists entirely of regular points. Suppose $f \in \mathscr{A}(X)$ vanishes on E. Let $\mathscr{U}_0$ be given. The condition (WD) implies (cf. § 5.1): there is a $v_1 \in \mathscr{A}(X)$ vanishing on an open nd. V_a of a in X and such that $v_1 f \in f+\frac{1}{2}\mathscr{U}_0$. The closed set $F = U_a \cap \complement V_a$ is distinguished relative to E, by (*), or even empty, so there is a $v_2 \in \mathscr{A}(X)$ vanishing near F and such that $v_2(v_1 f) \in v_1 f+\frac{1}{2}\mathscr{U}_0$. Since $v = v_1 v_2$ vanishes near U_a and $vf \in f+\mathscr{U}_0$, the nd. U_a is distinguished relative to E, that is, the point a is regular. Thus (**) is proved and the proof of (i) is complete.

(ii) Suppose $E \subset X$ is closed and $\operatorname{Bdr} E$ is a WD-set for $\mathscr{A}(X)$. Let $f \in \mathscr{A}(X)$ vanish on E and let a nd. $\mathscr{U}_0$ be given. There is a $v_1 \in \mathscr{A}(X)$ vanishing near $\operatorname{Bdr} E$ such that $v_1 f \in f+\mathscr{U}_0$ and clearly we may take v_1 also with compact support, by 4.1. Now define v by $v(x) = 0$ for $x \in E$, $v(x) = v_1(x)$ for $x \in X-E$. Then $v \in \mathscr{A}(X)$ (cf. § 1.3), v vanishes near E, and $vf = v_1 f \in f+\mathscr{U}_0$.

(iii) Suppose $E \subset X$ is closed and $E = \bigcup_{n \geqslant 1} E_n$, each E_n being closed and a WD-set for $\mathscr{A}(X)$. Let $f \in \mathscr{A}(X)$ vanish on E and let $\mathscr{U}_0$ be given. Take $v_0 \in \mathscr{A}(X)$ with compact support and such that $v_0 f \in f+\frac{1}{2}\mathscr{U}_0$. There is a $v_1 \in \mathscr{A}(X)$ vanishing on some open nd. U_1 of E_1 and such that $v_1(v_0 f) \in v_0 f+(\frac{1}{2})^2\mathscr{U}_0$. By induction, there is for each $n \geqslant 2$ a $v_n \in \mathscr{A}(X)$ vanishing on an open nd. U_n of E_n and such that

$$v_n(v_0 v_1 \dots v_{n-1} f) \in v_0 v_1 \dots v_{n-1} f+(\tfrac{1}{2})^{n+1}\mathscr{U}_0.$$

Hence $(v_0 v_1 \dots v_n) f \in f+\mathscr{U}_0$ for every $n \geqslant 1$. Put $C = E \cap \operatorname{Supp} v_0$; then C is compact (possibly empty) and $C \subset \bigcup_{n \geqslant 1} U_n$, hence $C \subset \bigcup_{1 \leqslant n \leqslant N} U_n$ for some N, $1 \leqslant N < \infty$. Put $v = v_0 v_1 \dots v_N$; then v vanishes near E and $vf \in f+\mathscr{U}_0$, so E is a WD-set for $\mathscr{A}(X)$. This proof resembles that of (*) in (i).

The results above are contained, essentially, in the writings of

Calderón [19], Herz [65, Theorem 6.2], and Rudin [120, Theorem 7.5.2]; the terminology there is somewhat different. See also Chapter 7, § 4.8.

5.4. The concept of Wiener–Ditkin set is a local one in the following sense. Let $\mathscr{A}(X)$ be a Wiener algebra with approximate units. Let E be a closed subset of X such that each point of E has a closed relative neighbourhood in E which is a Wiener–Ditkin set for $\mathscr{A}(X)$. Then E itself is a Wiener–Ditkin set for $\mathscr{A}(X)$.

This is a particular case of proposition (*) established in the proof of 5.3 (i).

6. Ideals in function algebras ; examples

6.1. Let $\mathscr{A}(X)$ be a Wiener algebra and I a closed proper ideal of $\mathscr{A}(X)$ which is maximal in the sense that there is no *closed* ideal I' of $\mathscr{A}(X)$ such that $I \subset I'$, $I \neq I' \neq \mathscr{A}(X)$. Then I coincides with an ideal I_a given by

$$I_a = \{f \mid f \in \mathscr{A}(X), f(a) = 0\},$$

where a is some point of X.

By Wiener's theorem (§ 2.4) cosp $I \neq \emptyset$. Since I is maximal in the sense above, cosp I must be a single point and I must consist of all $f \in \mathscr{A}(X)$ vanishing there.

An ideal I_a is also maximal in the purely algebraic sense, since it is the kernel of an algebraic morphism (i.e. a linear and multiplicative map) of $\mathscr{A}(X)$ onto **C**. Thus we can simply use the terminology 'closed maximal ideal' for Wiener algebras, without any ambiguity.†

Wiener's theorem 2.4 can be rephrased in a more algebraic terminology: a closed proper ideal of a Wiener algebra is contained in a closed maximal ideal. Likewise, if $\mathscr{A}(X)$ is a Wiener algebra such that every closed subset of X is a Wiener set for $\mathscr{A}(X)$, we can say that every closed ideal of $\mathscr{A}(X)$ is the intersection of the closed maximal ideals containing it.

6.2. A closed ideal I in a Wiener algebra $\mathscr{A}(X)$ is said to be *primary* if it is contained in a single closed maximal ideal. This means (§ 6.1) that cosp I consists of a single point and the property that every closed primary ideal of $\mathscr{A}(X)$ is maximal is equivalent to each point of X being a Wiener set for $\mathscr{A}(X)$.

Example 1. Consider the Wiener algebras $\mathscr{F}_{(p)}$ (§ 3.3 (i)). For $p = \infty$ we have: $\mathscr{F}_{(\infty)}$ contains closed primary ideals that are not maximal.

† A maximal ideal (in the purely algebraic sense) of a Wiener algebra is not necessarily closed. For example, let X be a l.c. space that is not compact. Let $\mathscr{A}(X)$ be the algebra consisting of $\mathscr{C}^0(X)$ and the constants, with the topology of uniform convergence on compact sets. Then $\mathscr{C}^0(X)$ is, algebraically, a maximal ideal of $\mathscr{A}(X)$ and is dense in $\mathscr{A}(X)$.

For any $a \in \mathbf{R}$ let I_a^+ and I_a^- consist of all functions $\hat{f} \in \mathscr{F}_{(\infty)}$ such that $\hat{f}(a) = 0$ and the right and left-hand derivatives respectively of $\hat{f}$ at a exist and are zero. It can be verified that I_a^+ and I_a^- are closed primary ideals of $\mathscr{F}_{(\infty)}$. Put $I_a^0 = I_a^+ \cap I_a^-$. The ideals I_a^+, I_a^-, I_a^0 are distinct, closed, primary but not maximal, and contained in the same maximal ideal.

Example 2. Let $\mathscr{C}^{(m)}$, $m \geqslant 1$, be the Banach algebra of all complex-valued continuous functions on $[0, 1]$ having continuous derivatives up to the mth order; the norm is $\|f\| = \|f\|_\infty + \sum_{1 \leqslant n \leqslant m} (1/n!)\, \|f^{(n)}\|_\infty$. In $\mathscr{C}^{(m)}$ the closed primary ideals are precisely the obvious ones ([125, Chap. 1, § 5], [51, § 37]). For general methods of investigating closed primary ideals see [101, § 11, No. 9]. Every closed ideal I of $\mathscr{C}^{(m)}$ is the intersection of the closed primary ideals containing I [125, Chap. 1, § 5, Theorem 14]; for a generalization to functions of several variables see [141].

Example 3. Let $\mathscr{A}(X)$ be a Wiener algebra with approximate units. If $\mathscr{A}(X)$ satisfies the condition of Wiener–Ditkin, then every closed primary ideal is maximal; whether the converse holds, has neither been proved nor disproved up to now (cf. also § 5.2 (ii) and (iii)).

6.3. A function f on $\mathbf{R}^\nu$, $\nu \geqslant 2$, is said to be a *radial function* if there is a function f^+ on $\mathbf{R}_+ = [0, \infty[$ such that $f(x) = f^+(|x|)$ for all $x \in \mathbf{R}^\nu$. If f is Lebesgue measurable, then so is f^+, and conversely.

If f is in $L^1(\mathbf{R}^\nu)$ and coincides with a radial function $x \to f^+(|x|)$ a.e., then the F.t. of f is again a radial function [12, p. 69 et seq.]:

$$\hat{f}(t) = \hat{f}^+(|t|) \qquad t \in \mathbf{R}^\nu,$$

where

$$\hat{f}^+(\rho) = a(S_{\nu-2}) \int_0^\infty f^+(\xi)\xi^{\nu-1} K_{\frac{1}{2}(\nu-2)}(2\pi\rho\xi)\, d\xi, \qquad \rho \in \mathbf{R}_+, \tag{1}$$

for $\nu \geqslant 2$. Here

$$K_m(\xi) = \int_0^\pi e^{-i\xi\cos\theta}(\sin\theta)^{2m}\, d\theta = c_m \frac{J_m(\xi)}{\xi^m},$$

where $c_m = 2^m\, \Gamma(m+\frac{1}{2})\, \Gamma(\frac{1}{2})$, J_m is the Bessel function of order $m \geqslant 0$, and $a(S_{j-1})$ denotes the $(j-1)$-dimensional area of the unit sphere S_{j-1} in $\mathbf{R}^j$, $j \geqslant 1$ $[a(S_0) = 2]$. We observe that $\int_0^\infty |f^+(\xi)|\xi^{\nu-1}\, d\xi < \infty$, since

$$\|f\|_1 = a(S_{\nu-1}) \int_0^\infty |f^+(\xi)|\xi^{\nu-1}\, d\xi. \tag{2}$$

Conversely, if f is in $L^1(\mathbf{R}^\nu)$ and the F.t. $\hat{f}$ is a radial function, then f coincides a.e. with a radial function (cf. e.g. [111, IV, Lemma 1]).

It is a remarkable fact that, if $f \in L^1(\mathbf{R}^\nu)$ coincides a.e. with a radial function $x \to f^+(|x|)$ and if $\nu \geqslant 3$, then the function (1) possesses a continuous derivative for $\rho > 0$:

$$(3) \qquad (\hat{f}^+)'(\rho) = a(S_{\nu-2}) \int_0^\infty f^+(\xi)\xi^{\nu-1}\, 2\pi\xi K'_{\frac{1}{2}(\nu-2)}(2\pi\rho\xi)\, d\xi \qquad (\nu \geqslant 3)$$

and there is a number $C_\nu(\rho)$ such that for *all* such functions f

$$(4) \qquad |(\hat{f}^+)'(\rho)| \leqslant C_\nu(\rho)\|f\|_1, \quad \rho > 0 \qquad (\nu \geqslant 3).$$

This follows from properties of the Bessel functions:

$$\frac{d}{d\xi}\{J_m(\xi)/\xi^m\} = -\xi\{J_{m+1}(\xi)/\xi^{m+1}\}$$

and $J_m(\xi)/\xi^m = O(\xi^{-m-\frac{1}{2}})$ as $\xi \to \infty$. Hence $\xi K'_{\frac{1}{2}(\nu-2)}(\xi) = O(\xi^{\frac{1}{2}(3-\nu)}) = O(1)$ as $\xi \to \infty$, for $\nu \geqslant 3$. Thus the integral in (3) converges absolutely (cf. (2)), even uniformly for $\rho \geqslant c$, $c > 0$. (4) then follows from (3).

The functions in $L^1(\mathbf{R}^\nu)$ coinciding a.e. with radial functions form a complete normed algebra under the L^1-norm. Let $\mathscr{F}^\circ_\nu(\mathbf{R}_+)$ be the isomorphic algebra of their Fourier transforms, considered as functions on $\mathbf{R}_+$, with the norm carried over from $L^1(\mathbf{R}^\nu)$.

$\mathscr{F}^\circ_\nu(\mathbf{R}_+)$ *is a Wiener algebra*, so Wiener's theorem (§ 2.4) holds for $\mathscr{F}^\circ_\nu(\mathbf{R}_+)$. Moreover, $\mathscr{F}^\circ_\nu(\mathbf{R}_+)$ *has approximate units.*

It is not difficult to show that, if $0 \leqslant a < b$, there is a function in $\mathscr{F}^\circ_\nu(\mathbf{R}_+)$ which is 1 for $0 \leqslant \rho \leqslant a$ and 0 for $b \leqslant \rho < \infty$ (cf. e.g. [111, IV, p. 472]. Hence, given any $\rho \geqslant 0$ and any nd. U_ρ of ρ in $\mathbf{R}_+$, there is a function in $\mathscr{F}^\circ_\nu(\mathbf{R}_+)$ that is 1 near ρ and vanishes outside U_ρ. The fact that $\mathscr{F}^\circ_\nu(\mathbf{R}_+)$ is a standard algebra now follows readily from the corresponding fact for $\mathscr{F}^1(\mathbf{R}^\nu)$, if we consider 'spherical shells' $a \leqslant |x| \leqslant b$ $(0 \leqslant a < b)$ in $\mathbf{R}^\nu$ and use 1.3, Corollary. It is also simple to show that the functions with compact supports are dense in $\mathscr{F}^\circ_\nu(\mathbf{R}_+)$ and that $\mathscr{F}^\circ_\nu(\mathbf{R}_+)$ has approximate units (see e.g. [111, IV, Lemma 2]).

If $\nu \geqslant 3$, then $\mathscr{F}^\circ_\nu(\mathbf{R}_+)$ contains closed primary ideals which are not maximal; more precisely, *if* $\nu \geqslant 3$, *then* $\{a\}$ *is not a Wiener set for* $\mathscr{F}^\circ_\nu(\mathbf{R}_+)$ *for* $a > 0$.

If $\nu \geqslant 3$, then for $a > 0$ the functions $\hat{f}^+ \in \mathscr{F}^\circ_\nu(\mathbf{R}_+)$ such that $\hat{f}^+(a) = 0$ *and* $(\hat{f}^+)'(a) = 0$ form a *closed* primary ideal (cf. (4)) which is clearly not maximal.

For $a = 0$, however, the situation is quite different: $\{0\}$ *is a Wiener set, even a Wiener–Ditkin set, for* $\mathscr{F}^\circ_\nu(\mathbf{R}_+)$, *for* ALL ν $(\nu \geqslant 2)$.

Cf. Chapter 1, § 2.4: in the proof there we now take for τ_1 any *radial* function in $L^1(\mathbf{R}^\nu)$ having a F.t. equal to 1 near the origin.

The case $\nu = 2$ will be discussed further in 7.4 (iii).

6.4. *Let* $\mathscr{A}(X)$ *be a Wiener algebra with approximate units. If* $\mathscr{A}(X)$ *satisfies the condition of Wiener–Ditkin, then the following holds. Whenever* I, I' *are distinct closed ideals of* $\mathscr{A}(X)$ *with the same cospectrum and such that* $I \subset I'$, *then there are at least continuum many closed ideals* I_0 *of* $\mathscr{A}(X)$

such that $I \subset I_0 \subset I'$, $I \neq I_0 \neq I'$. A particular case of this was proved by Helson [61]; the result given here is due, in essentially the generality above, to Y. Katznelson.† The proof follows.

First we introduce the following *definition* which extends that in 1.2: *the ideal* I' *belongs locally to* I *at the point* $a \in X$ if there is a nd. U_a of a such that every $f' \in I'$ coincides on U_a with some function in I.

Next let $P(I', I)$ be the set of all points in X where I' does *not* belong locally to I. Using the condition of Wiener–Ditkin, one can prove: $P(I', I)$ is a *perfect* subset of Bdr cosp I. See 4.4; the proof there still goes through (note that the function τ_a used in 4.4. is independent of f').

Since $I' \neq I$, we have $P(I', I) \neq \emptyset$ (cf. § 4.2) *Let* A_0 *be any perfect proper subset of* $P(I', I)$ *which is the closure of its relative interior* (i.e. relative to $P(I', I)$); put $B_0 = P(I', I) - A_0$.

Consider any *relatively interior* point $a \in A_0$. There is an open nd. Ω_a of a in X disjoint from B_0; let U_a be any compact nd. of a in X contained in Ω_a. There is a function $\tau_a \in \mathscr{A}(X)$ such that τ_a is 1 on U_a and $\operatorname{Supp} \tau_a \subset \Omega_a$ (cf. § 1.3, Remark), so $\operatorname{Supp} \tau_a \cap B_0 = \emptyset$. There is also an $f' \in I'$ which does *not* coincide on U_a with a function in I. Then $f'\tau_a \in I'$ does not coincide on U_a with a function in I, but vanishes near B_0 (and thus obviously belongs locally to I, in the sense of 1.2, at every point of B_0).

Now define I_0 as the family of all functions in I' that belong locally to I, in the sense of 1.2, at every point of B_0. Clearly I_0 is an ideal of $\mathscr{A}(X)$. Moreover, I_0 is closed; this is seen as follows. Let $h' \in I'$ be in the closure of I_0 and consider any point $b \in B_0$. Take $\tau_b \in \mathscr{A}(X)$ such that τ_b is 1 near b and vanishes near A_0. Then $h'\tau_b$ is in the closure of $I_0 . \tau_b$; but every function $f_0 \tau_b$ ($f_0 \in I_0$) is actually in I (§ 4.2), hence also $h'\tau_b \in I$, that is, h' belongs locally to I at b. Thus $h' \in I_0$, since $b \in B_0$ was arbitrary.

The preceding arguments also show that $P(I_0, I)$ is precisely A_0 (note that the nd. U_a of the relatively interior point $a \in A_0$ considered earlier was arbitrarily small, and that A_0 is the closure of its relative interior).

Thus *there are at least as many closed ideals* I_0 *of* $\mathscr{A}(X)$ *such that* $I \subset I_0 \subset I'$, $I \neq I_0 \neq I'$, *as there are perfect proper subsets of* $P(I', I)$ *which are the closure of their relative interior*. There are at least continuum many such subsets: this holds for any (non-empty) Hausdorff space that is locally compact (or, more generally, regular) and has no isolated points, as can readily be shown.

7. Wiener sets ; examples and counter-examples

7.1. The first example of a set in $\mathbf{R}^\nu$ which is not a Wiener set for the algebra $\mathscr{F}^1(\mathbf{R}^\nu)$ was given in 1948 by L. Schwartz [121]: he showed that a $(\nu-1)$-dimensional sphere is not a Wiener set for $\mathscr{F}^1(\mathbf{R}^\nu)$ if $\nu \geqslant 3$. Schwartz's result can be stated in a fairly general way as follows.

Let M *be a smooth manifold*‡ *in* $\mathbf{R}^\nu$, $\nu \geqslant 2$. *If* M *carries a bounded*

† It has not been published so far, but was communicated to the author in July 1966 by Professor Katznelson, by whose kind permission it is published here. For some previous extensions of Helson's result cf. [117] and [128].

‡ By 'smooth' we mean that locally on M one (at least) of the ν coordinates of a point can be expressed in terms of the $\nu-1$ others as a function of class $\mathscr{C}^{(\infty)}$ (i.e.

real measure μ ($\neq 0$) *such that for* $x \in \mathbf{R}^\nu$

(1) $$\phi(x) = \int \langle x, t\rangle \, d\mu(t) = O\left(\frac{1}{|x|}\right) \quad \textit{as } |x| \to \infty,$$

then M *is not a Wiener set for* $\mathscr{F}^1(\mathbf{R}^\nu)$.

On the unit sphere $S_{\nu-1}$ there is such a measure, if $\nu \geqslant 3$: we can take for μ the $(\nu-1)$-dimensional Lebesgue measure on $S_{\nu-1}$.

We have in this case (cf. e.g. [111, IV, p. 469])

(*) $$\phi(x) = A_\nu J_{\frac{1}{2}(\nu-2)}(2\pi|x|)/|x|^{\frac{1}{2}(\nu-2)} = O(|x|^{-\frac{1}{2}(\nu-1)}),$$

where A_ν is a constant; J_m is the Bessel function of order m.

More generally such a measure exists on the boundary of a smooth convex body in $\mathbf{R}^3$ if the Gaussian curvature is everywhere strictly positive (Herz [65, p. 212, Example]); likewise in $\mathbf{R}^\nu$, $\nu \geqslant 3$ [67, Theorem 3]; cf. also [68, Addendum]). This holds even for a wider class of smooth manifolds: see [90]. Some other examples are given in [37].

The proof of the result stated at the beginning follows.

With ϕ as in (1), put

(#) $$\psi_j(x) = 2\pi i x_j \phi(x) \qquad 1 \leqslant j \leqslant \nu.$$

Each ψ_j is *bounded*. Let f be any function in $L^1(\mathbf{R}^\nu)$ such that $\int |f(x)|\,.\,|x| \, dx < \infty$. Then we have by an interchange of integrals

$$\int f(x)\overline{\psi_j(x)} \, dx = \int \frac{\partial \hat{f}(t)}{\partial t_j} \, d\mu(t)$$

and, if $\hat{f}$ vanishes on M,

(##) $$\int f(x+y)\overline{\psi_j(x)} \, dx = \int \langle y, t\rangle \frac{\partial \hat{f}(t)}{\partial t_j} \, d\mu(t) \qquad y \in \mathbf{R}^\nu.$$

Let I_0 be the linear subspace of all $f_0 \in L^1(\mathbf{R}^\nu)$ such that $\int |f_0(x)|\,.\,|x| \, dx < \infty$ and $\hat{f}_0$ vanishes *near* M; I_0 is invariant under translations. Let I be the closure of I_0 in $L^1(\mathbf{R}^\nu)$; I is a closed ideal and clearly cosp $I = M$. We shall show: $I \neq I_M$, where I_M is the ideal of *all* $g \in L^1(\mathbf{R}^\nu)$ such that $\hat{g}$ vanishes on M.

If $f_0 \in I_0$, then $\partial \hat{f}_0/\partial t_j$ vanishes on M and (##) shows: $\int f_0(x+y)\overline{\psi_j(x)} \, dx = 0$, $1 \leqslant j \leqslant \nu$. Thus the continuous linear functionals

$$f \to \int f(x)\overline{\psi_j(x)} \, dx \qquad f \in L^1(\mathbf{R}^\nu)$$

all vanish on I. We shall show that they do not all vanish on I_M, hence $I \neq I_M$.

Let P be a point of M contained in the support of μ. We may assume that P is the origin and M is representable in the form $t_\nu = F(t_1,\ldots,t_{\nu-1})$ in a small box, say for $|t_1| < \delta,\ldots, |t_{\nu-1}| < \delta$, $|t_\nu| < \epsilon$, where F is of class $\mathscr{C}^{(\infty)}$; there are no other points of M in this box. Choose any c such that $0 < c < \min(\delta, \epsilon)$ and let $\hat{\tau}$ be a $\mathscr{C}^{(\infty)}$-function such that $\hat{\tau}(t) = 1$ if $|t| \leqslant \frac{1}{2}c$, $\hat{\tau}(t) = 0$ if $|t| \geqslant c$. Define for $t = (t_1,\ldots,t_\nu) \in \mathbf{R}^\nu$

$$\hat{g}(t) = \hat{\tau}(t)\{t_\nu - F(t_1,\ldots,t_{\nu-1})\} \quad \text{if } |t| \leqslant c,$$

$$\hat{g}(t) = 0 \quad \text{if } |t| \geqslant c.$$

indefinitely differentiable). Much weaker assumptions about the order of differentiability would actually suffice.

Then $\hat{g}$ is of class $\mathscr{C}^{(\infty)}$ and has compact support; hence $\hat{g}$ is the F.t. of a function $g \in L^1(\mathbf{R}^\nu)$ for which $\int |g(x)|.|x|\,dx < \infty$ (cf. e.g. [11, § 44.4]). Moreover, $\hat{g}$ vanishes on M, but $\partial\hat{g}(t)/\partial t_\nu = 1$ for $|t| \leqslant \frac{1}{2}c$. Thus g and its translates are in I_M, but (cf. (##)) $\int g(x+y)\overline{\psi_\nu(x)}\,dx \neq 0$ for some $y \in \mathbf{R}^\nu$, since the measure $\{\partial\hat{g}(t)/\partial t_\nu\}\,d\mu(t)$ is clearly not zero. Hence $I_M \not\subset I$ and the proof is complete.

7.2. Consider now the radial functions in $L^1(\mathbf{R}^\nu)$ (cf. § 6.3, also for the notation). *Let f be any radial function in $L^1(\mathbf{R}^\nu)$, $\nu \geqslant 3$, such that $\hat{f}^+(\rho_1) = 0$, but $(\hat{f}^+)'(\rho_1) \neq 0$ for some $\rho_1 > 0$. Then the closed ideal generated in $L^1(\mathbf{R}^\nu)$ by $f \star f$ does not contain f.*

Suppose first $\int |f(x)|.|x|\,dx < \infty$. Then, if $\hat{f}(t) = 0$ on some sphere $|t| = \rho_1$ ($\rho_1 > 0$), that is, if $\hat{f}^+(\rho_1) = 0$, we can apply the formula (##) in 7.1: it becomes here

$$(*) \qquad \int f(x+y)\overline{\psi_j(x)}\,dx = \frac{(\hat{f}^+)'(\rho_1)}{\rho_1}\int \langle y,t\rangle t_j\,d\mu(t),$$

where ψ_j is defined by 7.1 (*) and (#) and μ is the Lebesgue measure (area) of the $(\nu-1)$-dimensional sphere of radius ρ_1.

We prove now: the relation (*) above holds for *all* radial functions $f \in L^1(\mathbf{R}^\nu)$ for which $\hat{f}^+(\rho_1) = 0$. Choose some radial function $h \in L^1(\mathbf{R}^\nu)$ such that

$$\int |h(x)|.|x|\,dx < \infty \quad \text{and} \quad \hat{h}^+(\rho_1) = 1.$$

Let f be any radial function in $L^1(\mathbf{R}^\nu)$ with $\hat{f}^+(\rho_1) = 0$. Given $\epsilon > 0$, there is a radial function $g_\epsilon \in L^1(\mathbf{R}^\nu)$ such that $\operatorname{Supp} g_\epsilon$ is compact and

$$\|f-g_\epsilon\|_1 < \epsilon/(1+\|h\|_1).$$

Put $f_\epsilon = g_\epsilon - \hat{g}_\epsilon^+(\rho_1)h$; then $\|f-f_\epsilon\|_1 < \epsilon$, since

$$|\hat{g}_\epsilon^+(\rho_1)| = |\hat{g}_\epsilon^+(\rho_1)-\hat{f}^+(\rho_1)| \leqslant \|g_\epsilon - f\|_1.$$

The relation (*) applies to f_ϵ: hence (*) follows for f, when $\epsilon \to 0$, since then $(\hat{f}_\epsilon^+)'(\rho_1) \to (\hat{f}^+)'(\rho_1)$ (cf. § 6.3 (4)).

Next we observe: the right-hand side of (*) does not vanish for *all* $y \in \mathbf{R}^\nu$ if $(\hat{f}^+)'(\rho_1) \neq 0$, but it clearly does so if f is replaced by $f \star f$. Since $g \to \int g(x)\overline{\psi_j(x)}\,dx$, $g \in L^1(\mathbf{R}^\nu)$, is a continuous linear functional, the result to be proved now follows immediately. It is an extension of [111, IV, Theorem 1].

There are very simple examples of functions satisfying the above conditions. Taking, for instance, $\hat{f}^+(\rho) = e^{-\pi\rho^2} - e^{\pi}\,e^{-2\pi\rho^2}$, we obtain once more that the unit sphere $S_{\nu-1}$ is not a Wiener set for $\mathscr{F}^1(\mathbf{R}^\nu)$, if $\nu \geqslant 3$.†

7.3. Consider now a circle in $\mathbf{R}^2$ and the algebras $\mathscr{F}^1_\alpha(\mathbf{R}^2) \cong L^1_\alpha(\mathbf{R}^2)$ (Chap. 1, § 6.5). Here a remarkable situation obtains: *a circle is a Wiener set for $\mathscr{F}^1_\alpha(\mathbf{R}^2)$ if $0 \leqslant \alpha < \frac{1}{2}$, but not if $\alpha \geqslant \frac{1}{2}$.*

That a circle is a Wiener set for $\mathscr{F}^1_\alpha(\mathbf{R}^2)$ when $\alpha = 0$ was discovered by Herz [64] and his method still applies when $0 < \alpha < \frac{1}{2}$, with only trifling changes; this will be discussed in Chapter 7, § 3.9. For the case

† For a third method of proof see [51, § 42]; see also [134 *b*].

$\alpha \geqslant \frac{1}{2}$ we can apply the method of Schwartz above. We add that the same method shows: *the sphere* $S_{\nu-1}$ *is not a Wiener set for* $\mathscr{F}^1_\alpha(\mathbf{R}^\nu)$, $\alpha \geqslant 0$, *if* $\nu \geqslant 3$.

If $\alpha \geqslant \frac{1}{2}$ and f is any radial function in $L^1_\alpha(\mathbf{R}^2)$, then $\rho \to \hat{f}^+(\rho)$ has a derivative for $\rho > 0$ and

$$|(\hat{f}^+)'(\rho)| \leqslant c_\alpha(\rho)\|f\|_{1,\alpha},$$

where $c_\alpha(\rho)$ is independent of f; this is proved as for 6.3 (4). Thus 7.2 (*) remains applicable and the proof in 7.2 goes through as before: the function ϕ in 7.1 (*) is $O(|x|^{-\frac{1}{2}})$ as $|x| \to \infty$, for $\nu = 2$, hence the functions ψ_j in 7.1 (#) define *continuous* linear functionals on $L^1_\alpha(\mathbf{R}^2)$, since $\alpha \geqslant \frac{1}{2}$.

REMARK. More generally, a circle is not a Wiener set for any algebra $\mathscr{F}^1_w(\mathbf{R}^2) \cong L^1_w(\mathbf{R}^2)$ (Chap. 1, §§ 6.1 and 6.5) with w such that $w(x) \geqslant C(1+|x|)^{\frac{1}{2}}$, $x \in \mathbf{R}^2$, for some constant C. The proof above still applies.

7.4. (i) *The algebras* $\mathscr{F}^1_\alpha(\mathbf{R}^\nu) \cong L^1_\alpha(\mathbf{R}^\nu)$ (Chap. 1, §§ 6.1 and 6.5) *satisfy the condition of Wiener–Ditkin if* $0 \leqslant \alpha < 1$, *for all* $\nu \geqslant 1$. In particular (§ 5.2 (ii)) single points of $\mathbf{R}^\nu$ are Wiener sets for $\mathscr{F}^1_\alpha(\mathbf{R}^\nu)$ if† $0 \leqslant \alpha < 1$. The proof requires somewhat refined methods and will be given in Chapter 6, § 3.3.

(ii) The following, however, is very simple to prove: *a closed ball in* $\mathbf{R}^\nu$ *is a Wiener set for* $\mathscr{F}^1_\alpha(\mathbf{R}^\nu)$ *for all* $\alpha \geqslant 0$ $(\nu = 1, 2,...)$ *and so is the complement of an open ball in* $\mathbf{R}^\nu$.

Let K_0 be any closed ball $\{x \mid x \in \mathbf{R}^\nu,\ |x| \leqslant r\}$ $(r > 0)$. Suppose $\hat{f}_0 \in \mathscr{F}^1_\alpha(\mathbf{R}^\nu)$ vanishes on K_0. We shall show that for every $f \in L^1_\alpha(\mathbf{R}^\nu)$

$$(*) \qquad \|M_\rho f - f\|_{1,\alpha} \to 0 \qquad \text{as } \rho \to 1.$$

This, combined with 4.2, will imply that K_0 is a Wiener set, since

$$[M_\rho f_0]\hat{}\,(t) = \hat{f}_0(t/\rho)$$

vanishes near K_0, if $\rho > 1$.

To verify (*), observe that

$$\|M_\rho f - f\|_{1,\alpha} \leqslant A_\alpha(\rho) + B_\alpha(\rho),$$

where we put for $w_\alpha(x) = (1+|x|)^\alpha$

$$A_\alpha(\rho) = \|M_\rho(f.w_\alpha) - f.w_\alpha\|_1, \qquad B_\alpha(\rho) = \int |f(x)|.|w_\alpha(x) - w_\alpha(x/\rho)|\,dx.$$

Clearly $A_\alpha(\rho) \to 0$ as $\rho \to 1$, since $f.w_\alpha \in L^1(\mathbf{R}^\nu)$. This also holds for $B_\alpha(\rho)$, since

$$|w_\alpha(x) - w_\alpha(\lambda x)| \leqslant \alpha\, 2^{\alpha+1} w_\alpha(x)|1-\lambda| \qquad \text{for } \tfrac{1}{2} \leqslant \lambda \leqslant 2.$$

(Put $F(\lambda) = (1+\lambda|x|)^\alpha$, $\lambda > 0$. Then for some λ^* between 1 and λ

$$|F(1) - F(\lambda)| = \alpha(1+\lambda^*|x|)^{\alpha-1}.|x|.|1-\lambda| \leqslant \alpha \frac{2^\alpha(1+|x|)^\alpha}{1+(|x|/2)}|x|.|1-\lambda|,$$

if $\frac{1}{2} \leqslant \lambda \leqslant 2$.) Likewise for the complement of an open ball in $\mathbf{R}^\nu$.

More generally, (ii) is true for any star-shaped body in $\mathbf{R}^\nu$, that is, any closed set S with the following property: there is some interior point

† If $\alpha \geqslant 1$, single points are clearly not Wiener sets for $\mathscr{F}^1_\alpha(\mathbf{R}^\nu)$.

P_0 of S such that for every boundary point P of S the points of the segment P_0P other than P are all interior to S. The proof is the same.

REMARK. It is not known whether closed balls in $\mathbf{R}^\nu$, $\nu \geqslant 2$, are also Wiener–Ditkin sets for $\mathscr{F}^1_\alpha(\mathbf{R}^\nu)$; it is also not known whether the circle is a Wiener–Ditkin set for $\mathscr{F}^1_\alpha(\mathbf{R}^2)$, $0 \leqslant \alpha < \frac{1}{2}$.

(iii) Consider the algebras $\mathscr{F}^\circ_\nu(\mathbf{R}_+)$ (§ 6.3). Here the case $\nu = 2$ contrasts sharply with the others: *the algebra* $\mathscr{F}^\circ_2(\mathbf{R}_+)$ *satisfies the condition of Wiener–Ditkin.*

For the point $\rho = 0$ this has already been discussed in 6.3. For points $\rho > 0$ we obtain (iii) by combining the result in (i) above, for $\nu = 2$ and $\alpha = \frac{1}{2}$, with a recent theorem of M. Gatesoupe [47].† This theorem says (in particular) that *the algebras* $\mathscr{F}^1_{\frac{1}{2}}(\mathbf{R}^2)$ *and* $\mathscr{F}^\circ_2(\mathbf{R}_+)$ *are locally isomorphic in* $]0, \infty[$, that is, on any compact interval $[a, b]$, $0 < a < b < \infty$, they have the same restriction. The functions in those algebras having support in $[a, b]$ form an algebra complete in each of the two norms and an application of 3.6 finishes the proof.

7.5. We conclude with a discussion of some general properties of Wiener sets. Let $\mathscr{A}(X)$ be any Wiener algebra with approximate units.

(i) A closed ideal I of $\mathscr{A}(X)$ contains every function in $\mathscr{A}(X)$ that vanishes on a Wiener set for $\mathscr{A}(X)$ containing $\operatorname{cosp} I$.

(ii) The notion of a Wiener set is a local one in the following sense. Let $E \subset X$ be a closed set such that each point $a \in E$ has a closed relative neighbourhood V_a in E which is a Wiener set for $\mathscr{A}(X)$. Then E itself is a Wiener set for $\mathscr{A}(X)$. This is analogous to 5.4 and implies, in particular, that the union of two *disjoint* Wiener sets is a Wiener set.

(iii) We can generalize (ii) as follows. Let E be a closed set in X containing a closed subset F which is a Wiener–Ditkin set for $\mathscr{A}(X)$. If each point of $E - F$ has a closed relative neighbourhood in E which is a Wiener set for $\mathscr{A}(X)$, then E is a Wiener set for $\mathscr{A}(X)$. In particular, the union of a Wiener set and a Wiener–Ditkin set is a Wiener set.

(i) If $f \in \mathscr{A}(X)$ vanishes on a Wiener set $E \supset \operatorname{cosp} I$, then f can be approximated by functions in $\mathscr{A}(X)$ vanishing near E and thus near $\operatorname{cosp} I$; such functions belong to I (§ 4.2). Hence $f \in I$, since I is closed.

(ii) Let I be a closed ideal with $\operatorname{cosp} I = E$. We shall show that I contains every $f \in \mathscr{A}(X)$ vanishing on E. It is enough to verify that f belongs locally to I at every point $a \in E$ (§ 4.2). Let V_a be as stated in (ii); there is an open nd. U_a of a in X such that $E \cap U_a \subset V_a$. Take $\tau_a \in \mathscr{A}(X)$ such that τ_a is 1 near a and $\operatorname{Supp} \tau_a \subset U_a$. Given $\mathscr{U}_0$ in $\mathscr{A}(X)$, choose $\mathscr{V}_0$ so that $\mathscr{V}_0 \tau_a \subset \mathscr{U}_0$. Since V_a is a W-set, there is a $g \in \mathscr{A}(X)$ vanishing near V_a such that $g \in f + \mathscr{V}_0$, hence $g\tau_a \in f\tau_a + \mathscr{U}_0$.

† The author is indebted to Mr. Gatesoupe for a copy of this paper prior to publication.

Now $g\tau_a$ vanishes near E and hence is in I. Since $\mathscr{U}_0$ is arbitrarily small and I is closed, $f\tau_a$ is in I.

(iii) Let I be a closed ideal with cosp $I = E$. If $f \in \mathscr{A}(X)$ vanishes on E, then f can be approximated by functions vf, where $v \in \mathscr{A}(X)$ vanishes near F and has compact support. The argument in the proof of (ii) shows that vf belongs locally to I at every point of $(E-F) \cap \operatorname{Supp} vf$ and it follows that $vf \in I$; hence also $f \in I$.

It is not known whether, in general, the union of two Wiener sets is a Wiener set. For the intersection this need certainly not hold: as example consider $\mathscr{F}^1_\alpha(\mathbf{R}^2)$ and the Wiener sets $\{x \mid |x| \leqslant 1\}$ and $\{x \mid |x| \geqslant 1\}$ in $\mathbf{R}^2$ (cf. §§ 7.3 and 7.4 (ii)). In general, the properties of Wiener sets are, at present, less clearly understood than those of Wiener–Ditkin sets (cf. §§ 5.2 and 5.3).

References. [65, Theorem 6.4 and p. 227], [111, VI, § 4].†

† We remark that the lemma in the second reference clearly extends to Wiener algebras with approximate units.

3

LOCALLY COMPACT GROUPS AND HAAR MEASURE

1. Locally compact groups

In this section some notation and terminology to be used later is outlined, mainly in those cases where there are several choices in the literature; some of the relevant facts are also mentioned, for later reference. For complete expositions see [100], [138], [13 *e*, Chap. III], [106]. See also [26, pp. 26–32] and the exposition in [83, Chaps. IV and VIII]; an historical discussion is given in [46].

1.1. The general notation and the axioms for a *group* G are described in A. Weil's book [138]. Following Bourbaki, we call the element $e \in G$ such that $ex = xe = x$ for all $x \in G$ the *neutral element* of G. If we have several groups, say G_α, $\alpha \in \mathrm{A}$, then e_α denotes the neutral element of G_α.

We use multiplicative notation also for abelian groups, generally, but in special cases additive notation is adopted.

Subgroups of G are usually denoted by H; the *left cosets* of H are xH $(x \in G)$. We retain the term '*normal subgroup*' and call G/H the *quotient group* if H is normal.

1.2. By a *topological space* X is meant a non-empty set X in which certain subsets, called *open sets*, are given subject to the condition that the union of any number of open sets, the intersection of finitely many open sets, the whole space X and the *empty set* $\emptyset$ are open. Open sets will usually be denoted by Ω.

A *neighbourhood* (nd.) U of a point $x \in X$ is any subset of X such that $x \in \Omega \subset U$ for some open set Ω; likewise we define a nd. of a subset $E \subset X$. The use of nds. in this wider sense (i.e. not necessarily open) is convenient in practice. We often say '*near* x' or '*near* E' instead of the longer expression 'in a nd. of x (or E)'.

Topological spaces satisfying the Hausdorff separation axiom (Hausdorff spaces) are also called *separated topological spaces*.

A family of nds. of the same point $x \in X$ is a *basis at* x if to every nd. U of x there is a V in the family such that $V \subset U$.

The *product of two topological spaces* X_1, X_2 is a topological space (with the usual product topology) denoted by $X_1 \times X_2$. For arbitrarily many topological spaces X_α, $\alpha \in \mathrm{A}$, we denote the product by $\prod_{\alpha \in \mathrm{A}} X_\alpha$; its points are $(x_\alpha)_{\alpha \in \mathrm{A}}$, with $x_\alpha \in X_\alpha$. By '*almost all*' $\alpha \in \mathrm{A}$ we shall mean: all $\alpha \in \mathrm{A}$ except at most finitely many.

1.3. A *topological group* G is a topological space and a group, the two structures being related by continuity of the group operations: xy is a continuous function of $(x,y) \in G \times G$, x^{-1} a continuous function of $x \in G$. An equivalent 'local' definition can be given in terms of neighbourhoods of e. We shall consider only topological groups or spaces satisfying the Hausdorff separation axiom.

The term '*compact*' will be used in the sense of A. Weil and Bourbaki ('bicompact' in the Russian literature). We simply say '*locally compact* (*l.c.*) *group*' (or space), omitting the adjective 'topological' (likewise we say 'compact group', 'separated group', etc.). G is locally compact if there is a compact neighbourhood of e.

We call a locally compact space *countable at infinity* if it can be represented as a union of countably many compact sets. A locally compact group generated by a compact subset is said to be *compactly generated*.

1.4. A subgroup H of a topological group G is itself a topological group in the *induced topology* (i.e. induced by that of G). In case a different topology is used for H, it will be mentioned explicitly.

1.5. A mapping (or map) of a set G_1 into a set G_2 is called *injective* or an *injection* if different points of G_1 have different images in G_2. A map which is both injective and *surjective* (i.e. which maps G_1 onto G_2) is called *bijective* or a *bijection*.

Two topological groups G_1 and G_2 are said to be *isomorphic* if there is a bijection of G_1 onto G_2 which is an algebraic isomorphism and a homeomorphism.†

An *open map* of a topological space into another is a map such that the image of an open set is again an open set.

A *representation* of a group G_1 in a group G_2 is a map f of G_1 into G_2 such that $f(xy) = f(x)f(y)$ for all $x, y \in G_1$. If G_1 and G_2 are topological groups, the representation is called a *morphism* if it is continuous, and a *strict morphism* if it is continuous and open.‡

1.6. Let G be a topological group, H any subgroup of G. The left cosets of H form the points of a new topological space, the *quotient space*§ G/H, the topology being defined as follows. We denote the *canonical map* of G onto G/H by $\pi_H: x \to \dot{x} = \pi_H(x) = xH$ (this will be used as standard notation throughout). A set $E \subset G/H$ is defined to be open if $\pi_H^{-1}(E)$ is open in G—in other words, the open sets in G/H correspond to the sets

† There are non-trivial examples where one but not the other of these conditions is fulfilled.

‡ This is essentially the terminology of [13 *e*, Chap. III, § 2, n° 8] and differs slightly from that of [138, p. 11] and [100, p. 23].

§ 'Espace homogène' [138, p. 10], [13 *e*, Chap. III, § 2, n° 5], 'coset space' [100, § 1.16].

$\Omega H \subset G$ where Ω is open in G. If H is a normal subgroup, then G/H is a topological group. Since we shall consider only the case where the Hausdorff separation axiom holds, we shall restrict H to be a *closed* subgroup; this also entails that G/H is locally compact if G is locally compact.

The map π_H is not only continuous, but also open (and thus a strict morphism if H is normal).

G/H is, of course, a topological group (or space) in its own right. In practice it is sometimes convenient not to distinguish between the points of G/H and the left cosets of H.

1.7. The product of any family of topological groups is again a topological group, multiplication being defined in the obvious way.

REMARK. The product of discrete groups G_α, $\alpha \in A$, is *not* discrete if A is infinite and each G_α has at least two elements. Similarly for the product of locally compact, but not compact, groups.

1.8. We collect here some useful facts for reference. Let G be a topological group.

(i) If $K \subset G$ is compact and Ω is an open set in G containing K, then there is a neighbourhood V of e such that $KV \subset \Omega$ (cf. e.g. [138, p. 16] or [100, § 2.1, Lemma]).

(ii) If G is locally compact and H a closed subgroup of G, then for any compact set $\dot{K} \subset G/H$ there is a compact set $K \subset G$ such that $\pi_H(K) = \dot{K}$ [138, p. 19].

(iii) The map $(x,y) \to xy$ of $G \times G$ onto G is *open*; likewise for the maps $(x,y) \to y^{-1}x$, $(x,y) \to x$. This is an immediate consequence of the definition of the product topology.

(iv) A locally compact group contains open (and hence closed) subgroups which are countable at infinity and even compactly generated, viz. the subgroups $G_V = \bigcup_{n \geq 1} V^n$, where V is any compact symmetric neighbourhood of e. Clearly G is the union of all such subgroups G_V.

(v) A locally compact group or quotient space which is countable is discrete. This follows from a Baire category argument (cf. e.g. [100, § 1.7.3]).

(vi) A complex-valued function f on G is said to be *right* [left] *uniformly continuous* if, given $\epsilon > 0$, there is a nd. U of e such that†

$$|f(yx)-f(x)| < \epsilon \quad [|f(xy)-f(x)| < \epsilon]$$

† This definition of right and left uniform continuity, used in [13 e, Chap. III, § 3, n° 1], reverses that in [138, p. 14].

if $y \in U$, for all $x \in G$. There are simple examples of l.c. groups for which these two 'uniform structures' are distinct (cf. e.g. [13 *e*, Chap. III, § 3, Exercise 4] or [74, § 4.2.4]). A (complex-valued) continuous function on G with compact support is right and left uniformly continuous ([138, p. 19], [91, § 28B]).

(vii) Let H be a closed subgroup of G. A (complex-valued) function F on G is of the form $F' \circ \pi_H$, where F' is a function on G/H and '$\circ$' the usual composition of mappings, if and only if F is constant on each left coset of H (i.e. $F(x\xi) = F(x)$ for all $x \in G$, $\xi \in H$), and F is continuous if and only if F' is continuous; this follows directly from the definitions. If H is a closed *normal* subgroup, then a function F on G constant on each coset of H is said to be *H-periodic*.

1.9. We use the terms '*positive*' and '*increasing*' in the wide sense (that is, for $a \geqslant 0$ and $a_n \leqslant a_{n+1}$, $n \geqslant 1$) and we say '*strictly positive*' if $a > 0$, '*strictly increasing*' if $a_n < a_{n+1}$, $n \geqslant 1$; likewise for functions and sequences of functions, and for 'negative' and 'decreasing'. The notation '$f \leqslant g$' for two real-valued functions on a space X means: $f(x) \leqslant g(x)$ for all $x \in X$; a *positive function* f is one such that $f(x) \geqslant 0$ for all $x \in X$ and f is a *strictly positive function* if $f(x) > 0$ for all $x \in X$. These notations and conventions are useful in practice; they will be used also for functions with values in $\overline{\mathbf{R}}_+ = [0, \infty]$, the one-point compactification of the positive reals $\mathbf{R}_+ = [0, \infty[$.

The *characteristic function* of a subset $A \subset X$ is denoted by ϕ_A, where $\phi_A(x) = 1$ for $x \in A$, $\phi_A(x) = 0$ for $x \in \complement A$. The letters j, m, n, ν, *used as subscripts or superscripts, will always denote integers.*

2. Integration on locally compact spaces

The approach to integration theory emphasized in A. Weil's book [138, p. 32] has made an ever-growing impact on the development of analysis, especially harmonic analysis, on groups. This section contains a short survey of the principal notations, definitions, and results, in that form in which they will be used; it will serve as reference for the later work.

A detailed treatment is given by Bourbaki [13 *c*, Chaps. III, IV, and V];† see also [130]. The theory expounded by Bourbaki is not required, however, in its entire generality: the selection given below is sufficient for the applications to groups.

† References are to the second edition. The notation there—which is also used here, essentially—differs slightly from that of the first.

2.1. *Measures: general definitions and properties* (cf. [13 *c*, Chap. III, §§ 1 and 2]).

(i) Let X be a locally compact space; let $\mathscr{K}_{\mathbf{R}}(X)$ and $\mathscr{K}_{\mathbf{C}}(X)$ denote respectively the space of real and complex-valued continuous functions on X with compact support. The subset of positive functions in $\mathscr{K}_{\mathbf{R}}(X)$ is denoted by $\mathscr{K}_{+}(X)$. The real [complex] continuous functions on X vanishing at infinity form a real [complex] Banach space, denoted by $\mathscr{C}^0_{\mathbf{R}}(X)$ [$\mathscr{C}^0_{\mathbf{C}}(X)$], with the norm $\|f\|_\infty = \max_{x \in X} |f(x)|$; this is the completion of $\mathscr{K}_{\mathbf{R}}(X)$ [$(\mathscr{K}_{\mathbf{C}}(X)$] if X is not compact. When X is fixed, we often write simply $\mathscr{K}_{\mathbf{R}}$, $\mathscr{K}_{+}$,.... These spaces have already been considered in Chapter 2, § 3.1 (i).

(ii) A real [complex] *measure* μ on X is defined as a real [complex] linear functional on $\mathscr{K}_{\mathbf{R}}(X)$ [$\mathscr{K}_{\mathbf{C}}(X)$] with the following property: to every compact set $K \subset X$ there is a constant M_K such that

$$|\mu(f)| \leqslant M_K \|f\|_\infty$$

for all $f \in \mathscr{K}_{\mathbf{R}}(X)$ [$\mathscr{K}_{\mathbf{C}}(X)$] with support in K. A real measure μ is said to be *positive* if $\mu(f) \geqslant 0$ for all $f \in \mathscr{K}_{+}(X)$. Every real measure can be represented as the difference of two positive measures (cf. (viii) below).

REMARK. This terminology differs from that of A. Weil: to 'measure' in [138] corresponds 'positive measure' here.

The number $\mu(f)$ is also called the *integral* of f (with respect to μ) and is denoted, in analogy with the classical notation, by $\int_X f(x)\, d\mu(x)$, or simply by $\int f(x)\, d\mu(x)$ or $\int f\, d\mu$ [$f \in \mathscr{K}_{\mathbf{R}}(X)$ or $\mathscr{K}_{\mathbf{C}}(X)$].†

(iii) A complex measure is uniquely determined by its restriction to $\mathscr{K}_{\mathbf{R}}(X)$. Every real measure can be uniquely extended, in the obvious way, to a complex measure. It is convenient to call a complex measure positive if it is the extension of a real positive measure.

From now on we shall just speak of measures, meaning either real or complex measures. Correspondingly, $\mathscr{K}(X)$, or simply $\mathscr{K}$, will denote either $\mathscr{K}_{\mathbf{R}}(X)$ or $\mathscr{K}_{\mathbf{C}}(X)$; likewise for $\mathscr{C}^0(X)$.

(iv) If μ is a positive measure, then

$$(1) \qquad \left|\int f(x)\, d\mu(x)\right| \leqslant \int |f(x)|\, d\mu(x) \qquad f \in \mathscr{K}(X).$$

We may assume $\mu(f) \neq 0$: put $|\mu(f)| = \lambda\mu(f)$. Then $\mu(\lambda f)$ is real and

$$0 < \mu(\lambda f) = \mu(\mathrm{Re}\{\lambda f\}) \leqslant \mu(|\lambda f|) = \mu(|f|),$$

since $|\lambda| = 1$.

† In the literature the term 'integral' is sometimes used in the sense of 'positive measure'.

(v) Any positive linear functional on $\mathscr{K}(X)$ (i.e. any linear functional μ on $\mathscr{K}(X)$ such that $\mu(f) \geqslant 0$ if $f \in \mathscr{K}_+(X)$) is a (positive) measure.

The relation $|\mu(f)| \leqslant \mu(|f|)$, $f \in \mathscr{K}$, is proved as in (iv). Given any compact set $K \subset X$, there is a $g_K \in \mathscr{K}_+$ which is 1 on K. If $f \in \mathscr{K}$ and $\operatorname{Supp} f \subset K$, then $f = f.g_K$, hence $|\mu(f)| \leqslant \mu(|f|.g_K) \leqslant M_K\|f\|_\infty$, with $M_K = \mu(g_K)$.

(vi) A *bounded measure* μ is one for which there is a number M such that $|\mu(f)| \leqslant M\|f\|_\infty$ for all $f \in \mathscr{K}(X)$. The smallest such M is called the *norm* of μ, denoted by $\|\mu\|$:

$$\|\mu\| = \sup_{f \in \mathscr{K}(X), \|f\|_\infty \leqslant 1} |\mu(f)|.$$

The bounded measures can be extended, by continuity, to $\mathscr{C}^0(X)$, for non-compact X. They form a Banach space, denoted by† $M^1(X)$, the dual of $\mathscr{C}^0(X)$, with the norm above.

(vii) The *support of a measure* μ, $\operatorname{Supp}\mu$, is the smallest closed set $E \subset X$ such that $\mu(f) = 0$ for all $f \in \mathscr{K}(X)$ with support in $X—E$: the existence of a smallest such set—possibly X itself—is readily shown (cf. Chap. 2, § 3.1 (i), Remark). The measures with compact support are dense in $M^1(X)$, but we shall not need this fact.

(viii) Given any measure μ, we define a positive measure $|\mu|$ as follows. For $f \in \mathscr{K}_+$ we put

$$|\mu|(f) = \sup_{g \in \mathscr{K}, |g| \leqslant f} |\mu(g)|.$$

Then $|\mu|(\alpha f) = \alpha|\mu|(f)$ for $\alpha \in \mathbf{R}_+$ and it can readily be shown that

$$|\mu|(f_1+f_2) = |\mu|(f_1)+|\mu|(f_2) \quad \text{for } f_1, f_2 \in \mathscr{K}_+$$

[13 *c*, Chap. III, § 1, n° 6]; we can extend $|\mu|$ to $\mathscr{K}_{\mathbf{R}}$ by putting

$$|\mu|(f) = |\mu|(f_1)-|\mu|(f_2) \quad \text{if } f = f_1-f_2.$$

Thus we obtain a positive measure $|\mu|$ (cf. (iii) and (v)) and obviously

$$|\mu(f)| \leqslant |\mu|(|f|) \qquad f \in \mathscr{K}.$$

If μ is real, we put $\mu_+ = \frac{1}{2}(|\mu|+\mu)$, $\mu_- = \frac{1}{2}(|\mu|-\mu)$: then μ_+, μ_- are real, positive measures and $\mu = \mu_+-\mu_-$; also $\mu = |\mu| \Leftrightarrow \mu$ is positive.

(ix) Let Ω be an open set in X. The restriction of a measure μ on X to the linear subspace of $\mathscr{K}(X)$ consisting of all $f \in \mathscr{K}(X)$ such that $\operatorname{Supp} f \subset \Omega$ is called the *restriction of* μ *to* Ω, since there is an obvious isomorphism between this subspace and $\mathscr{K}(\Omega)$.

2.2. *Extension of a positive measure* (cf. [13 *c*, Chap. IV, § 1 and § 2, n°s 1–5]; there complex measures are considered).

(i) Let X be a locally compact space. We denote by $\mathscr{I}_+(X)$, or simply $\mathscr{I}_+$, the class of all lower semi-continuous functions F on X with values

† This is a slight deviation from Bourbaki's notation.

in $\overline{\mathbf{R}}_+$ (thus F may assume the value ∞, 'plus infinity'). $\mathscr{I}_+$ is closed under addition and multiplication.†

The functions in $\mathscr{I}_+$ are precisely those representable in the form $\sup_{f \in \mathfrak{F}} f$, where $\mathfrak{F}$ is a '*filtering family*' in $\mathscr{K}_+$.‡

(ii) Let μ be a *positive* measure on X. We define a functional $\mu^\times$ on $\mathscr{I}_+$ by $\mu^\times(F) = \sup_{k \in \mathscr{K}_+, k \leqslant F} \mu(k)$, $F \in \mathscr{I}_+$. Then we have:

$$0 \leqslant \mu^\times(F) \leqslant \infty, \qquad \mu^\times(aF) = a\mu^\times(F) \quad (a \in \mathbf{R}_+),$$

and
$$\mu^\times(F_1) \leqslant \mu^\times(F_2) \quad \text{if } F_1 \leqslant F_2;$$

also
$$\mu^\times(F_1+F_2) = \mu^\times(F_1)+\mu^\times(F_2)$$

for all $F_1, F_2 \in \mathscr{I}_+$ (the proof is based on relation (2) below). Clearly $\mu^\times$ coincides with μ on $\mathscr{K}_+$.

(iii) For arbitrary functions f on X with values in $\overline{\mathbf{R}}_+$ we define $\mu^\times(f) = \inf_{F \in \mathscr{I}_+, F \geqslant f} \mu^\times(F)$. If $f \in \mathscr{I}_+$, this is still the same number as in (ii). Instead of $\mu^\times(f)$ we usually write $\int_X^\times f(x)\, d\mu(x)$ or simply $\int^\times f\, d\mu$ and call it the *upper integral* of f (with respect to μ). We have

$$0 \leqslant \int^\times f\, d\mu \leqslant \infty, \qquad \int^\times af\, d\mu = a \int^\times f\, d\mu \quad (a \in \mathbf{R}_+),$$

$$\int^\times f\, d\mu \leqslant \int^\times g\, d\mu \quad \text{if } f \leqslant g,$$

and
$$\int^\times (f_1+f_2)\, d\mu \leqslant \int^\times f_1\, d\mu + \int^\times f_2\, d\mu.$$

(iv) The following fact is of fundamental importance in the applications: if $\mathfrak{F}$ is any *filtering family in* $\mathscr{I}_+$ (cf. (i)), then

$$\int^\times \sup_{F \in \mathfrak{F}} F\, d\mu = \sup_{F \in \mathfrak{F}} \int^\times F\, d\mu. \tag{2}$$

Likewise, if $(f_n)_{n \geqslant 1}$ is an *increasing sequence of arbitrary functions with values in* $\overline{\mathbf{R}}_+$, then

$$\int^\times \sup_{n \geqslant 1} f_n\, d\mu = \sup_{n \geqslant 1} \int^\times f_n\, d\mu. \tag{3}$$

† We define $\alpha+\infty = \infty+\alpha = \infty$ for all $\alpha \in \overline{\mathbf{R}}_+$, $\infty.\alpha = \alpha.\infty = \infty$, if $0 < \alpha \leqslant \infty$, and $\infty.0 = 0.\infty = 0$.

‡ The function $\sup_{f \in \mathfrak{F}} f$ is defined for each $x \in X$ as $\sup_{f \in \mathfrak{F}} f(x)$. A family $\mathfrak{F}$ of functions with values in $\mathbf{R}$ or $\overline{\mathbf{R}}_+$ is said to be *filtering with respect to* '$\leqslant$', or simply *filtering*, if for every pair $f, g \in \mathfrak{F}$ there is an $h \in \mathfrak{F}$ such that $f \leqslant h$, $g \leqslant h$. This is a useful extension of the concept of 'increasing sequence' of functions [cf. § 1.9]. Analogously one can consider a family *filtering with respect to* '$\geqslant$'.

We have (cf. (ii))

$$\int^{\times} \sum_{n\geqslant 1} F_n \, d\mu = \sum_{n\geqslant 1} \int^{\times} F_n \, d\mu \quad \text{for } F_n \in \mathscr{I}_+, \tag{4}$$

and (cf. (iii))

$$\int^{\times} \sum_{n\geqslant 1} f_n \, d\mu \leqslant \sum_{n\geqslant 1} \int^{\times} f_n \, d\mu \quad \text{for arbitrary } f_n \geqslant 0. \tag{5}$$

(v) For arbitrary sets $A \subset X$ we define the *outer measure* $\mu^{\times}(A)$ as $\mu^{\times}(\phi_A)$, where ϕ_A is the characteristic function of A (§ 1.9). We have $0 \leqslant \mu^{\times}(A) \leqslant \infty$, $\mu^{\times}(A) \leqslant \mu^{\times}(B)$ if $A \subset B$ and (cf. (5))

$$\mu^{\times}\Big(\bigcup_{n\geqslant 1} A_n\Big) \leqslant \sum_{n\geqslant 1} \mu^{\times}(A_n) \qquad A_n \subset X. \tag{6}$$

(vi) We say that $A \subset X$ is a *negligible set* (relative to μ) if $\mu^{\times}(A) = 0$. Subsets and countable unions of negligible sets are negligible. We use the expression '*almost everywhere*' (a.e.) in the sense of 'for all points of X except for a negligible set' (which may, of course, be empty).

(vii) Let f, g be any functions on X with values in $\overline{\mathbf{R}}_+$. We have

$$\int^{\times} f \, d\mu = 0 \Leftrightarrow f(x) = 0 \text{ a.e.}; \qquad \int^{\times} f \, d\mu < \infty \Rightarrow f(x) < \infty \text{ a.e.};$$

$$f(x) = g(x) \text{ a.e.} \Rightarrow \int^{\times} f \, d\mu = \int^{\times} g \, d\mu;$$

$$f(x) \leqslant g(x) \text{ a.e.} \Rightarrow \int^{\times} f \, d\mu \leqslant \int^{\times} g \, d\mu.$$

(viii) Let $(f_n)_{n\geqslant 1}$ be any sequence of functions on X with values in $\overline{\mathbf{R}}_+$. If $\sum_{n\geqslant 1} \int^{\times} f_n \, d\mu < \infty$, then $\sum_{n\geqslant 1} f_n(x) < \infty$ a.e. and in particular $f_n(x) \to 0$ a.e. $(n \to \infty)$. This is a corollary of (5) and (vii), useful in the applications.

2.3. *The spaces* $\mathfrak{L}^1_B(X, \mu)$ *and* $L^1_B(X, \mu)$ (cf. [13 *c*, Chap. IV, § 3, n°s 1–7, § 4, n°s 1–4]).

Let μ be a positive measure on X (Bourbaki (loc. cit.) considers also complex measures). Let B be a real or complex Banach space (with norm $\|.\|$) or, in particular, $\mathbf{R}$ or $\mathbf{C}$.

(i) We define $\mathfrak{F}^1_B(X, \mu)$, or $\mathfrak{F}^1_B$ for short, as the linear space of all functions $\mathbf{f}$: $x \to \mathbf{f}(x)$, $x \in X$, with values in B and such that

$$N_1(\mathbf{f}) = \int^{\times} \|\mathbf{f}(x)\| \, d\mu(x) < \infty. \tag{7}$$

N_1 is a semi-norm on $\mathfrak{F}^1_B$ and $\mathfrak{F}^1_B$ is complete.

(ii) We define the space $\mathfrak{L}^1_B = \mathfrak{L}^1_B(X, \mu)$ as the closure of $\mathscr{K}_B$ in $\mathfrak{F}^1_B$, where $\mathscr{K}_B = \mathscr{K}_B(X)$ is the space of continuous functions on X with values in B and compact support (the definition of support being analogous to the complex case).

If $\mathbf{f} \in \mathfrak{L}^1_B$, then $x \to \|\mathbf{f}(x)\|$ is in $\mathfrak{L}^1_{\mathbf{R}}$. If f, $g \in \mathfrak{L}^1_{\mathbf{R}}$, then $\sup(f, g)$ and $\inf(f, g)$ are also in $\mathfrak{L}^1_{\mathbf{R}}$.

Remark. Every $\mathbf{f} \in \mathfrak{L}^1_B$ vanishes a.e. outside a set which is a union of countably many compact sets.

(iii) Let $\mathfrak{N}_B = \mathfrak{N}_B(X, \mu)$ be the closed linear subspace of all $\mathbf{f} \in \mathfrak{L}^1_B$ with $N_1(\mathbf{f}) = 0$. The quotient space $\mathfrak{L}^1_B/\mathfrak{N}_B$ is a Banach space, denoted by $L^1_B(X, \mu)$ or L^1_B, with a norm derived from (7) and denoted by $\|.\|_1$.

Two functions in $\mathfrak{L}^1_B$ represent the same element of L^1_B if and only if they coincide a.e.; this follows from 2.2 (vii). In practice it is often convenient to disregard the distinction between L^1_B and $\mathfrak{L}^1_B$ and to write, say, $\mathbf{f} \in L^1_B$ rather than $\mathbf{f} \in \mathfrak{L}^1_B$, or $\|\mathbf{f}\|_1$ instead of $N_1(\mathbf{f})$ for $\mathbf{f} \in \mathfrak{L}^1_B$.

Remark. Analogously we define for $1 < p < \infty$ the spaces $\mathfrak{L}^p_B$, with semi-norm N_p, and the Banach spaces L^p_B with norm $\|.\|_p$.

(iv) When $B = \mathbf{R}$ or $\mathbf{C}$, we usually write $\mathfrak{L}^1$ and L^1 for either $\mathfrak{L}^1_{\mathbf{R}}$, $\mathfrak{L}^1_{\mathbf{C}}$ and $L^1_{\mathbf{R}}$, $L^1_{\mathbf{C}}$. The given integral can be extended from $\mathscr{K} = \mathscr{K}(X)$ to $\mathfrak{L}^1$ by continuity (cf. § 2.1 (iv)); it is again denoted by $\int f\,d\mu$ for $f \in \mathfrak{L}^1$ and satisfies the fundamental inequality

$$\left|\int f(x)\,d\mu(x)\right| \leqslant \int |f(x)|\,d\mu(x) \qquad f \in \mathfrak{L}^1 \tag{8}$$

(cf. also (v) below).

For two functions in $\mathfrak{L}^1$ that coincide a.e. the integral has the same value: it is thus a linear functional on L^1.

The integral can also be extended, in the obvious way, to those functions with values in $\mathbf{R}$, $\mathbf{C}$, or $\overline{\mathbf{R}}_+$ which are defined a.e. and coincide a.e. with a function in $\mathfrak{L}^1$. These are called *integrable functions* (with respect to μ). This is a slightly more general term than 'function in $\mathfrak{L}^1$': $\mathfrak{L}^1$ consists precisely of the integrable functions with values in $\mathbf{R}$ or $\mathbf{C}$ and defined everywhere. Similarly we define *p*th-*power integrable functions*.

Remark. We can also consider a real or complex measure μ. Let $|\mu|$ be the corresponding positive measure (§ 2.1 (viii)). Then $\int f\,d\mu$ can be extended from $\mathscr{K}(X)$ to $\mathfrak{L}^1(X, |\mu|)$ by continuity, the inequality (8) being replaced by

$$\left|\int f(x)\,d\mu(x)\right| \leqslant \int |f(x)|\,d|\mu|(x) \qquad f \in \mathfrak{L}^1(X, |\mu|).$$

We can again consider integrable functions for μ: they are those which are integrable for $|\mu|$.

(v) *If* $f \geqslant 0$ and $f \in \mathfrak{L}^1_{\mathbf{R}}$, then $\int f\,d\mu = \int^{\times} f\,d\mu$. If $F \in \mathscr{I}_+$ and $\int^{\times} F\,d\mu < \infty$, then F coincides a.e. with a function in $\mathfrak{L}^1_{\mathbf{R}}$. Likewise, if f is a real, positive, upper semi-continuous function and if $\int^{\times} f\,d\mu < \infty$, then $f \in \mathfrak{L}^1_{\mathbf{R}}$.

(vi) Let A be an open or closed set in X. Then for any $f \in \mathfrak{L}^1$ the function $f\phi_A$ is in $\mathfrak{L}^1$. We define

$$\int_A f\,d\mu = \int_X f\phi_A\,d\mu.$$

Given any $f \in \mathfrak{L}^1$ and $\epsilon > 0$, there is a compact set $K \subset X$ such that $\int_{X-K} |f|\,d\mu < \epsilon$.

(It is enough to consider an open set A, so that $\phi_A \in \mathscr{I}_+$. If $f \in \mathscr{K}$, then

$$\int^{\times} |f|\phi_A\,d\mu = \sup_{k \in \mathscr{K}_+, k \leqslant \phi_A} \int |f| k\,d\mu,$$

by (2). Hence for any $\epsilon > 0$ there is a $k \in \mathscr{K}_+$ such that $\int^{\times} |f\phi_A - fk|\,d\mu < \epsilon$, so $f\phi_A \in \mathfrak{L}^1$. The assertion above now follows readily, since $\mathscr{K}$ is dense in $\mathfrak{L}^1$.)

(vii) Let f be in $L^1(X,\mu)$. Then $k \to \int k(x)f(x)\,d\mu(x)$, $k \in \mathscr{K}$, is a bounded measure on X, denoted by $f\mu$. Also

$$(*) \qquad \|f\mu\| = \int |f(x)|\,d\mu(x).$$

Thus $f \to f\mu$ *is an algebraic and isometric injection of* $L^1(X,\mu)$ *into* $M^1(X)$.

(For $f \in \mathscr{K}$ this is immediate: put $f_n(x) = |f(x)/\|f\|_\infty|^{1/n}.\mathrm{sgn}\,\overline{f(x)}$, $n \geqslant 1$ ($\mathrm{sgn}\,a = a/|a|$ if $a \neq 0$, $\mathrm{sgn}\,0 = 0$), then $f_n \in \mathscr{K}$, $\|f_n\|_\infty \leqslant 1$, $f_n f \geqslant 0$ and $f_n f \uparrow |f|$ ($n \to \infty$). Thus clearly $\|f\mu\| \geqslant \int |f|\,d\mu$ and, since the opposite inequality is obvious, (*) holds for all $f \in \mathscr{K}$. Then (*) follows for all $f \in L^1$ by approximation.)

(viii) In the case of $\mathfrak{L}^1_B$, where B is any Banach space, we first *define* $\int \mathbf{f}\,d\mu$ for $\mathbf{f} \in \mathscr{K}_B$ by the relation

$$(9) \qquad \left\langle \int \mathbf{f}\,d\mu, \mathbf{z}' \right\rangle = \int \langle \mathbf{f}(x), \mathbf{z}' \rangle\,d\mu(x) \qquad \text{for } every\ \mathbf{z}' \in B',$$

where B' is the dual of B; we have $\int \mathbf{f}\,d\mu \in B$ (cf. [13 *c*, Chap. III, § 3]). We can then extend this integral to $\mathfrak{L}^1_B$ as before. The relation (8) becomes here

$$(10) \qquad \left\| \int \mathbf{f}(x)\,d\mu(x) \right\| \leqslant \int \|\mathbf{f}(x)\|\,d\mu(x) \qquad \mathbf{f} \in \mathfrak{L}^1_B,$$

and (9) holds, in fact, for all $\mathbf{f} \in \mathfrak{L}^1_B$.

The integral carries over to L^1_B and also to functions (with values in B) defined a.e. and coinciding a.e. with a function in $\mathfrak{L}^1_B$; these are again called *integrable functions* (cf. (iv)).

2.4. We discuss here *sequences of functions in* $\mathfrak{L}^1_B$ and the *interchange of integral and limit*; μ is a positive measure, except in (iv).

(i) If $\mathbf{f}_n \to \mathbf{f}$ in $\mathfrak{L}^1_B$ ($n \to \infty$), then $\mathbf{f}_{n_j}(x) \to \mathbf{f}(x)$ a.e. ($j \to \infty$) for some subsequence $(n_j)_{j \geqslant 1}$ [13 *c*, Chap. IV, § 3, n° 4, Theorem 3]. If $\sum_{n \geqslant 1} N_1(\mathbf{f} - \mathbf{f}_n) < \infty$, then already $\mathbf{f}_n(x) \to \mathbf{f}(x)$ a.e. ($n \to \infty$) (cf. § 2.2 (vii)).

(ii) Suppose $f_n \in \mathfrak{L}^1_{\mathbf{R}}$, $f_n \geqslant 0$ and $f_n \leqslant f_{n+1}$ ($n \geqslant 1$). Then $\sup_{n \geqslant 1} f_n$ is integrable if and only if $\sup_{n\geqslant 1} \int f_n \, d\mu < \infty$ and in this case $\int \sup_{n\geqslant 1} f_n \, d\mu = \sup_{n\geqslant 1} \int f_n \, d\mu$ [13 *c*, Chap. IV, § 3, nº 6, Theorem 5 and § 4, nº 3, Prop. 4].

(iii) Suppose $\mathbf{f}_n \in \mathfrak{L}^1_B$, $\|\mathbf{f}_n(x)\| \leqslant g(x)$ a.e. $(n \geqslant 1)$, where $\int^{\times} g \, d\mu < \infty$, and $\lim \mathbf{f}_n(x)$ exists a.e. $(n \to \infty)$. Then $\lim \mathbf{f}_n$ is integrable and, if $\mathbf{f} \in \mathfrak{L}^1_B$ coincides a.e. with $\lim \mathbf{f}_n$, then $\mathbf{f}_n \to \mathbf{f}$ in $\mathfrak{L}^1_B$ $(n \to \infty)$; thus $\int \lim \mathbf{f}_n \, d\mu = \lim \int \mathbf{f}_n \, d\mu$ [13 *c*, Chap. IV, § 3, nº 7, Theorem 6 and § 4, nº 3, Theorem 2]. Similarly for $\mathfrak{L}^p_B$, $1 < p < \infty$.

Remark. (ii) and (iii) can also be stated, in an obvious way, for integrable functions.

(iv) In the case of a complex measure μ, let $|\mu|$ be the corresponding positive measure (§ 2.1 (viii)). If a sequence $(f_n)_{n\geqslant 1}$ in $\mathfrak{L}^1(X, |\mu|)$ is such that $|f_n(x)| \leqslant g(x)$ a.e., where $\int^{\times} g \, d|\mu| < \infty$, and if $\lim f_n(x)$ exists a.e. ('a.e.' with respect to $|\mu|$), as $n \to \infty$, then $\int \lim f_n \, d\mu = \lim \int f_n \, d\mu$ (cf. § 2.3 (iv), Remark). Bourbaki (loc. cit. (iii)) combines (iii) and (iv) in a single formulation.

(v) Let (F_α) be a family in $\mathscr{I}_+ \cap \mathfrak{L}^1_{\mathbf{R}}$ filtering with respect to '$\leqslant$' and such that $\sup_\alpha \int F_\alpha \, d\mu < \infty$. Then $\sup_\alpha F_\alpha$ is integrable and $\int \sup_\alpha F_\alpha \, d\mu = \sup_\alpha \int F_\alpha \, d\mu$. This follows at once from (2) and 2.3 (v). As a corollary we have: if (f_α) is a family of positive, upper semi-continuous functions, in $\mathfrak{L}^1_{\mathbf{R}}$ filtering with respect to '$\geqslant$', then $\inf_\alpha f_\alpha \in \mathfrak{L}^1_{\mathbf{R}}$ and $\int \inf_\alpha f_\alpha \, d\mu = \inf_\alpha \int f_\alpha \, d\mu$ (cf. [13 *c*, Chap. IV, § 4, nº 4, Cor. 2 of Prop. 5]).

2.5. (Cf. [13 *c*, Chap. IV, § 4, nºs 5 and 6].) Let μ be a positive measure on X. If a set $A \subset X$ is such that ϕ_A is in $\mathfrak{L}^1(X, \mu)$, we call A an *integrable set* (with respect to μ) and put $\mu(A) = \int \phi_A \, d\mu$: this is called the *measure* of A. We note that $0 \leqslant \mu(A) < \infty$. All open or closed sets of finite outer measure are integrable, in particular all compact sets.

A very important general criterion is this: a set $A \subset X$ is integrable if and only if for every $\epsilon > 0$ there is an integrable open set $\Omega \supset A$ and a compact set $K \subset A$ such that $\mu(\Omega) - \mu(K) < \epsilon$. That is, for every integrable set A we have:

(11 a) $$\mu(A) = \inf_{\Omega \supset A} \mu(\Omega) \qquad (\Omega \text{ open, integrable}),$$

(11 b) $$\mu(A) = \sup_{K \subset A} \mu(K) \qquad (K \text{ compact}).$$

These relations are quite important and also receive special attention in other approaches to measure theory.

2.6. We mention here *product measures* (cf. [13 *c*, Chap. III, § 4, nºs 1–4]). Let X, Y be locally compact spaces, $X \times Y$ their product. Let μ_1, μ_2 be (real or complex) measures on X, Y respectively. It is very simple to show: if $k \in \mathscr{K}(X \times Y)$, then the functions

$$x \to \int_Y k(x,y) \, d\mu_2(y), \qquad y \to \int_X k(x,y) \, d\mu_1(x), \qquad x \in X, \ y \in Y$$

are in $\mathscr{K}(X)$, $\mathscr{K}(Y)$ respectively and

$$\int_X \left\{ \int_Y k(x,y)\, d\mu_2(y) \right\} d\mu_1(x) = \int_Y \left\{ \int_X k(x,y)\, d\mu_1(x) \right\} d\mu_2(y).$$

By making correspond to $k \in \mathscr{K}(X \times Y)$ this common value, we obtain a measure μ on $X \times Y$, as is readily seen; μ is called the *product of* μ_1 *and* μ_2 and we write $\mu = \mu_1 \otimes \mu_2$. If the measures μ_1, μ_2 are positive, then so is $\mu_1 \otimes \mu_2$. If μ_1 and μ_2 are bounded, then $\mu_1 \otimes \mu_2$ is bounded and $\|\mu_1 \otimes \mu_2\| = \|\mu_1\| . \|\mu_2\|$. The product $\otimes$ has the usual algebraic properties.

2.7. *Measurable functions and measurable sets* (cf. [13 *c*, Chap. IV, § 5, n°s 1–6]).

Let μ be a positive measure on X.

(i) Let Φ be a function on X with values in a metric or, more generally, topological space. We say that Φ is a *measurable function* (with respect to μ) if it has the following property: given any compact set $K \subset X$, there is a partition of K into a negligible set and countably many compact sets K_n such that the restriction of Φ to each K_n is continuous. Measurability is thus a local property and is related to continuity: a function is measurable if it is not 'too badly discontinuous'.

This definition, somewhat different from the traditional one, is extremely convenient in the applications. There are various familiar criteria for measurability (cf. also (ii)).

If f, g are real- or complex-valued functions which are measurable, then so are $c.f$ ($c \in \mathbf{R}$ or $\mathbf{C}$), $f+g$, fg, and $1/f$ if $f(x) \neq 0$ for all x. Similarly for functions $\mathbf{f}$, $\mathbf{g}$ with values in a Banach space or algebra. If $\mathbf{f}$ is measurable, then so is $x \to \|\mathbf{f}(x)\|$.

(ii) $A \subset X$ is said to be a *measurable set* if its characteristic function is measurable. It is then shown that A is measurable if and only if $A \cap K$ is integrable for every compact set $K \subset X$. All open or closed sets, in particular X and $\emptyset$, are measurable. If A, B are measurable, then so are $A \cup B$, $A \cap B$ and $X - A$.

For functions with values in $\mathbf{R}$ or $\overline{\mathbf{R}}_+$ there is a familiar criterion for measurability, based on measurable sets; similarly for functions with values in a metric space. Functions in $\mathscr{I}_+$ are measurable.

(iii) In connexion with measurable functions we shall need an extension of the concept of 'almost everywhere'. We call $A \subset X$ a *locally negligible set* if $A \cap K$ is negligible for every compact set $K \subset X$; correspondingly we say '*locally almost everywhere*' (l.a.e.) in the sense of 'for all points of X except for a locally negligible set' (which may, of course, be empty). If X is countable at infinity, these definitions amount to the same as those in 2.2 (vi). Subsets and countable unions of locally negligible sets are locally negligible.

(iv) A function on X coinciding l.a.e. with a measurable function is measurable. On account of this fact one can also call a function defined l.a.e. measurable if it coincides l.a.e. with a measurable function in the sense of (i).

(v) If a sequence of measurable functions on X converges l.a.e., then the limit function is measurable (cf. (iv)). It follows that countable unions and intersections of measurable sets are measurable.

(vi) The following *integrability criterion* is of fundamental importance: a function $\mathbf{f}$ on X, with values in B, is integrable (and hence in $\mathfrak{L}^1_B$) if and only if $\mathbf{f}$ is measurable and $\int^\times \|\mathbf{f}(x)\|\, d\mu(x) < \infty$. Likewise for functions with values in $\overline{\mathbf{R}}_+$, except for the statement in parentheses.

(vii) If Φ is '*locally integrable*' (i.e. if $k\Phi$ is integrable for every $k \in \mathscr{K}$) and if $\int k\Phi\, d\mu = 0$ for all $k \in \mathscr{K}$, then $\Phi(x) = 0$ l.a.e. (For every $k_1 \in \mathscr{K}$ we have $\int |k_1 \Phi|\, d\mu = 0$, by 2.3 (vii), hence $k_1(x)\Phi(x) = 0$ a.e.)

2.8. *The spaces* $\mathfrak{L}^\infty_B(X,\mu)$ *and* $L^\infty_B(X,\mu)$ (cf. [13 *c*, Chap. IV, § 6 and Chap. V, § 5, n° 8]).

Let μ be a positive measure on X.

(i) Let f be any function on X with values in $\overline{\mathbf{R}}_+$. There is a smallest M $(0 \leqslant M \leqslant \infty)$ such that $f(x) \leqslant M$ l.a.e.; we denote it by $\operatorname{ess.sup}_{x \in X} f(x)$.

A function $\boldsymbol{\varphi}$ on X with values in a Banach space B is said to be *essentially bounded* (with respect to μ) if

$$N_\infty(\boldsymbol{\varphi}) = \operatorname*{ess.sup}_{x \in X} \|\boldsymbol{\varphi}(x)\| < \infty. \tag{12}$$

(ii) The measurable, essentially bounded functions $\boldsymbol{\varphi}$ on X with values in B form a complete space, denoted by $\mathfrak{L}^\infty_B(X,\mu)$ or $\mathfrak{L}^\infty_B$, with (12) as semi-norm. It should be noted that $\mathfrak{L}^\infty_B$ consists of functions defined *everywhere*, just like $\mathfrak{L}^p_B$, $1 \leqslant p < \infty$.

Let $\mathfrak{N}^\infty_B = \mathfrak{N}^\infty_B(X,\mu)$ be the closed linear subspace of all $\boldsymbol{\varphi} \in \mathfrak{L}^\infty_B$ such that $N_\infty(\boldsymbol{\varphi}) = 0$. Then $\mathfrak{L}^\infty_B/\mathfrak{N}^\infty_B$ is a Banach space, denoted by $L^\infty_B(X,\mu)$ or L^∞_B, with a norm $\|.\|_\infty$ derived from (12).

Two functions in $\mathfrak{L}^\infty_B$ represent the same element of L^∞_B if and only if they coincide l.a.e. In practice it is often convenient not to distinguish between L^∞_B and $\mathfrak{L}^\infty_B$ (cf. § 2.3 (iii)).

The continuous functions in $\mathfrak{L}^\infty_B$ do not form a *dense* subspace, in general; this contrasts sharply with $\mathfrak{L}^p_B$, $1 \leqslant p < \infty$. The distinction between $\mathfrak{N}^\infty_B$ and $\mathfrak{N}_B$ (cf. § 2.3 (iii)) should also be noted: it corresponds exactly to that between 'l.a.e.' and 'a.e.'

The above applies in particular when $B = \mathbf{R}$ or $\mathbf{C}$; in these cases we simply write $\mathfrak{L}^\infty$ and L^∞.

REMARK. If $f_1, f_2 \in \mathfrak{L}^1$ coincide a.e. and $\phi_1, \phi_2 \in \mathfrak{L}^\infty$ coincide l.a.e., then the functions $f_1\phi_1, f_2\phi_2 \in \mathfrak{L}^1$ coincide a.e. (cf. § 2.3 (ii), Remark). We can thus multiply 'functions in L^1' by 'functions in L^∞' without ambiguity; likewise for L^1_B and L^∞.

(iii) *The dual space of the Banach space $L^1(X,\mu)$ is (isomorphic to) $L^\infty(X,\mu)$*: every continuous linear functional on $L^1(X,\mu)$ can be written in the form

$$f \to \langle f,\phi\rangle \equiv \int f(x)\overline{\phi(x)}\,d\mu(x), \tag{13}$$

where $\phi \in L^\infty(X,\mu)$, and the norm of the functional is precisely $\|\phi\|_\infty$. The use of the complex conjugate in (13), for the complex case, is convenient in the applications.

This fundamental result follows from the Lebesgue–Radon–Nikodym theorem. For the most general and thorough treatment of that theorem see [13 *c*, Chap. V, § 5, n° 5], but for our purposes we require it only for *bounded* positive measures, in which case it is quite simple to prove (loc. cit.). Then one can readily derive the result stated above if X and μ satisfy a certain condition (cf. [101, § 6, No. 16]); this condition holds for a locally compact group and the Haar measure ([101, § 27, No. 5 (IX)]; see [92, § 3] for a reference). The general case is treated in [13 *c*, Chap. V, § 5, n° 8] for $L^1_{\mathbf{R}}$ and then follows for $L^1_{\mathbf{C}}$; but it is not needed here.

Similarly, *the dual of L^p, $1 < p < \infty$, is $L^{p'}$, where $1/p+1/p' = 1$.* The proof is somewhat easier than for $p = 1$.

(iv) The family of all functionals $f \to \langle f,\phi\rangle$, $\phi \in L^\infty$, defines in L^1 a 'weak' topology, the topology $\sigma(L^1, L^\infty)$; similarly the family of all functionals $\phi \to \langle f,\phi\rangle$, $f \in L^1$, defines in L^∞ the topology $\sigma(L^\infty, L^1)$ [13 *b*, Chap. II, § 6, n° 2].

(v) If a sequence $(f_n)_{n\geqslant 1}$ in L^1 is such that $\lim\langle f_n,\phi\rangle$ exists for every $\phi \in L^\infty$, then there is an $f \in L^1$ such that $\lim\langle f_n,\phi\rangle = \langle f,\phi\rangle$ for all $\phi \in L^\infty$ ($n \to \infty$). See e.g. [13 *c*, Chap. V, § 5, Exercise 16], [32, p. 92, Cor.], and [58, Chap. V, § 4, n° 1, Cor. 3 of Theorem 1].

3. Haar measure

3.1. (i) A measure μ on a locally compact group G is said to be *left invariant* if $\mu(L_a f) = \mu(f)$, $f \in \mathscr{K}(G)$, for all $a \in G$, where L_a is the *left translation operator*:

$$L_a f(x) = f(a^{-1}x).$$

On a locally compact group there exists a left invariant positive measure, not identically zero and uniquely determined up to a constant factor. It is called left Haar measure; for shortness we usually say '*Haar measure*' for 'left Haar measure' and speak of 'the' Haar measure, assuming that some Haar measure has been chosen. In analogy with the classical notation we denote the Haar measure by dx and write $\int_G f(x)\,dx$ or $\int f(x)\,dx$, or merely $\int f$. The left invariance is then expressed by

$$\int f(a^{-1}x)\,dx = \int f(x)\,dx \quad \text{or} \quad d(ax) = dx \qquad a \in G.$$

The fundamental theorem above will be taken for granted here.

Historical remarks up to 1938 are given by A. Weil [138, p. 38] who introduced a new method, now classical, into the proof (cf. the review of the first edition of [138] in *Math. Rev.* **3**, 198–9 (1942)). This method was modified by H. Cartan [21] (cf. also [138, p. 159]), and expounded again, essentially, by Bourbaki [13 c, Chap. VII, § 1, n° 2].† A very illuminating proof, close to classical analysis and to Weil's method, is in [100, p. 85]; for another interesting proof see G. E. Bredon [16]. H. F. Davis [28] has shown that the most general Haar measure is a combination of three familiar types: Lebesgue measure on $\mathbf{R}^\nu$, the mean value on compact groups, summation on discrete groups.

(ii) It is useful to note that, if $f \in \mathscr{K}_+(G)$ and f is not identically zero, then $\int f(x)\,dx > 0$. Thus the support of the Haar measure is G itself and any compact neighbourhood of e has strictly positive Haar measure.

Suppose $f_0 \in \mathscr{K}_+$, $f_0(x_0) > 0$ for some $x_0 \in G$, and $\int f_0 = 0$. Given any $f \in \mathscr{K}_+$, there are finitely many $a_n \in G$ such that $\sum_n L_{a_n} f_0$ is strictly positive on $\operatorname{Supp} f$. Then $\lambda \sum_n L_{a_n} f_0 \geqslant f$ for some $\lambda \in \mathbf{R}_+$, and hence $0 \geqslant \int f$, by the left invariance of the Haar measure. Thus $\int f = 0$ for all $f \in \mathscr{K}_+$ and therefore for all $f \in \mathscr{K}$, a contradiction.

(iii) One can also consider a *right Haar measure* $d_R x$ on G satisfying

$$\int f(xa^{-1})\,d_R x = \int f(x)\,d_R x \quad \text{or} \quad d_R(xa) = d_R x \qquad a \in G.$$

Left and right Haar measures are of course related: put

$$\check{f}(x) = f(x^{-1}),$$

then $f \to \int \check{f}(x)\,dx$ is a right Haar measure, $f \to \int \check{f}(x)\,d_R x$ a left Haar measure ($f \in \mathscr{K}(G)$). The relation between these two measures will be studied in more detail in 3.5.

(iv) If G has finite Haar measure, then G is compact (and conversely).

Take $f \in \mathscr{K}_+, f \neq 0, f \leqslant 1$; put $C = \operatorname{Supp} f$, a compact set. If G is not compact, there is a sequence $(a_n)_{n \geqslant 1}$ in G such that the sets $a_n C$ are mutually disjoint for $n \geqslant 1$: choose a_1 arbitrarily and, having chosen $a_1, \ldots, a_n$, take any a_{n+1} outside

† Bourbaki's uniqueness proof is particularly elegant and applies directly to *complex* left invariant measures: such measures are of the form $c.dx$, $c \in \mathbf{C}$.

$\Big(\bigcup_{1\leqslant m\leqslant n} a_m C\Big)C^{-1}$. Then $\int \sum_{1\leqslant n\leqslant N} L_{a_n}f = N\int f$ and $\sum_{1\leqslant n\leqslant N} L_{a_n}f \leqslant 1$ for all integers $N \geqslant 1$. Thus the Haar measure is unbounded.

(v) In practice, the following situation often arises: in some open neighbourhood V of e in G a positive measure $d_V x$ ($\neq 0$) is given which is '*locally left invariant*', that is, for all $f \in \mathscr{K}(G)$ such that $\operatorname{Supp} f \subset V$

$$\int_V L_a f(x)\, d_V x = \int_V f(x)\, d_V x \textit{ if } a \in V \textit{ and } \operatorname{Supp} L_a f \subset V.$$

Then $d_V x$ is the restriction to V of a Haar measure dx on G and dx can be defined explicitly in terms of $d_V x$.

Let U be a symmetric open nd. of e such that $U^2 \subset V$. Then, whenever $f \in \mathscr{K}(G)$ and $\operatorname{Supp} f \subset U$, we have

$$(*) \qquad \int_V L_y f(x)\, d_V x = \int_V f(x)\, d_V x \textit{ for all } y \in G \textit{ such that } \operatorname{Supp} L_y f \subset U.$$

Now let $f \in \mathscr{K}$ be such that $\operatorname{Supp} L_s f \subset U$ for *some* $s \in G$. Then we say that f has 'small' support and *define* $\int f(x)\, dx$ by $\int_V L_s f(x)\, d_V x$; this number depends only on f, not on s, by (*). Thus we obtain a left invariant integral for each $f \in \mathscr{K}$ with small support. We shall show:

(*a*) Every $f \in \mathscr{K}$ can be represented as a (finite) sum $f = \sum_n f_n$, $f_n \in \mathscr{K}$, where each f_n has small support; if $f \in \mathscr{K}_+$, we can take $f_n \in \mathscr{K}_+$.

(*b*) If $\sum_n f_n = \sum_j g_j$, where f_n, $g_j \in \mathscr{K}$ and each f_n, g_j has small support, then

$$\sum_n \int f_n(x)\, dx = \sum_j \int g_j(x)\, dx.$$

Clearly (*a*) and (*b*) yield the result to be proved: for arbitrary $f \in \mathscr{K}$ we put $\int f(x)\, dx = \sum_n \int f_n(x)\, dx$, whenever $f = \sum_n f_n$, $f_n \in \mathscr{K}$ and f_n has small support.

Proof of (*a*). We can cover $\operatorname{Supp} f$ by finitely many translates $s_n^{-1}U$, $1 \leqslant n \leqslant N$. There are N functions $e_n \in \mathscr{K}_+$ such that $\operatorname{Supp} e_n \subset s_n^{-1}U$ and $\sum_n e_n$ is 1 on $\operatorname{Supp} f$ (Chap. 2, § 3.1 (i), Remark). Then $f = \sum_n e_n f$ and each $e_n f$ has small support; also $e_n f \in \mathscr{K}_+$ if $f \in \mathscr{K}_+$.

Proof of (*b*). Let C be a compact set in G so large that all f_n, g_j vanish outside C. There are finitely many functions $e_m \in \mathscr{K}_+$, each with small support, such that $\sum_m e_m$ is 1 on C. Putting

$$F = \sum_n f_n = \sum_j g_j,$$

we can write

$$F = \sum_m \sum_n e_m f_n = \sum_m \sum_j e_m g_j.$$

Since each f_n and each g_j has small support, we have

$$(\alpha) \qquad \int f_n = \sum_m \int e_m f_n, \qquad \int g_j = \sum_m \int e_m g_j \quad \text{for each } n, j.$$

Since each e_m has small support, we also have

$$(\beta) \qquad \int e_m F = \sum_n \int e_m f_n, \qquad \int e_m F = \sum_j \int e_m g_j \quad \text{for each } m.$$

From (α) and (β) we get $\sum_n \int f_n = \sum_m \int e_m F = \sum_j \int g_j$, which proves ($b$).

References. [26, pp. 161–4 and 167–8], [13 c, Chap. VII, § 1, nº 7].

(vi) For the Haar measure we have: if a continuous function vanishes l.a.e., then it vanishes everywhere; more generally, if ϕ is continuous, then $\operatorname{ess.sup}_{x\in G}|\phi(x)|$ coincides with $\sup_{x\in G}|\phi(x)|$ (cf. (ii)). Similarly, if $F \in \mathscr{I}_+(G)$ and $\int^\times F(x)\,dx = 0$, then $F(x) = 0$ for *all* $x \in G$.

(vii) From now on the terms 'a.e.', 'l.a.e.', 'measurable', 'integrable', etc. will refer to Haar measure unless stated otherwise.† The Haar measure of an integrable set $A \subset G$ is denoted by $m_G(A)$ or $m(A)$.

3.2. Let H be a closed subgroup of G. Since H is locally compact, it has a Haar measure $d\xi$. Given $f \in \mathscr{K}(G)$, consider the function

$$x \to \int_H f(x\xi)\,d\xi \qquad x \in G.$$

Since this function is constant on each left coset of H, it is of the form $\dot{f} \circ \pi_H$, where $\dot{f}$ is a function on the quotient space G/H (§ 1.8 (vii)). We put $\dot{f} = T_H f$ and also write by *abus de notation*

(1)
$$T_H f(\dot{x}) = \int_H f(x\xi)\,d\xi \qquad \dot{x} \in G/H,$$

where $\dot{x} = \pi_H(x)$. Clearly $T_H f$ *vanishes outside* $\pi_H(\operatorname{Supp} f)$. Moreover, $T_H f$ *is continuous*; we show actually a little more, for later applications.

LEMMA. Let f be in $\mathscr{K}(G)$. Then, given $\epsilon > 0$, there is a neighbourhood U_ϵ of the neutral element such that for all $x \in G$

(2)
$$\int_H |f(y^{-1}x\xi) - f(x\xi)|\,d\xi < \epsilon \qquad \text{for all } y \in U_\epsilon.$$

In particular, if $H = G$, there is a U_ϵ such that $\int |f(y^{-1}x) - f(x)|\,dx < \epsilon$ for all $y \in U_\epsilon$.

Let V be a compact nd. of e; take $g \in \mathscr{K}_+(G)$ such that g is 1 on $V.\operatorname{Supp} f$. Then

$$|f(y^{-1}x) - f(x)| \leqslant |f(y^{-1}x) - f(x)|g(x) \qquad x \in G,\ y \in V.$$

Put $M_x = 1 + \int_H g(x\xi)\,d\xi$. Given $\epsilon > 0$, there is a nd. $U(\epsilon, M_x)$ of e, contained in V, such that

$$|f(y^{-1}z) - f(z)| < \epsilon/M_x \qquad y \in U(\epsilon, M_x),\ z \in G$$

(it is, of course, enough to consider $z \in V.\operatorname{Supp} f$). Hence

$$|f(y^{-1}x\xi) - f(x\xi)| \leqslant (\epsilon/M_x)g(x\xi) \quad \text{for } y \in U(\epsilon, M_x),\ x \in G,\ \xi \in H,$$

† In Chapter 8, where various measures are considered, the precise meaning will be clear from the context.

which yields (2) with $U(\epsilon, M_x)$ instead of U_ϵ. This already shows that $x \to \int_H f(x\xi)\, d\xi$ is continuous on G, so $T_H f$ is continuous on G/H, that is, *$T_H f$ is in $\mathscr{K}(G/H)$ for every $f \in \mathscr{K}(G)$.*

In particular we have for the auxiliary function g above: $T_H g \in \mathscr{K}_+(G/H)$. Hence we can replace M_x in the preceding proof by $M = \max_{\dot{x} \in G/H} T_H g(\dot{x}) < \infty$ and thus obtain (2).

REMARK. Let μ be any measure on G. If $f \in \mathscr{K}(G)$, then $\int f(xt)\, d\mu(t)$ and $\int f(tx)\, d\mu(t)$ are continuous functions of $x \in G$. This is shown as in the first part of the proof above, if μ is positive, and then follows for real or complex μ (§ 2.1 (ii)).

Thus *the mapping T_H defined by* (1) *carries functions in $\mathscr{K}(G)$ into functions in $\mathscr{K}(G/H)$.*

3.3. (i) Let H be a closed *normal* subgroup of G. Then $\dot{G} = G/H$ is a locally compact group with its own Haar measure $d\dot{x}$. Thus we can consider the functional $f \to \int_{G/H} T_H f(\dot{x})\, d\dot{x}$, $f \in \mathscr{K}(G)$. This is a positive linear functional on $\mathscr{K}(G)$, hence a measure on G (§ 2.1 (v)); moreover, it does not vanish identically and is left invariant.

If $f \in \mathscr{K}_+(G)$ and $f \neq 0$, then $T_H f \in \mathscr{K}_+(G/H)$ and $T_H f \neq 0$, so $\int_{G/H} T_H f(\dot{x})\, d\dot{x} > 0$ (§ 3.1 (ii)). The left invariance follows from the relation $T_H[L_a f] = L_{\pi_H(a)}[T_H f]$ and from the left invariance of $d\dot{x}$.

Thus there is a constant $c > 0$ such that $\int_{G/H} T_H f(\dot{x})\, d\dot{x} = c \int_G f(x)\, dx$ for all $f \in \mathscr{K}(G)$. Now let two of the Haar measures on G, H and G/H be given: then we can normalize the third so that $c = 1$. We shall always use this *canonical relation* and write, in a convenient notation,

$$(3) \qquad \int_{G/H} \left\{ \int_H f(x\xi)\, d\xi \right\} d\dot{x} = \int_G f(x)\, dx \qquad f \in \mathscr{K}(G)$$

or simply

$$d\xi\, d\dot{x} = dx.$$

(3) will be called *Weil's formula*: it is a special case, for normal subgroups, of a result of A. Weil [138, pp. 42–45] and is of fundamental importance.

(ii) Let G_1, G_2 be locally compact groups with Haar measures dx, dy respectively. Then $f \to \int_{G_2} \left\{ \int_{G_1} f(x, y)\, dx \right\} dy$, $f \in \mathscr{K}(G_1 \times G_2)$, is obviously a Haar measure on $G_1 \times G_2$. Thus *the Haar measure of the product $G_1 \times G_2$ is the product of the Haar measures of G_1 and G_2.* Weil's formula may be considered as a generalization of this result. We write

$$dx\, dy = d(x, y).$$

This is the canonical relation of the Haar measures for G_1, G_2, $G_1 \times G_2$.

(iii) Weil's formula (3) extends, of course, to other classes of functions. In the applications the following case is very useful. Let H be a closed normal subgroup of G. *For every* $F \in \mathscr{I}_+(G)$ *the function* $\dot{x} \to \int_H^\times F(x\xi)\,d\xi$ $(\dot{x} = \pi_H(x))$ *is in* $\mathscr{I}_+(G/H)$ *and*

$$\int_{G/H}^\times \left\{ \int_H^\times F(x\xi)\,d\xi \right\} d\dot{x} = \int_G^\times F(x)\,dx. \tag{4}$$

This follows from Weil's formula: consider the filtering family

$$\mathfrak{F}_F = \{k \mid k \in \mathscr{K}_+(G),\ k \leqslant F\},$$

then (cf. § 2.2 (iv))

$$\int_G^\times F(x)\,dx = \sup_{k \in \mathfrak{F}_F} \int_G k(x)\,dx = \sup_{k \in \mathfrak{F}_F} \int_{G/H} \left\{ \int_H k(x\xi)\,d\xi \right\} d\dot{x}$$

$$= \int_{G/H}^\times \left\{ \sup_{k \in \mathfrak{F}_F} \int_H k(x\xi)\,d\xi \right\} d\dot{x} = \int_{G/H}^\times \left\{ \int_H^\times F(x\xi)\,d\xi \right\} d\dot{x}.$$

For this method of proof, and related results, see [136, I, n°s 2 and 3].

Another extension will be given in § 4.5.

(iv) Let G_1, G_2 be locally compact groups; let the Haar measure on $G_1 \times G_2$ be determined as in (ii). Then *for* $F \in \mathscr{I}_+(G_1 \times G_2)$ *the functions* $x \to \int_{G_2}^\times F(x,y)\,dy$, $y \to \int_{G_1}^\times F(x,y)\,dx$ *are* in $\mathscr{I}_+(G_1)$, $\mathscr{I}_+(G_2)$, *respectively, and*

$$\int_{G_1}^\times \left\{ \int_{G_2}^\times F(x,y)\,dy \right\} dx = \int_{G_2}^\times \left\{ \int_{G_1}^\times F(x,y)\,dx \right\} dy = \int_{G_1 \times G_2}^\times F(x,y)\,d(x,y). \tag{5}$$

This follows from (iii). It is also contained in § 4.8, Lemma.

3.4. A complex measure μ on G, not identically zero, is said to be *relatively invariant* if there is a function D_μ on G such that for all $f \in \mathscr{K}(G)$

$$\int f(a^{-1}x)\,d\mu(x) = D_\mu(a) \int f(x)\,d\mu(x) \qquad a \in G \tag{6}$$

(or $d\mu(ax) = D_\mu(a)\,d\mu(x)$). Then $D_\mu(e) = 1$ and

$$D_\mu(ab) = D_\mu(a)D_\mu(b) \qquad a, b \in G, \tag{7}$$

hence $D_\mu(a) \neq 0$ for all a; also $D_\mu > 0$ if μ is positive. We note that D_μ is continuous.†

The relatively invariant measures considered above are, more precisely, relatively left invariant. We can also consider relatively right

† Observe that the left-hand side of (6) is a continuous function of $a \in G$ (cf. § 3.2, Remark).

invariant measures μ_R, characterized by the property that for all $f \in \mathscr{K}(G)$

(8) $$\int f(xa^{-1})\, d\mu_R(x) = D_{\mu_R}(a) \int f(x)\, d\mu_R(x) \qquad a \in G.$$

Here D_{μ_R} again satisfies (7) and is continuous.

If we know a relatively left invariant measure μ on G, we can easily obtain the Haar measure: in fact,

(9) $$f \to \int f(x) \frac{1}{D_\mu(x)}\, d\mu(x) \qquad f \in \mathscr{K}(G)$$

is a left invariant measure and hence up to a constant (possibly complex) a Haar measure (cf. § 3.1 (i), footnote). Likewise, if μ_R is a relatively right invariant measure on G, we can obtain a right Haar measure in the same way.

3.5. Let dx be a Haar measure on G. For every $a \in G$ the measure $f \to \int f(xa^{-1})\, dx, f \in \mathscr{K}(G)$, is left invariant, hence there is a number $\Delta(a)$ such that

(10) $$\int f(xa^{-1})\, dx = \Delta(a) \int f(x)\, dx \qquad f \in \mathscr{K}(G).$$

Thus a (left) Haar measure is relatively right invariant (cf. (8)), Δ is a morphism of G into $\mathbf{R}_+^*$, the multiplicative group of strictly positive real numbers, and $f \to \int f(x)\Delta(x)^{-1}\, dx$ is a right Haar measure, or

(11) $$d_R x = \Delta(x^{-1})\, dx$$

(and likewise $dx = \Delta(x) d_R x$). Then $f \to \int \check{f}(x)\Delta(x^{-1})\, dx$ is a (left) Haar measure (§ 3.1 (iii)) and thus for some constant $c > 0$

$$\int \check{f}(x)\Delta(x^{-1})\, dx = c \int f(x)\, dx \qquad f \in \mathscr{K}(G).$$

Replacing here f by $(f.\Delta)^\vee \in \mathscr{K}(G)$ we obtain $c = 1$ or

(12) $$\int f(x^{-1})\Delta(x^{-1})\, dx = \int f(x)\, dx \qquad f \in \mathscr{K}(G).$$

In (12) we may interchange $f(x)$ and $f(x^{-1})$, hence

(13) $$\int f(x^{-1})\, dx = \int f(x)\Delta(x^{-1})\, dx \quad \text{or} \quad d(x^{-1}) = \Delta(x^{-1})\, dx.$$

The function Δ defined by (10) is called the *Haar modulus* of G. If $\Delta = 1$, G is said to be a *unimodular group* and the left and right Haar measures of G coalesce (as for abelian groups).

REMARK 1. The elements $x \in G$ for which $\Delta(x) = 1$ form a closed normal subgroup G_1 of G. It can happen that G_1 is abelian and open. An example is the multiplicative group of all matrices $\begin{bmatrix} a^n & t \\ 0 & 1 \end{bmatrix}$ ($n \in \mathbf{Z}$, $t \in \mathbf{R}$) with $a > 1$ fixed; here G_1 consists of the matrices $\begin{bmatrix} 1 & t \\ 0 & 1 \end{bmatrix}$.

REMARK 2. The negligible sets, the locally negligible sets, and the measurable sets in G are the same for left and right Haar measures (cf. (11)). But in any non-unimodular l.c. group there are open sets integrable for the left Haar measure but not for the right Haar measure (cf. e.g. [74, § 20.29]).

3.6. Each closed subgroup H of G, being a locally compact group itself, has its own Haar measure; we write Δ_G, Δ_H,..., for the Haar moduli when considering different groups.

(i) If H is a closed *normal* subgroup of G, then $\Delta_H(\xi) = \Delta_G(\xi)$ for all $\xi \in H$. In particular, if a closed normal subgroup H is such that Δ_G is 1 on H, then H is unimodular.†

(ii) If H is a *compact* subgroup of G, then $\Delta_H(\xi) = \Delta_G(\xi) = 1$ for all $\xi \in H$. In particular, compact groups are unimodular.

(i) Fix $\eta \in H$ and put $f_\eta(x) = f(x\eta^{-1})$ for $f \in \mathscr{K}(G)$. Then by Weil's formula

$$\int_{G/H}\left\{\int_H f_\eta(x\xi)\,d\xi\right\}d\dot{x} = \int_G f_\eta(x)\,dx = \int_G f(x\eta^{-1})\,dx$$

or by (10), applied once on H and once on G,

$$\Delta_H(\eta)\int_{G/H}\left\{\int_H f(x\xi)\,d\xi\right\}d\dot{x} = \Delta_G(\eta)\int_G f(x)\,dx,$$

whence $\Delta_H(\eta) = \Delta_G(\eta)$.

(ii) The image of H under Δ_H or Δ_G is a *compact* subgroup of $\mathbf{R}_+^*$ and thus must reduce to $\{1\}$.

3.7. Let μ be a measure on G such that $\int f\,d\mu > 0$ for all $f \in \mathscr{K}_+(G)$, $f \neq 0$. Suppose that μ has the following property: for all $f \in \mathscr{K}(G)$ the relation

$$\int f(a^{-1}x)\,d\mu(x) = \int f(x)\lambda(a,x)\,d\mu(x) \qquad a \in G \tag{14}$$

(or $d\mu(ax) = \lambda(a,x)\,d\mu(x)$) holds, where λ is a continuous function on $G \times G$, independent of f.

Then λ necessarily satisfies

$$\lambda(ab,c) = \lambda(a,bc)\lambda(b,c) \qquad a, b, c \in G, \tag{15}$$

as is readily seen by replacing a in (14) by ab and using the fact that λ is continuous (and that μ satisfies the condition stated at the beginning). Now (14) shows that $\lambda(e,x) = 1$ for all $x \in G$ and hence we get from (15): $1 = \lambda(x^{-1},xy)\lambda(x,y)$ for all x, $y \in G$. Thus λ never vanishes; moreover, (14) also implies that λ is positive. Now put

$$\vartheta(x) = \lambda(x,e) \qquad x \in G. \tag{16}$$

† Thus the subgroup G_1 of 3.5, Remark 1, is necessarily unimodular.

Then we obtain from (15)

$$\vartheta(ax) = \lambda(a,x)\vartheta(x) \qquad a, x \in G. \tag{17}$$

Thus $f \to \int f(x)\vartheta(x)^{-1}\,d\mu(x)$ is a Haar measure on G, as is easy to verify.

Any (complex) measure on G satisfying (14)—with λ possibly complex —is said to be *quasi-invariant*.† A relatively invariant measure corresponds to the special case when $\lambda(a,x)$ is independent of x.

REMARK. In practice, it usually happens that a measure μ is given in some open neighbourhood W of e and has the properties: (i) $\int_W f\,d\mu > 0$ if $f \in \mathscr{K}_+(G)$, $f \neq 0$, and $\operatorname{Supp} f \subset W$; (ii) the relation (14) holds locally in W, i.e. there is a continuous function λ, defined on $W \times W$, such that for all $f \in \mathscr{K}(G)$ with $\operatorname{Supp} f \subset W$

$$\int_W L_a f(x)\,d\mu(x) = \int_W f(x)\lambda(a,x)\,d\mu(x) \quad \text{if } a \in W \text{ and } \operatorname{Supp} L_a f \subset W. \tag{18}$$

Then the considerations above remain valid in some neighbourhood $V \subset W$ of e and yield a 'locally left invariant' positive measure there which is the restriction of the Haar measure of G (cf. § 3.1 (v)).

More precisely, let V be a symmetric open nd. of e such that $V^2 \subset W$. Applying (18) to all $f \in \mathscr{K}(G)$ with $\operatorname{Supp} f \subset V$, we obtain (15) for all a, b in V such that $ab \in V$, and all $c \in V$. Thus λ is strictly positive in $V \times V$ and (17) holds (at least) for all a, x in V such that $ax \in V$; hence we obtain the situation described in 3.1 (v), with $d_V x = \vartheta(x)^{-1}\,d\mu_V(x)$, where μ_V is the restriction of μ to V.

The preceding can be applied to calculate the Haar measure of concrete matrix groups, or Lie groups, as shown below.

3.8. Let G be a multiplicative group of real $n \times n$ matrices which is locally compact in the 'natural' topology induced by that of $\mathbf{R}^{n^2}$. Suppose that we can inject G topologically into $\mathbf{R}^m$ as an *open* subset, for some $m \leqslant n^2$, by expressing the n^2 elements of a general matrix $x \in G$ as functions of m parameters. We can then identify $x \in G$ with a point $(x_1,\ldots,x_m)$ in $\mathbf{R}^m$ and multiplication in G, say $z = xy$, is expressed by

$$z_j = F_j(x_1,\ldots,x_m; y_1,\ldots,y_m) = F_j(x,y) \qquad 1 \leqslant j \leqslant m. \tag{19}$$

We assume that the functions F_j have continuous partial derivatives. Examples are given below. Put‡

$$\lambda(x,y) = |\det[\partial F_j(x,y)/\partial y_k]| \qquad x, y \in G \tag{20}$$

† Quasi-invariant measures can be defined in a much more general way, but this is not needed here (see also Chap. 8, § 1.3).

‡ If G is connected we may omit the absolute value sign.

and define $\vartheta(x)$ by (16). Then *the Haar measure of* G *is* $\vartheta(x)^{-1}d_{\mathbf{R}^m}x$, *where* $d_{\mathbf{R}^m}x$ *is the restriction of the Lebesgue measure of* $\mathbf{R}^m$ *to* G. To see this, it is enough to observe that, if λ is defined by (20), then (14) holds for $d_{\mathbf{R}^m}x$, since $\lambda(a,x)$ is the absolute value of the Jacobian of the transformation $x \to ax$ $(a, x \in G)$.† Quite analogously the *right* Haar measure is $\vartheta_R(x)^{-1}d_{\mathbf{R}^m}x$, where $\vartheta_R(x) = \lambda_R(e,x)$, and λ_R is defined by

$$\lambda_R(x,y) = |\det[\partial F_j(x,y)/\partial x_k]| \qquad x, y \in G.$$

In general not the whole group but only some neighbourhood of e can be injected as an *open* set into $\mathbf{R}^m$ (for some m). Then the remark in 3.7 applies and the calculation of the 'local' Haar measure is the same.

The preceding considerations apply to all Lie groups; thus the Haar measure of a Lie group can be obtained by more or less simple computations.

References. [79], [4, especially pp. 572–6].

Examples

(i) $GL(n,\mathbf{R})$, $n \geqslant 2$, the group of all $n\times n$ real matrices x with $\det x \neq 0$; this is, topologically, an open set in $\mathbf{R}^{n^2}$. It is readily seen that the Lebesgue measure $d_{\mathbf{R}^{n^2}}x$, restricted to this set, is relatively left and right invariant; the Haar measure is $|\det x|^{-n}d_{\mathbf{R}^{n^2}}x$ and is left and right invariant, i.e. $GL(n,\mathbf{R})$ is unimodular.

(ii) $SL(n,\mathbf{R})$, $n \geqslant 2$, the group of all $n\times n$ real matrices x with $\det x = 1$. This is a closed normal subgroup of $GL(n,\mathbf{R})$ and hence (§ 3.6 (i)) unimodular.

To obtain the Haar measure explicitly we inject an open nd. of e topologically into $\mathbf{R}^{n^2-1}$. If $n = 2$, $x = [x_{ij}]_{1\leqslant i,j\leqslant 2}$, we can take $u = x_{11}$, $v = x_{12}$, $w = x_{21}$ as parameters in a nd. of the unit matrix: this nd. is then injected topologically into the half-space $u > 0$ of $\mathbf{R}^3$ and the (restriction of the) Haar measure is $u^{-1}\,du dv dw$. A remarkable geometrical method for obtaining the Haar measure of $SL(n,\mathbf{R})$ has been given by C. L. Siegel [126].

(iii) $ST_+(n,\mathbf{R})$, $n \geqslant 2$, the group of all $n\times n$ real matrices $x = [x_{ij}]$ with $\det x = 1$ and such that $x_{ij} = 0, 1 \leqslant j < i \leqslant n, x_{ii} > 0, 1 \leqslant i \leqslant n$. This group is not unimodular. It can be injected as an open set into $\mathbf{R}^{\frac{1}{2}n(n+1)-1}$; the left and right Haar measures can then be computed by the method above.

(iv) $ST_1(n,\mathbf{R})$, $n \geqslant 2$, the group of all $n\times n$ real matrices $[x_{ij}]$ with $x_{ii} = 1$, $1 \leqslant i \leqslant n$, $x_{ij} = 0$, $1 \leqslant j < i \leqslant n$; this is a closed *normal*

† The Jacobian is 1 for $a = e$ and hence always positive for connected G. We remark that here (15) can also be verified by the 'chain rule' of the differential calculus.

subgroup of $ST_+(n, \mathbf{R})$. Topologically it is the whole space $\mathbf{R}^{\frac{1}{2}n(n-1)}$ and the Lebesgue measure of $\mathbf{R}^{\frac{1}{2}n(n-1)}$ is left (and right) invariant, hence it is also the Haar measure of $ST_1(n, \mathbf{R})$.†

(v) The Haar measures of the corresponding complex groups $GL(n, \mathbf{C})$, $SL(n, \mathbf{C})$ etc. can be obtained in an analogous way.

References. [49, Chap. I, §§ 2 and 4], [13 *c*, Chap. VII, § 3, n° 3].

(vi) A quite different example of a Haar measure is the following. Let G be the product $\prod_{n \geqslant 1} G_n$, each G_n being a group of order K_n, $2 \leqslant K_n < \infty$. The group G can be topologically injected into the interval $[0,1]$ as a Cantor set C of Lebesgue measure zero. The Haar measure of G can be represented as a Cantor function associated with C; it depends only on the sequence $(K_n)_{n \geqslant 1}$.

References. [76, p. 368]; cf. also the reference to L. E. J. Brouwer in [100, § 1.11].

We represent the elements of each G_n by 0, 1,..., $K_n - 1$, with 0 corresponding to e_n. Every $x = (x_n) \in G$ corresponds to a (unique) sequence of integers (a_n), $0 \leqslant a_n \leqslant K_n - 1$, and conversely. Put $P_n = \prod_{1 \leqslant j \leqslant n} (2K_j - 1)$ and

$$x' = \sum_{n \geqslant 1} \frac{2a_n}{P_n}. \qquad (*)$$

The set C of all x' thus obtained lies in $[0,1]$ and $x \to x'$ is a topological bijection of G onto C. (Since $\sum_{n \geqslant 1} (2K_n - 2)/P_n = 1$, we have $0 \leqslant x' \leqslant 1$. Also

$$\sum_{n \geqslant N+1} (2K_n - 2)/P_n = 1/P_N, \quad N \geqslant 1,$$

hence $x_1' = x_2' \Leftrightarrow x_1 = x_2$. Moreover, the elements $(x_n) \in G$ such that $x_n = e_n$ for $1 \leqslant n \leqslant N$ form a nd. U_N of e in G, $(U_N)_{N \geqslant 1}$ is a fundamental sequence of nds. of e, and for any $x, y \in G$ we have: $y^{-1}x \in U_N \Leftrightarrow |x' - y'| \leqslant 1/P_N$.)

C is a Cantor set of Lebesgue measure zero. Now put $Q_n = \prod_{1 \leqslant j \leqslant n} K_j$, $n \geqslant 1$, and define a function F_C on C by

$$F_C(x') = \sum_{n \geqslant 1} \frac{a_n}{Q_n} \quad \text{if } x' \text{ is as in } (*).$$

Then $F_C(x')$ is the Haar measure of the open subset of G corresponding to $C \cap [0, x'[$.

(The sets $C \cap [0, 2/P_n[$ correspond to the nds. U_n introduced above which are actually open normal subgroups of G, of index Q_n. For general x' as in (*), the set $C \cap [0, x'[$ corresponds to a union of disjoint cosets of the subgroups U_n, containing precisely a_n cosets of U_n for each $n \geqslant 1$ for which $a_n > 0$.)

The function F_C is the restriction to C of a positive, continuous function on $[0,1]$ increasing from 0 to 1, but constant on each interval contiguous to C: this is called a Cantor function attached to C.

† Thus there are very simple non-discrete l.c. groups which are quite different algebraically, but are isomorphic as topological spaces and also have the same Haar measure.

3.9. We consider now measurable sets and measurable functions on l.c. groups (cf. also § 3.1 (vii)).

Let H be a closed normal subgroup of a l.c. group G. If E is a locally negligible set in G/H, then $\pi_H^{-1}(E)$ is locally negligible in G, and conversely $[\pi_H\colon G \to G/H]$.

COROLLARY. *If Φ is a (complex-valued) measurable function on G/H, then $\Phi \circ \pi_H$ is measurable on G.*† *Also*

$$\operatorname*{ess.\,sup}_{x \in G} |\Phi \circ \pi_H(x)| = \operatorname*{ess.\,sup}_{\dot{x} \in G/H} |\Phi(\dot{x})|.$$

First we show: for every compact set $K \subset G$ there is a constant C_K such that

$$(*) \qquad m_G^{\times}(\Omega) \leqslant C_K m_{G/H}^{\times}(\pi_H(\Omega)) \quad \text{for all open sets } \Omega \subset K,$$

where $m^{\times}$ denotes outer Haar measure (Ω and $\pi_H(\Omega)$ are, in fact, integrable, but we do not need this here).

Let ϕ' be the characteristic function of the *open* set $\pi_H(\Omega)$; thus $\phi' \in \mathscr{I}_+(G/H)$ and $\phi' \circ \pi_H \in \mathscr{I}_+(G)$. Take an $f_1 \in \mathscr{K}_+(G)$ such that f_1 is 1 on K. Then (cf. (1) and (4))

$$m_G^{\times}(\Omega) \leqslant \int_G^{\times} \phi'(\pi_H(x)) f_1(x)\, dx = \int_{G/H}^{\times} \phi'(\dot{x}) T_H f_1(\dot{x})\, d\dot{x}.$$

This yields (*) with $C_K = \max_{\dot{x} \in G/H} T_H f_1(\dot{x})$.

Now suppose $E \subset G/H$ is loc. negl. Let K_1 be any compact set in G and put $E_1 = K_1 \cap \pi_H^{-1}(E)$: we want to show that E_1 is *negligible*. Choose a compact set $K \subset G$ such that $E_1 \subset K^{\circ}$, the interior of K. The set $\pi_H(E_1)$ is negligible, hence for any $\epsilon > 0$ there is an open set $\Omega' \supset \pi_H(E_1)$ such that $m_{G/H}^{\times}(\Omega') < \epsilon$; put

$$\Omega = K^{\circ} \cap \pi_H^{-1}(\Omega').$$

By (*) we have $m_G^{\times}(\Omega) < C_K \epsilon$; it follows that $m_G^{\times}(E_1) = 0$. Thus $\pi_H^{-1}(E)$ is loc. negl.

Conversely, suppose $\pi_H^{-1}(E)$ is loc. negl. Let K' be any compact set in G/H: it must be shown that $A' = K' \cap E$ is negligible. We can take an $f \in \mathscr{K}_+(G)$ such that $T_H f(\dot{x}) \geqslant 1$, $\dot{x} \in K'$. The set $A = \pi_H^{-1}(A') \cap \operatorname{Supp} f$ is negligible; hence, given $\epsilon > 0$, there is an open set $\Omega \subset G$ such that $\Omega \supset A$ and $m_G^{\times}(\Omega) < \epsilon/M$, where $M = \max_{x \in G} f(x)$. The characteristic function ϕ_Ω is in $\mathscr{I}_+(G)$ and

$$\int_G^{\times} \phi_\Omega(x) f(x)\, dx < \epsilon;$$

hence, putting

$$\dot{F}(\dot{x}) = \int_H^{\times} \phi_\Omega(x\xi) f(x\xi)\, d\xi, \qquad \dot{x} = \pi_H(x),$$

we have: $\dot{F} \in \mathscr{I}_+(G/H)$ and $\int_{G/H}^{\times} \dot{F}(\dot{x})\, d\dot{x} < \epsilon$, by (4). Now $\dot{F}(\dot{x}) \geqslant 1$ if $\dot{x} \in A'$ (observe that $\phi_\Omega(x) f(x) \geqslant \phi_{A'}(\pi_H(x)) f(x)$ for *all* $x \in G$). Thus A' is negligible, and E is loc. negl. A shorter, but less elementary, proof of the converse part can be given by means of 4.5.

† The converse is also true (and can be proved by means of 4.5 (ii)); but we shall establish a stronger result in 6.5. Compare also 1.8 (vii).

The Corollary follows immediately, by the definition of measurability and 'ess. sup'.

For an extension of the results above to the case that H is not normal, and references, see Chapter 8, § 2.2.

3.10. A result analogous to 3.9 is the following.

Let π_o be the mapping $(x,y) \to xy$ of $G \times G$ onto G. If E is a locally negligible set in G, then $\pi_o^{-1}(E)$ is locally negligible in $G \times G$, and conversely.

COROLLARY. *If Φ is a (complex-valued) measurable function on G, then $\Phi \circ \pi_o$ is measurable on $G \times G$.*† *Also*

$$\operatorname*{ess.\,sup}_{(x,y)\in G\times G} |\Phi(xy)| = \operatorname*{ess.\,sup}_{x \in G} |\Phi(x)|.$$

REMARK. These results hold likewise for the mappings

$$(x,y) \to y^{-1}x, \quad (x,y) \to xy^{-1}, \quad (x,y) \to x, \quad \ldots.$$

First we show in analogy to 3.9 (*): given any compact set $K \subset G \times G$, there is a constant C_K such that

$$(*) \qquad m^{\times}_{G\times G}(\Omega) \leqslant C_K\, m^{\times}_G(\pi_o(\Omega)) \quad \text{for all open sets } \Omega \subset K.$$

The map π_o is open (§ 1.8 (iii)), so $\pi_o(\Omega)$ is an open set and hence the characteristic function ϕ' of $\pi_o(\Omega)$ is in $\mathscr{I}_+(G)$. Then the function $\phi' \circ \pi_o$ is in $\mathscr{I}_+(G \times G)$. Now take $f,\ g \in \mathscr{K}_+(G)$ such that $g \leqslant 1$ and $f(x)g(y) = 1$ for all $(x,\ y) \in K$, the given compact set. Then for any open set $\Omega \subset K$ we have (cf. (5))

$$m^{\times}_{G\times G}(\Omega) \leqslant \int^{\times}_{G\times G} f(x)g(y)\phi'(xy)\, d(x,y) \leqslant \int^{\times}_{G\times G} f(x)\phi'(xy)\, d(x,y) = \int_G f(x)\, dx \int^{\times}_G \phi'(y)\, dy,$$

which yields (*). We can then show in the same way as in 3.9, merely with a slight change in notation, that $\pi_o^{-1}(E)$ is loc. negl. in $G \times G$ if E is loc. negl. in G.

The proof of the converse is less elementary, using the Lebesgue–Fubini theorem (§ 4.8). Let ϕ_E be the characteristic function of $E \subset G$. If $\pi_o^{-1}(E) \subset G \times G$ is loc. negl., then $\phi_E \circ \pi_o$ is measurable on $G \times G$ and for any $f,\ g \in \mathscr{K}_+(G)$ we have

$$\int_{G\times G} f(x)g(xy)\phi_E(xy)\, d(x,y) = 0.$$

The Lebesgue–Fubini theorem now implies that $g\,\phi_E$ is integrable on G and $\int_G g(y)\phi_E(y)\, dy = 0$. Since $g \in \mathscr{K}_+(G)$ was arbitrary, E is loc. negl.

Likewise for the other mappings mentioned in the remark above.

References. [11, Anhang, No. 4], [138, p. 49].

4. L^1-spaces on groups

4.1. Let G be a locally compact group. We denote the space of *complex-valued* functions on G integrable with respect to the Haar measure by $\mathfrak{L}^1(G)$ and the corresponding L^1-space by $L^1(G)$ (cf. § 2.3).

† The converse is also true: this results, essentially, from the converse part of the proof that follows.

We also write from now on $\mathscr{K}(G)$ for $\mathscr{K}_{\mathbf{C}}(G)$. We observe that $k \to \int |k(x)|\, dx$, $k \in \mathscr{K}(G)$, is a norm on $\mathscr{K}(G)$, not only a semi-norm (cf. § 3.1 (ii)). Thus $L^1(G)$ may be considered, abstractly, as the completion of $\mathscr{K}(G)$ in this norm and the properties of $L^1(G)$ can often be obtained most simply by a *reduction to* $\mathscr{K}(G)$, as pointed out by A. Weil [138, p. 32].

Likewise for $\mathfrak{L}^p(G)$ and $L^p(G)$, $1 < p < \infty$.

REMARK. The support of a function in $\mathfrak{L}^p(G)$, $1 \leqslant p < \infty$, is a union of countably many compact sets. For $p = 1$ this follows from 2.3 (7), 1.8 (iv), and 3.1 (ii) (cf. [13 *c*, Chap. VII, § 1, Exercise 10]); likewise for $p > 1$.

4.2. Let H be a closed normal subgroup of G; put $\dot{G} = G/H$. We want to investigate the relation between $L^1(G)$ and $L^1(\dot{G})$.

Let dx, $d\xi$, $d\dot{x}$ be Haar measures on G, H, $\dot{G}$ respectively which are canonically related and consider the mapping T_H defined in 3.2:

$$(1) \qquad f \to T_H f\colon\ T_H f(\dot{x}) = \int_H f(x\xi)\, d\xi \qquad f \in \mathscr{K}(G),$$

where $\dot{x} = \pi_H(x)$. As shown in 3.2, T_H maps $\mathscr{K}(G)$ into $\mathscr{K}(\dot{G})$ and from Weil's formula it follows that

$$(2) \qquad \|T_H f\|_1 \leqslant \|f\|_1,$$

the norms being, respectively, those in $L^1(\dot{G})$ and in $L^1(G)$. Thus T_H is a *bounded* linear mapping of $\mathscr{K}(G)$ into $\mathscr{K}(\dot{G})$. We assert that T_H *maps* $\mathscr{K}(G)$ *onto*† $\mathscr{K}(\dot{G})$. Indeed, let any $\dot{f} \in \mathscr{K}(\dot{G})$ be given; then we can obtain an $f_1 \in \mathscr{K}(G)$ such that $T_H f_1 = \dot{f}$ as follows. Put $\dot{K} = \operatorname{Supp}\dot{f}$; there is a compact set $K \subset G$ such that $\pi_H(K) = \dot{K}$ (§ 1.8 (ii)). Take a $k \in \mathscr{K}_+(G)$ strictly positive on K; then $T_H k(\dot{x}) > 0$ for $\dot{x} \in \dot{K}$. Now define for $x \in G$

$$(3) \qquad f_1(x) = \begin{cases} \dot{f} \circ \pi_H(x)\, \dfrac{k(x)}{T_H k \circ \pi_H(x)} & x \in \pi_H^{-1}(\dot{K}) \\ 0 & x \notin \pi_H^{-1}(\dot{K}). \end{cases}$$

Then f_1 is continuous (this is readily seen for $x \in \pi_H^{-1}(\dot{K})$ and is trivial for the complement, an open set); also $\operatorname{Supp} f_1$ is compact. Thus $f_1 \in \mathscr{K}(G)$; moreover, $T_H f_1 = \dot{f}$.

We even have by Weil's formula, since k in (3) is positive,

$$(4) \qquad \|\dot{f}\|_1 = \|f_1\|_1.$$

Thus we may write, on comparing (4) with (2),

$$(5) \qquad \|\dot{f}\|_1 = \inf_{f \in \mathscr{K}(G),\, T_H f = \dot{f}} \|f\|_1.$$

† This is also true if H is an *arbitrary* closed subgroup of G and $\dot{G}$ the quotient *space* G/H. In fact, the proof does not depend on H being normal.

Let $\mathscr{J}(G,H)$ be the kernel of T_H in $\mathscr{K}(G)$:

(6) $$\mathscr{J}(G,H) = \{k_0 \mid k_0 \in \mathscr{K}(G),\ T_H k_0 = 0\}.$$

$\mathscr{J}(G,H)$ is a *closed* linear subspace of $\mathscr{K}(G)$, considered as a normed linear space (with norm $\|.\|_1$): indeed, T_H is continuous ((2)). We have

(7) $$\mathscr{K}(G/H) \simeq \mathscr{K}(G)/\mathscr{J}(G,H)$$

and this isomorphism is not only algebraic, but also isometric, if we provide $\mathscr{K}(G)/\mathscr{J}(G,H)$ with the usual quotient norm: this is the significance of relation (5).

4.3. Now let, quite generally, $\mathscr{S}$ be a normed linear space and $\mathscr{S}_0$ a closed linear subspace of $\mathscr{S}$. Then

(8) $$(\mathscr{S}/\mathscr{S}_0)^{-} \simeq \bar{\mathscr{S}}/\bar{\mathscr{S}}_0,$$

where bars denote completion and '$\simeq$' is an *algebraic and isometric* isomorphism, the two quotient spaces being provided with the usual quotient norms. $\bar{\mathscr{S}}_0$, of course, is the closure of $\mathscr{S}_0$ in $\bar{\mathscr{S}}$.

Consider $\mathscr{S}_0$ as a linear subspace of $\bar{\mathscr{S}}$. Then it is simple to verify that

(*) $$\mathscr{S}/\mathscr{S}_0 \cong (\mathscr{S}+\bar{\mathscr{S}}_0)/\bar{\mathscr{S}}_0,$$

where $\mathscr{S}+\bar{\mathscr{S}}_0$ is the usual sum of two subspaces. $(\mathscr{S}+\bar{\mathscr{S}}_0)/\bar{\mathscr{S}}_0$ is dense in $\bar{\mathscr{S}}/\bar{\mathscr{S}}_0$, hence its completion is $\bar{\mathscr{S}}/\bar{\mathscr{S}}_0$, which gives (8).

We add that *the canonical map* $\bar{\mathscr{S}} \to \bar{\mathscr{S}}/\bar{\mathscr{S}}_0$ *is the* (uniquely determined) *continuous extension of the canonical map* $\mathscr{S} \to \mathscr{S}/\mathscr{S}_0$, with $\mathscr{S}/\mathscr{S}_0$ injected into $\bar{\mathscr{S}}/\bar{\mathscr{S}}_0$ via the isomorphism (*).

We remark that of course $\mathscr{S}/\mathscr{S}_0$ can be complete even when $\mathscr{S}$ itself is not.

4.4. If we apply (8) to (7), we obtain (cf. § 4.1):

(9) $$L^1(G/H) \simeq L^1(G)/J^1(G,H),$$

where $J^1(G,H)$ is the closure of $\mathscr{J}(G,H)$ in $L^1(G)$, $\mathscr{J}(G,H)$ being defined by (6). *The isomorphism* (9) *is algebraic and isometric*, $L^1(G)/J^1(G,H)$ being provided with the quotient norm, and is defined via the extension of the map T_H (cf. (1)), by continuity, to the whole space $L^1(G)$. We denote the extended map still by T_H and also write

(10) $$T_H f(\dot{x}) = \int_H f(x\xi)\, d\xi \qquad f \in L^1(G).$$

T_H maps $L^1(G)$ *onto* $L^1(G/H)$. The notation (10) has not only a formal meaning: it will be justified in 4.5.

References. [110, I]; for a generalization cf. Chapter 8, § 2.3 (6).

4.5. Let H be a closed normal subgroup of G and let π_H be the canonical map $G \to G/H$. Let f be in $\mathfrak{L}^1(G)$. Then the following holds.

(i) There is a negligible set A_0 in G/H such that $\xi \to f(x\xi)$, $\xi \in H$, is in $\mathfrak{L}^1(H)$ for every $x \in G$ such that $\pi_H(x) \notin A_0$.

(ii) The function $\dot{x} \to \int_H f(x\xi)\, d\xi$, $\dot{x} = \pi_H(x)$, defined almost everywhere on G/H, is integrable.

(iii)
$$\int_{G/H} \left\{ \int_H f(x\xi)\, d\xi \right\} d\dot{x} = \int_G f(x)\, dx,$$

if the Haar measures $d\xi$, $d\dot{x}$, dx are canonically related.

We call (iii) the *extended Weil formula* (cf. § 3.3 (i)); the proof, and references, will be given in 4.7 (cf. also Chapter 8, § 2.3).

4.6. The notation (10) is now justified by 4.5: the mapping T_H *can be defined* DIRECTLY *by* (10) *for all* $f \in L^1(G)$ and is the extension of (1) by continuity. Thus its kernel in $L^1(G)$ is precisely $J^1(G, H)$, as defined in 4.4; more explicitly: any $f \in L^1(G)$ such that $T_H f$, defined by (10), is (a.e. equal to) 0 can be approximated in $L^1(G)$ by functions $k_0 \in \mathscr{K}(G)$ such that $T_H k_0 = 0$. This fact is not trivial from the viewpoint of classical analysis, as already the case $G = \mathbf{R}$, $H = \mathbf{Z}$ shows.

4.7. We give here the proof of 4.5.

Given $f \in \mathfrak{L}^1(G)$, we have for every $F \in \mathscr{I}_+(G)$ such that $|f| \leqslant F$ (cf. § 3.3 (iii))
$$\int_{G/H}^{\times} \left\{ \int_H^{\times} |f(x\xi)|\, d\xi \right\} d\dot{x} \leqslant \int_{G/H}^{\times} \left\{ \int_H^{\times} F(x\xi)\, d\xi \right\} d\dot{x} = \int_G^{\times} F(x)\, dx.$$

Hence

(*)
$$\int_{G/H}^{\times} \left\{ \int_H^{\times} |f(x\xi)|\, d\xi \right\} d\dot{x} \leqslant N_1(f).$$

Take $f_n \in \mathscr{K}(G)$ such that $N_1(f - f_n) < 2^{-n}$, $n \geqslant 1$. Then by (*)

(**)
$$\int_{G/H}^{\times} \left\{ \int_H^{\times} |f(x\xi) - f_n(x\xi)|\, d\xi \right\} d\dot{x} < 2^{-n} \qquad n \geqslant 1,$$

and hence (§ 2.2 (viii)) there is a negligible set A_0 in G/H such that
$$\int_H^{\times} |f(x\xi) - f_n(x\xi)|\, d\xi \to 0 \quad (n \to \infty) \quad \text{if } \pi_H(x) \notin A_0,$$

which yields 4.5 (i).

Now put $\dot{f}(\dot{x}) = \int_H f(x\xi)\, d\xi$ if $\dot{x} = \pi_H(x) \notin A_0$, and $\dot{f}(\dot{x}) = 0$, say, if $\dot{x} \in A_0$. Then
$$|\dot{f}(\dot{x}) - T_H f_n(\dot{x})| \leqslant \int_H |f(x\xi) - f_n(x\xi)|\, d\xi \quad \text{for } \dot{x} \notin A_0,$$

hence (**) yields (cf. § 2.2 (vii))
$$\int_{G/H}^{\times} |\dot{f}(\dot{x}) - T_H f_n(\dot{x})|\, d\dot{x} < 2^{-n} \qquad n \geqslant 1.$$

This implies 4.5. (ii), since $T_H f_n \in \mathscr{K}(G/H)$, and also shows that

$$\int_{G/H} \dot{f}(\dot{x})\, d\dot{x} = \lim_{n\to\infty} \int_{G/H} T_H f_n(\dot{x})\, d\dot{x} = \lim_{n\to\infty} \int_G f_n(x)\, dx = \int_G f(x)\, dx,$$

which gives 4.5 (iii).

References. [130, III]; cf. also § 4.8.

4.8. The extended Weil formula in 4.5 is of course related to the theorem of Lebesgue–Fubini which we now discuss for the sake of completeness. Let μ_1, μ_2 be positive measures on locally compact spaces X, Y respectively; put $\mu = \mu_1 \otimes \mu_2$ (§ 2.6).

Lemma [13 *c*, Chap. V, § 8, n° 3, Prop. 5]. If $F \in \mathscr{I}_+(X\times Y)$, then the functions

$$x \to \int_Y^\times F(x,y)\, d\mu_2(y), \qquad y \to \int_X^\times F(x,y)\, d\mu_1(x), \qquad x\in X,\ y\in Y$$

are in $\mathscr{I}_+(X)$, $\mathscr{I}_+(Y)$ respectively and

$$\int_X^\times \left\{ \int_Y^\times F(x,y)\, d\mu_2(y)\right\} d\mu_1(x) = \int_Y^\times \left\{ \int_X^\times F(x,y)\, d\mu_1(x)\right\} d\mu_2(y) = \int_{X\times Y}^\times F(x,y)\, d\mu(x,y).$$

This lemma (and its proof) is entirely analogous to 3.3 (iii) and yields again 3.3 (iv). In fact, one can include both the lemma and 3.3 (iii) in a more general statement [13 *c*, Chap. V, § 3, n° 1, Prop. 2].

Theorem of Lebesgue–Fubini ([130, III]; [13 *c*, Chap. V, § 8, n° 4, Theorem 1]). Let μ_1, μ_2 and $\mu = \mu_1 \otimes \mu_2$ be as above. Let f be in $\mathfrak{L}^1(X\times Y, \mu)$.

(i) The function $y \to f(x,y)$ is in $\mathfrak{L}^1(Y,\mu_2)$ for all $x \in X$ outside a μ_1-negligible set in X.

(ii) The function $x \to \int_Y f(x,y)\, d\mu_2(y)$, defined a.e. on X, is μ_1-integrable.

(iii) $$\int_X \left\{ \int_Y f(x,y)\, d\mu_2(y)\right\} d\mu_1(x) = \int_{X\times Y} f(x,y)\, d\mu(x,y).$$

(iv) Since the roles of X and Y may be interchanged, it follows that

$$\int_X \left\{ \int_Y f(x,y)\, d\mu_2(y)\right\} d\mu_1(x) = \int_Y \left\{ \int_X f(x,y)\, d\mu_1(x)\right\} d\mu_2(y).$$

The proof is based on the lemma above and is entirely analogous to that in 4.7. The extended Weil formula and the Lebesgue–Fubini theorem can even be subsumed under a more general formulation ([130, III], [13 *c*, Chap. V, § 3, n° 3, Theorem 1]).

4.9. The following result—somewhat analogous to the extended Weil formula—will be useful later. *Let G be a locally compact group* countable at infinity *and H a closed normal subgroup of G. Let Φ be a function on G with values in $\overline{\mathbf{R}}_+$ and measurable. Then*

$$\operatorname*{ess.\,sup}_{\dot{x}\in G/H} \left\{ \operatorname*{ess.\,sup}_{\xi\in H} \Phi(x\xi)\right\} = \operatorname*{ess.\,sup}_{x\in G} \Phi(x). \tag{11}$$

We note that in (11) each 'ess. sup' refers to a different Haar measure.

First we show

$$(*)\qquad \operatorname*{ess.\,sup}_{\dot{x}\in G/H}\Big\{\operatorname*{ess.\,sup}_{\xi\in H}\Phi(x\xi)\Big\} \leqslant \operatorname*{ess.\,sup}_{x\in G}\Phi(x).$$

We may assume that the right-hand side equals $c_0 < \infty$. Since G is countable at infinity, there is a *negligible* set $M_0 \subset G$ such that $\Phi(x) \leqslant c_0$ if $x \in \complement M_0$. Let ϕ_0 be the characteristic function of M_0; thus $\int_G \phi_0(x)\,dx = 0$. Applying 4.5 to ϕ_0 we obtain, on observing that $\xi \to \phi_0(x\xi)$, $\xi \in H$, is the characteristic function of $H \cap x^{-1}M_0$ and using 2.2 (vii):

(*a*) $H \cap x^{-1}M_0$ is an integrable subset of H except when $\pi_H(x) \in M_0'$, a negligible set in G/H.

(*b*) There is a negligible set M_0'' in G/H containing M_0' such that

$$m_H(H \cap x^{-1}M_0) = 0$$

whenever $\pi_H(x) \in \complement M_0''$.

Now, if $x \in G$ is such that $\operatorname*{ess.\,sup}_{\xi\in H}\Phi(x\xi) > c_0$, then there is a set $A \subset H$ such that $m_H^{\times}(A) > 0$ and $\Phi(x\xi) > c_0$ for all $\xi \in A$. Thus $xA \subset M_0$ or $A \subset H \cap x^{-1}M_0$. But this implies $\pi_H(x) \in M_0''$ (cf. (*b*)). Hence $\operatorname*{ess.\,sup}_{\xi\in H}\Phi(x\xi) \leqslant c_0$ a.e. on G/H and (*) follows.

Now let us prove

$$(**)\qquad \operatorname*{ess.\,sup}_{\dot{x}\in G/H}\Big\{\operatorname*{ess.\,sup}_{\xi\in H}\Phi(x\xi)\Big\} \geqslant \operatorname*{ess.\,sup}_{x\in G}\Phi(x).$$

Take any $c < \operatorname*{ess.\,sup}_{x\in G}\Phi(x)$. Since Φ is measurable, there is an integrable set $M \subset G$ such that $m_G(M) > 0$ and $\Phi(x) > c$ for all $x \in M$ (e.g.

$$M = \{x \mid x \in G,\ \Phi(x) > c\} \cap K,$$

where K is *some* compact set in G). Then we have in a completely analogous way to that above:

(*a'*) $H \cap x^{-1}M$ is an integrable subset of H except when $\pi_H(x) \in N'$, a negligible set in G/H.

(*b'*) There is a set A' in G/H, disjoint from N', which is not negligible and such that $m_H(H \cap x^{-1}M) > 0$ whenever $\pi_H(x) \in A'$.

Hence, if $\pi_H(x) \in A'$, then $\operatorname*{ess.\,sup}_{\xi\in H}\Phi(x\xi) > c$; thus

$$\operatorname*{ess.\,sup}_{\dot{x}\in G/H}\Big\{\operatorname*{ess.\,sup}_{\xi\in H}\Phi(x\xi)\Big\} > c.$$

Since $c < \operatorname*{ess.\,sup}_{x\in G}\Phi(x)$ was arbitrary, (**) follows and the proof is complete.

Remark. The condition that G be countable at infinity cannot be omitted, as the following example, due to H. de Vries, shows. Let G' be any non-discrete l.c. group and put $G = G' \times G_d'$, where G_d' is the abstract group G' with the discrete topology. Since G' is not countable (§ 1.8 (v)), G is not countable at infinity. Define Φ on G by $\Phi(x, y) = 1$ or 0 according as $x = y$ or $x \neq y$. Taking $H = G_d'$, we readily see that (11) fails.

5. The algebra $L^1(G)$

5.1. $\mathscr{K}(G)$ is not only a normed linear space under the norm $\|.\|_1$, but also a normed algebra, multiplication being defined by the *convolution*† $f \star g$:

$$(1) \qquad f \star g(x) = \int f(y) g(y^{-1}x)\, dy.$$

The usual algebraic properties and the relation

$$(2) \qquad \|f \star g\|_1 \leqslant \|f\|_1 . \|g\|_1$$

are easy to verify. If G is abelian, then convolution is commutative (the converse can also readily be shown [123, Theorem 1.11]).

The convolution extends by continuity to $L^1(G)$. In this way the properties of the convolution in $L^1(G)$ can be obtained simply by studying them in $\mathscr{K}(G)$ (cf. also § 4.1).

We can write (1) also in the form

$$(3) \qquad f \star g(x) = \int f(xy^{-1}) \Delta(y^{-1}) g(y)\, dy.$$

(1) can of course be defined directly for all functions f, $g \in \mathfrak{L}^1(G)$; more precisely:

(i) The function $y \to f(y)g(y^{-1}x)$ is in $\mathfrak{L}^1(G)$ for all $x \in G$ outside a negligible subset of G.

(ii) The function $x \to \int f(y)g(y^{-1}x)\, dy$, defined a.e. on G, is integrable. Let $f \star g$ be any function in $\mathfrak{L}^1(G)$ coinciding with it a.e.; then

(iii) $\int |f \star g(x)|\, dx \leqslant \int |f(x)|\, dx . \int |g(x)|\, dx$.

The proof is analogous to that in 4.7. If F_1, $F_2 \in \mathscr{I}_+(G)$, then the function $(x, y) \to F_1(y) F_2(y^{-1}x)$ is in $\mathscr{I}_+(G \times G)$ (cf. § 2.2 (i)) and 3.3 (iv) yields

$$\int^\times \Big\{ \int^\times F_1(y) F_2(y^{-1}x)\, dy \Big\}\, dx = \int^\times F_1(y)\, dy . \int^\times F_2(x)\, dx.$$

(This holds even if one of the factors on the right is ∞.) It follows that, if f, $g \in \mathfrak{L}^1(G)$, then

$$(*) \qquad \int^\times \Big\{ \int^\times |f(y) g(y^{-1}x)|\, dy \Big\}\, dx \leqslant N_1(f) N_1(g).$$

Now choose f_n, $g_n \in \mathscr{K}(G)$ so that $N_1(f-f_n) \leqslant 2^{-n}$, $N_1(g-g_n) < 2^{-n}$, $n \geqslant 1$. Then (*) gives

$$(**) \qquad \int^\times \Big\{ \int^\times |f(y)g(y^{-1}x) - f_n(y) g_n(y^{-1}x)|\, dy \Big\}\, dx$$
$$\leqslant N_1(f-f_n) N_1(g) + N_1(f_n) N_1(g-g_n) \leqslant \text{const}. 2^{-n} \quad (n \geqslant 1).$$

Hence (§ 2.2 (viii)) there is a negligible set $A_0 \subset G$ such that

$$\int^\times |f(y)g(y^{-1}x) - f_n(y) g_n(y^{-1}x)|\, dy \to 0 \quad (n \to \infty) \text{ if } x \in \complement A_0,$$

† When several groups G, H,..., are considered, we write $\star^G$, $\star^H$,..., for clarity.

which yields (i). Now define $f \star g$ by

$$f \star g(x) = \int f(y)g(y^{-1}x)\,dy \quad \text{if } x \in \complement A_0, \qquad f \star g(x) = 0 \quad \text{if } x \in A_0.$$

Then

$$|f \star g(x) - f_n \star g_n(x)| \leqslant \int |f(y)g(y^{-1}x) - f_n(y)g_n(y^{-1}x)|\,dy \quad \text{if } x \in \complement A_0.$$

Combining this with (**) we obtain: $f \star g \in \mathfrak{L}^1(G)$ (since $f_n \star g_n \in \mathscr{K}(G)$). Thus (ii) is proved. Finally (iii) follows from (*).

REMARK. We can also show in almost the same way as above: *if $f, g \in \mathfrak{L}^1(G)$, then the function Φ: $\Phi(x,y) = f(y)g(y^{-1}x)$, is in $\mathfrak{L}^1(G \times G)$.* (We show first that $\int^{\times}_{G\times G} |\Phi(x,y)|\,d(x,y) \leqslant N_1(f)N_1(g)$ and then use approximation in $\mathfrak{F}^1(G \times G)$.) Thus the Lebesgue–Fubini theorem can be applied, yielding again (i), (ii), (iii). But the first method can be used in more general situations (cf. § 5.4, first footnote), p. 76.

5.2. In $\mathscr{K}(G)$ there is an involution $f \to f^*$, defined by

(4) $$f^*(x) = \overline{f(x^{-1})}\,\Delta(x^{-1}),$$

which is isometric,

(5) $$\|f^*\|_1 = \|f\|_1,$$

hence continuous, and satisfies the usual conditions:

$$f^{**} = f, \qquad (\alpha f + \beta g)^* = \bar{\alpha} f^* + \bar{\beta} g^* \quad (\alpha, \beta \in \mathbf{C})$$

and

(6) $$(f \star g)^* = g^* \star f^*.$$

This involution can be extended to $L^1(G)$ by continuity: then (4), (5), (6) apply to arbitrary $f, g \in L^1(G)$. (These formulae can also be considered directly for integrable functions, as in 5.1.)

REMARK. Analogously we can define an isometric involution in $L^p(G)$, $1 < p < \infty$, by $f^*(x) = \overline{f(x^{-1})}\,\Delta^{1/p}(x^{-1})$.

5.3. Let H be a closed *normal* subgroup of G. The mapping T_H studied in § 4 is not only linear and continuous: *T_H is a morphism of the Banach algebra $L^1(G)$ onto the Banach algebra $L^1(G/H)$*, i.e. T_H also has the property

(7) $$T_H(f \star^G g) = T_H f \star^{G/H} T_H g \qquad f, g \in L^1(G).$$

Thus *the subspace $J^1(G,H)$* (cf. § 4.4) *is a closed, two-sided ideal of $L^1(G)$.* Moreover, *T_H also satisfies*

(8) $$T_H(f^*) = (T_H f)^* \qquad f \in L^1(G),$$

where the star on the right denotes involution in $L^1(G/H)$.

(7) is readily verified for $f, g \in \mathscr{K}(G)$ and then follows for $L^1(G)$ by continuity: T_H is a continuous map and $f \star g$ is continuous in the two 'variables' $f, g \in L^1(G)$.

Moreover, $\mathscr{J}(G,H)$ (defined in § 4.2 (6)) is now seen to be a two-sided ideal of $\mathscr{K}(G)$, hence its closure $J^1(G,H)$ is a two-sided ideal of $L^1(G)$.

To prove (8), take $k \in \mathscr{K}(G/H)$, $f \in \mathscr{K}(G)$. Then we can write, using 3.5 (13) and the notation $\tilde{g}(x) = \overline{g(x^{-1})}$

$$\int_G k \circ \pi_H(x) f^*(x)\, dx = \left\{ \int_G [k \circ \pi_H]^\sim(x) f(x)\, dx \right\}^-.$$

Applying Weil's formula to both sides, we get

$$\int_{G/H} k(\dot{x}) T_H[f^*](\dot{x})\, d\dot{x} = \left\{ \int_{G/H} \tilde{k}(\dot{x}) T_H f(\dot{x})\, d\dot{x} \right\}^-,$$

where we have used the fact that $[k \circ \pi_H]^\sim = \tilde{k} \circ \pi_H$. Making the transformation $\dot{x} \to \dot{x}^{-1}$ on the right, we get

$$\int_{G/H} k(\dot{x}) T_H[f^*](\dot{x})\, d\dot{x} = \int_{G/H} k(\dot{x}) [T_H f]^*(\dot{x})\, d\dot{x}.$$

Since $k \in \mathscr{K}(G/H)$ was arbitrary, (8) is true for $f \in \mathscr{K}(G)$ and hence for all $f \in L^1(G)$ by continuity.

The image of a left (right, two-sided) ideal of $L^1(G)$ under the morphism T_H is an ideal of the same kind in $L^1(G/H)$, since T_H is surjective. But it is not known whether the image of a *closed* ideal of $L^1(G)$ is a *closed* ideal of $L^1(G/H)$: this will be discussed in Chapter 8, § 4.6.

References. [110, I], [112, p. 15].

5.4. The convolution in $L^1(G)$ can be generalized as follows. Let H be any closed subgroup of G. A function $h \in L^1(H)$ yields a bounded measure μ_h on G if we put $\mu_h(k) = \int_H k(\xi) h(\xi)\, d\xi$, $k \in \mathscr{K}(G)$, or

$$d\mu_h(x) = h(\xi)\, d\xi. \tag{9}$$

We define for $f \in \mathscr{K}(G)$ the convolution $\mu_h \star f$ by

$$\mu_h \star f(x) = \int_H h(\xi) f(\xi^{-1}x)\, d\xi \qquad x \in G. \tag{10}$$

Then $\mu_h \star f$ is a continuous function (e.g., since f is left uniformly continuous, cf. 1.8 (vi)) and

$$\|\mu_h \star f\|_1 \leqslant \|h\|_{L^1(H)} \cdot \|f\|_1. \tag{11}$$

Thus (10) can be extended, by continuity, to all $f \in L^1(G)$ and (11) still holds.

We can consider (10) also in $L^p(G)$, $1 < p < \infty$: for $f \in \mathscr{K}(G)$ we have

$$\|\mu_h \star f\|_p \leqslant \|h\|_{L^1(H)} \cdot \|f\|_p, \tag{12}$$

thus $\mu_h \star f$ is a (continuous) function in $L^p(G)$. Hence (10) and (12) can be extended to all $f \in L^p(G)$.

To prove (12), note that if $g \in \mathcal{K}(G)$, then

$$\int_G \mu_h \star f(x)g(x)\,dx = \int_H h(\xi)\left\{\int_G f(\xi^{-1}x)g(x)\,dx\right\}d\xi$$

and by Hölder's inequality

$$\left|\int_G \mu_h \star f(x)g(x)\,dx\right| \leqslant \|h\|_{L^1(H)}\cdot\|f\|_p\cdot\|g\|_{p'} \quad (1/p+1/p'=1)$$

which yields (12).

Actually, if $f \in \mathfrak{L}^p(G)$, $1 \leqslant p < \infty$, the integral (10), as a function of x, is defined a.e. on G, is pth-power integrable and (12) holds.†

More generally, if μ is any complex bounded measure on G, we can define the operators $f \to \mu \star f$ and $f \to f \star \mu$, $f \in \mathcal{K}(G)$, by

(13 a) $$\mu \star f(x) = \int_G f(y^{-1}x)\,d\mu(y)$$

(13 b) $$f \star \mu(x) = \int_G f(xy^{-1})\Delta(y^{-1})\,d\mu(y).$$

These operators can be extended to $L^1(G)$ by continuity. We can define analogous operators in $L^p(G)$, $1 < p < \infty$ (with Δ replaced by $\Delta^{1/p}$ in the case of (13 b)).‡

If in (13) $d\mu(y) = h(y)\,dy$, with $h \in L^1(G)$, then (13) reduces to the ordinary convolutions $h \star f$ and $f \star h$ respectively. Another special case of (13 a) is (10), μ being of the form (9); likewise for (13 b).

Remark. We mention here the *convolution* $\mu_1 \star \mu_2$ of two bounded complex measures μ_1, μ_2 on G: this is defined by (cf. § 2.6)

$$\mu_1 \star \mu_2(k) = \int_{G\times G} k(xy)\,d\mu_1\otimes\mu_2(x,y) \qquad k \in \mathcal{K}(G).$$

$\mu_1 \star \mu_2$ is again a bounded measure on G and $\|\mu_1 \star \mu_2\| \leqslant \|\mu_1\|\cdot\|\mu_2\|$. *The bounded complex measures on G form a Banach algebra $M^1(G)$ of* which $L^1(G)$ is a subalgebra, in fact a two-sided ideal, by the injection§ $f \to f(x)\,dx$. There is also an *involution in* $M^1(G)$, defined by

(14) $$\mu^*(k) = \overline{\mu(\tilde{k})} \qquad k \in \mathcal{K}(G),\ \tilde{k}(x) = \overline{k(x^{-1})}.$$

This has all the usual properties of an involution and is an extension

† The proof is entirely analogous to that in 5.1; the remark made there applies again, *if* $p = 1$, with $H \times G$ in the place of $G \times G$.

‡ We have $\|\mu \star f\|_p \leqslant \|\mu\|\,\|f\|_p$, $\|f \star \mu\|_p \leqslant \|\mu\|\,\|f\|_p$ for $f \in \mathcal{K}(G)$ and hence for $f \in L^p(G)$; the proof is the same as for (12).

§ See 2.3 (vii). Note also that, if $\mu_1 \in M^1(G)$ and $f \in L^1(G)$, then to $\mu_1 \star f$, $f \star \mu_1$, as defined by (13), there correspond $\mu_1 \star \mu_2$, $\mu_2 \star \mu_1$ respectively, where $d\mu_2(x) = f(x)\,dx$.

of that in $L^1(G)$, defined in 5.2. We can write for (13 b)

$$f\star\mu = [\mu^*\star f]^* \qquad f\in L^1(G)$$

and likewise for $f\in L^p(G)$, $1<p<\infty$ (cf. also § 5.2, Remark).

Reference. [138, pp. 48–49].

5.5. The Dirac measure δ_a $(a\in G)$ is defined by $\delta_a(k) = k(a)$, $k\in\mathscr{K}(G)$. It gives rise to the left and right *translation operators* respectively:

$$(15\,\lambda)\quad L_a f = \delta_a\star f \qquad (15\,\rho)\quad R_a f = f\star\delta_a \qquad f\in L^1(G).$$

We summarize below, for reference, the main properties of L_a and R_a.

(λ_1) $L_a f(x) = f(a^{-1}x)$, $\qquad(\rho_1)$ $R_a f(x) = f(xa^{-1})\Delta(a^{-1})$.

(λ_2) $L_{ab} = L_a L_b$, $\qquad(\rho_2)$ $R_{ab} = R_b R_a$.

(λ_3) $\|L_a f\|_1 = \|f\|_1$, $\qquad(\rho_3)$ $\|R_a f\|_1 = \|f\|_1$.

Given $f\in L^1(G)$ and $\epsilon>0$ there are neighbourhoods U, V of e such that

(λ_4) $\|L_y f - f\|_1 < \epsilon$ $(y\in U)$, $\qquad(\rho_4)$ $\|R_y f - f\|_1 < \epsilon$ $(y\in V)$.

(λ_5) $f\star g(x) = \int f(y)L_y g(x)\,dy$, $\qquad(\rho_5)$ $f\star g(x) = \int R_y f(x) g(y)\,dy$.

(λ_6) $L_a[f\star g] = [L_a f]\star g$, $\qquad(\rho_6)$ $R_a[f\star g] = f\star R_a g$.

(λ_7) $L_a[f^*] = [R_{a^{-1}} f]^*$.

(λ_2) means that $a\to L_a$ is a representation of G by operators in $L^1(G)$. By (λ_3) these operators are isometric. (λ_4) is contained in 3.2, Lemma, for $f\in\mathscr{K}(G)$ and follows for all $f\in L^1(G)$ by continuity; (ρ_4) results from (λ_4) via (λ_7) and (5) ([138, p. 41]; cf. also Chap. 1, § 2.1).

REMARK. We can consider the operator L_a of course also in $L^p(G)$, $1<p<\infty$; likewise we can define an operator $R_a^{(p)}$ in $L^p(G)$ by $R_a^{(p)} f(x) = f(xa^{-1})\Delta^{1/p}(a^{-1})$. (This amounts to transferring (15) to $L^p(G)$.) Then the analogues of $(\lambda_1-\lambda_4)$, $(\rho_1-\rho_4)$, and (λ_7) hold (cf. also § 5.2, Remark).

(λ_4) and (λ_2) imply that for fixed $f\in L^1(G)$ the function $y\to L_y f$, $y\in G$, is (left uniformly) continuous; likewise $y\to R_y f$ is (right uniformly) continuous, by (ρ_4) and (ρ_2).

The formulae (λ_5), (ρ_5) can be regarded simply as another notation for the convolution, but they can also be interpreted from the viewpoint of integration of functions with values in a Banach space: this will be shown in 6.6.

5.6. The algebra $L^1(G)$ possesses approximate left units and approximate right units. The proof is quite simple.

Given $f \in L^1(G)$ and $\epsilon > 0$, choose U according to 5.5 (λ_4). Let u be any function in $\mathscr{K}_+(G)$ such that $\int u = 1$ and $\operatorname{Supp} u \subset U$. Then (cf. § 5.5 ($\lambda_5$))

$$u \star f(x) - f(x) = \int \{L_y f(x) - f(x)\} u(y)\, dy$$

and hence

$$\|u \star f - f\|_1 \leqslant \int \|L_y f - f\|_1 u(y)\, dy < \epsilon.$$

The proof for approximate right units is the same, by 5.5 (ρ_4) and (ρ_5).

It is not difficult to show that $L^1(G)$ has a left (or right) unit only if G is discrete ([123, Theorem 1.10]; cf. also [101, § 28, No. 1] and Chap. 1, § 2.3).

5.7. The closed left [right] ideals of $L^1(G)$ coincide with the closed *left* [*right*] *invariant linear subspaces* (i.e. invariant under the operators L_y [R_y]); the proof is in two parts:

(i) Every closed left [right] ideal I is a left [right] invariant linear subspace.

Given $f \in I$, there are approximate left units u_n such that $\|u_n \star f - f\|_1 \to 0$ ($n \to \infty$), by 5.6. Hence, for any $a \in G$, $[L_a u_n] \star f \to L_a f$ by 5.5 (λ_6) and (λ_3); since $[L_a u_n] \star f \in I$ for each n and I is closed, $L_a f$ is in I. For R_a the proof is analogous, using 5.5 (ρ_6) and (ρ_3).

(ii) Every closed left [right] invariant linear subspace is a left [right] ideal.

We shall obtain (ii) as a corollary to Proposition 5.8 below which will be needed later for other applications.†

5.8. Let f be in $L^1(G)$ and $h \in L^1(H)$, where H is any closed subgroup of G (which may coincide with G). Then, given $\epsilon > 0$, there are finitely many elements $\xi_n \in H$ and coefficients $c_n \in \mathbf{C}$ such that

$$\int_G \left| \int_H f(\xi^{-1}x) h(\xi)\, d\xi - \sum_n c_n f(\xi_n^{-1}x) \right| dx < \epsilon. \tag{16}$$

The coefficients are given by

$$c_n = \int_{A_n} h(\xi)\, d\xi, \tag{17}$$

where (A_n) is a partition of H into measurable subsets, and hence

$$\sum_n c_n = \int_H h(\xi)\, d\xi. \tag{18}$$

Likewise there are elements $\xi_n \in H$ and coefficients $c_n \in \mathbf{C}$ given by (17) —and thus satisfying (18)—such that‡

$$\int_G \left| \int_H f(x\xi^{-1}) \Delta_G(\xi^{-1}) h(\xi)\, d\xi - \sum_n c_n f(x\xi_n^{-1}) \Delta_G(\xi_n^{-1}) \right| dx < \epsilon. \tag{19}$$

† Both (i) and (ii) can also be proved by the method of linear functionals.

‡ The elements ξ_n and the coefficients c_n in (16) and (19) are not necessarily the same.

There is a compact set $K \subset H$ such that $\int_{H-K} |h(\xi)|\, d\xi < \epsilon$. Since the function $\xi \to L_\xi f$, $\xi \in H$, is continuous (§ 5.5) and K is compact, there are finitely many points $\xi_n \in K$, say $1 \leqslant n \leqslant N$, and open nds. U_n of ξ_n in H covering K, such that $\|L_\xi f - L_{\xi_n} f\|_1 < \epsilon$ for $\xi \in U_n$. Define now a partition of H as follows. Let

$$A_1 = K \cap U_1, \qquad A_n = \Big(K - \bigcup_{1 \leqslant j < n} A_j\Big) \cap U_n$$

for $1 < n \leqslant N$ if $N > 1$; put $A_0 = H - K$ and $\xi_0 = e$. The sets A_n, $0 \leqslant n \leqslant N$, are measurable (for $d\xi$), mutually disjoint, and their union is H. Hence, if c_n is defined by (17),

$$\int_H L_\xi f(x) h(\xi)\, d\xi - \sum_{0 \leqslant n \leqslant N} c_n L_{\xi_n} f(x) = \sum_{0 \leqslant n \leqslant N} \int_{A_n} \{L_\xi f(x) - L_{\xi_n} f(x)\} h(\xi)\, d\xi$$

and thus

$$\int_G \Big| \int_H L_\xi f(x) h(\xi)\, d\xi - \sum_{0 \leqslant n \leqslant N} c_n L_{\xi_n} f(x) \Big|\, dx$$

$$\leqslant 2\|f\|_1 \int_{H-K} |h(\xi)|\, d\xi + \sum_{1 \leqslant n \leqslant N} \int_{A_n} \|L_\xi f - L_{\xi_n} f\|_1 \cdot |h(\xi)|\, d\xi$$

$$< \Big\{2\|f\|_1 + \sum_{1 \leqslant n \leqslant N} \int_{A_n} |h(\xi)|\, d\xi\Big\}\epsilon \leqslant \Big\{2\|f\|_1 + \int_H |h(\xi)|\, d\xi\Big\}\epsilon,$$

so that (16) follows after a change of ϵ. The proof of (19) is exactly the same, with L_ξ replaced by R_ξ.

REMARK. An entirely analogous approximation holds for measures $\mu \in M^1(G)$ and the convolutions $\mu \star f, f \star \mu$ (cf. (13 a, b)), the subgroup H being replaced by Supp μ. The proof is the same. The case above corresponds to $\mu = \mu_h$ (cf. (10)).

References. [142, relation (2.53)], [112, Lemma 3].

5.9. We can interpret Proposition 5.8 by means of vector-valued integrals.

Let B be a Banach space with norm $\|.\|$ and let $\mathbf{f}$ be a bounded continuous function on a l.c. group H, with values in B. Let $h \in L^1(H)$; thus $\mathbf{f}h$ is integrable over H. Then, given $\epsilon > 0$, there are finitely many points $\xi_n \in H$ and coefficients $c_n \in \mathbf{C}$ given by (17) such that

$$(20) \qquad \Big\| \int_H \mathbf{f}(\xi) h(\xi)\, d\xi - \sum_n c_n \mathbf{f}(\xi_n) \Big\| < \epsilon.$$

The proof is entirely analogous to that in 5.8. In fact, 5.8 is merely a special case of the above, with $B = L^1(G)$ and $\mathbf{f}(\xi) = L_\xi f$ or $R_\xi f$ (cf. in this respect also § 6.6).

Example. Consider the space $L^p(G)$, $1 < p < \infty$. Let h be in $L^1(H)$, H being a closed subgroup of G. We can prove the analogue of 5.8 for the operators L_y and $R_y^{(p)}$ (cf. § 5.5, Remark), by the same method as in 5.8. This is again a special case of the more abstract result above.

REMARK. We can likewise extend (20) to the case of bounded measures. Also 5.8, Remark, applies to functions with values in a Banach space. But we shall not need this.

6. The space $L^\infty(G)$

6.1. Let $L^\infty(G)$ be the space of *complex-valued* functions on G which are measurable and essentially bounded (with respect to the Haar measure of G); functions coinciding locally almost everywhere (l.a.e.) are identified. This is a Banach space with

(1) $$\|\phi\|_\infty = \operatorname*{ess.\,sup}_{x \in G} |\phi(x)|$$

as norm.† It is the dual of $L^1(G)$: every continuous linear functional on $L^1(G)$ has the form

$$f \to \langle f, \phi \rangle \equiv \int f(x)\overline{\phi(x)}\, dx \qquad \phi \in L^\infty(G),$$

and the norm of this functional is precisely $\|\phi\|_\infty$ (cf. § 2.8).

In $L^\infty(G)$ there is an *involution* $\phi \to \tilde{\phi}$, where $\tilde{\phi}(x) = \overline{\phi(x^{-1})}$. This is the precise analogue for $p = \infty$ of the involution in $L^p(G)$, $1 \leqslant p < \infty$ (§ 5.2, Remark).

We can form the convolution of $f \in L^1(G)$ with $\phi \in L^\infty(G)$:

$$f \star \phi(x) = \int f(y)\phi(y^{-1}x)\, dy.$$

For reasons that will become apparent in § 7 we shall use $f^* \star \phi$ rather than $f \star \phi$, where f^* is defined by 5.2 (4). The function $f^* \star \phi$ can be written in the form

(2) $$f^* \star \phi(x) = \int \overline{f(y)}\phi(yx)\, dy,$$

it satisfies

(3) $$\|f^* \star \phi\|_\infty \leqslant \|f\|_1 \cdot \|\phi\|_\infty$$

and is (right uniformly) continuous.‡

REMARK 1. If $f \in L^1(G)$ and $u \in \mathscr{K}(G)$, then $f \star u$ is a bounded continuous function in $L^1(G)$.

REMARK 2. If $\mu \in M^1(G)$, and $\phi \in L^\infty(G)$ is continuous, we can define $\mu^* \star \phi$ and $\mu^* \star \phi$ is a *continuous* function in $L^\infty(G)$. (If $\operatorname{Supp}\mu$ is compact, this results from the fact that ϕ is (left) uniformly continuous on compact sets; the general case then follows by approximation.)

We can write

(4) $$\langle f, \phi \rangle \equiv \int f(x)\overline{\phi(x)}\, dx = \{f^* \star \phi(e)\}^-.$$

† In [138] the notation $L^\infty(G)$ has a different meaning (cf. [138, p. 33]).

‡ The last assertion follows from 5.5 (ρ_4), since

$$|f^* \star \phi(yx) - f^* \star \phi(x)| \leqslant \|R_y f - f\|_1 \cdot \|\phi\|_\infty.$$

If $\langle f, \phi \rangle = 0$, we call ϕ *orthogonal* to f and write $\phi \perp f$. For a linear subspace $I \subset L^1(G)$ we write $\phi \perp I$ if $\phi \perp f$ for all $f \in I$.

If $f, g \in L^1(G)$ and $\phi \in L^\infty(G)$, then

$$(5) \qquad (g \star f)^* \star \phi = f^* \star (g^* \star \phi)$$

and in particular

$$(6) \qquad \langle g \star f, \phi \rangle = \langle f, g^* \star \phi \rangle.$$

Also

$$(7) \qquad \langle g \star f, \phi \rangle = \langle g, [f \star \tilde{\phi}]^{\sim} \rangle.$$

(5) expresses associativity, combined with 5.2 (6):

$$(g \star f)^* \star \phi = (f^* \star g^*) \star \phi = f^* \star (g^* \star \phi).$$

The proof of (5) and (7) depends, of course, on the Lebesgue–Fubini theorem; to verify that the functions concerned are in $\mathfrak{L}^1(G \times G)$ we can use 2.7 (vi), 3.10, and 5.1, Remark.

6.2. We state here some useful applications of (6) for later reference.

(i) Let I be a *left* ideal of $L^1(G)$. Then $\phi \perp I$ implies $g^* \star \phi \perp I$ for all $g \in L^1(G)$, and conversely.

(ii) Let I be a *right* ideal of $L^1(G)$. Then $\phi \perp I$ holds if and only if $f^* \star \phi = 0$ for all $f \in I$.

(i) follows directly from (6) (the converse part depends on the fact that $L^1(G)$ has approximate left units). To prove (ii) we write (6) in the form

$$\langle f \star g, \phi \rangle = \langle g, f^* \star \phi \rangle,$$

with $f \in I$ and general $g \in L^1(G)$. Thus, if $\phi \perp I$, then $\langle g, f^* \star \phi \rangle = 0$ for all $g \in L^1(G)$, hence $f^* \star \phi = 0$. Likewise for the converse, since $L^1(G)$ has approximate right units.

6.3. The formula (7) implies the following very useful fact. Let I be a closed left ideal of $L^1(G)$; let f be in $L^1(G)$. If every *continuous* (or only every right uniformly continuous) function $\phi \perp I$ satisfies $\phi \perp f$, then f is in I.

It is enough to show that the condition above implies: $\phi \perp f$ for *all* $\phi \perp I$. Suppose $\phi \perp I$, then for any $g \in L^1(G)$ we have $g^* \star \phi \perp I$ (§ 6.2 (i)) and $g^* \star \phi$ is (right uniformly) continuous (§ 6.1). By hypothesis this implies $\langle f, g^* \star \phi \rangle = 0$ or $\langle g \star f, \phi \rangle = 0$. Thus by (7) $\langle g, [f \star \tilde{\phi}]^{\sim} \rangle = 0$ for all $g \in L^1(G)$, hence the continuous function $[f \star \tilde{\phi}]^{\sim}$ vanishes identically and for $x = e$ we get $\langle f, \phi \rangle = 0$.

6.4. Let H be a closed normal subgroup of G and consider the mapping T_H (§ 4.4 (10)). We can now give a new characterization of its kernel $J^1(G, H)$ (cf. § 4.6).

$J^1(G, H)$ is the closed linear subspace of $L^1(G)$ generated by the family of all functions $L_\eta k - k$ with $k \in \mathscr{K}(G)$ and $\eta \in H$.

Denote this subspace by J, for the moment. Since H is normal, we have

$$T_H L_\eta k = T_H k$$

if $\eta \in H$; hence $J \subset J^1(G,H)$. To obtain the opposite inclusion, observe that J is left invariant, since for any $a \in G$ we have $L_a[L_\eta k - k] = L_{\eta'} k' - k'$, where $k' = L_a k \in \mathscr{K}(G)$ and $\eta' = a\eta a^{-1} \in H$. Thus J is a closed left ideal and we can apply 6.3. But every *continuous* $\phi \perp J$ is constant on each coset of H: if $\eta \in H$, we have $0 = \langle L_\eta k - k, \phi\rangle = \langle k, L_{\eta^{-1}}\phi - \phi\rangle$ for all $k \in \mathscr{K}(G)$, which implies $\phi(\eta x) = \phi(x)$ for *all* $x \in G$, since ϕ is continuous. Thus $\phi = \phi' \circ \pi_H$, where ϕ' is a continuous function in $L^\infty(G/H)$, and an application of Weil's formula shows that $\phi \perp J^1(G,H)$. Hence $J \supset J^1(G,H)$ by 6.3 and the proof is complete.

REMARK. The use of *continuous* functions ϕ avoids a certain difficulty mentioned in [111, I, p. 260, footnote]. See also 6.5 below.

6.5. Using the result in 6.4, we can now prove the following proposition, useful for applications.

Let H be a closed normal subgroup of G. Let Φ be a (complex-valued) measurable function on G such that for each $\xi \in H$ we have $\Phi(\xi x) = \Phi(x)$ locally almost everywhere on G; here the exceptional locally negligible set may depend on ξ. Then Φ coincides l.a.e. with a function of the form $\Phi' \circ \pi_H$, where Φ' is a measurable function on G/H.

First let ϕ be an essentially bounded function on G satisfying the hypotheses. Then the continuous linear functional $f \to \langle f, \phi\rangle$ on $L^1(G)$ has the property that $\langle L_\eta f, \phi\rangle = \langle f, \phi\rangle$ for all $\eta \in H$; thus $\langle L_\eta f - f, \phi\rangle = 0$ and hence $\phi \perp J^1(G,H)$ by 6.4. Therefore ϕ defines a continuous linear functional on $L^1(G/H)$ (§ 4.4 (9)): it is defined by $f' \to \langle f_1, \phi\rangle$, where f_1 is *any* function in $L^1(G)$ such that $T_H f_1 = f'$. Hence there is a function $\phi' \in L^\infty(G/H)$ such that for all $f \in L^1(G)$

$$\int_G f(x)\overline{\phi(x)}\,dx = \int_{G/H} T_H f(\dot{x})\overline{\phi'(\dot{x})}\,d\dot{x}.$$

The function $\phi' \circ \pi_H$ is in $L^\infty(G)$ (§ 3.9, Corollary). Hence we can apply the extended Weil formula (§ 4.5) to $f.\overline{\phi' \circ \pi_H}$ and obtain

$$\int_G f(x)\overline{\phi(x)}\,dx = \int_G f(x)\overline{\phi' \circ \pi_H(x)}\,dx,$$

for all $f \in L^1(G)$; hence $\phi(x) = \phi' \circ \pi_H(x)$ l.a.e. (§ 2.7 (vii)).

For general Φ as above put $\phi = \Phi/(1+|\Phi|)$. Then ϕ is measurable and $|\phi(x)| < 1$ for *all* $x \in G$. Now ϕ coincides l.a.e. with $\phi' \circ \pi_H$, where ϕ' is a measurable function on G/H; moreover, we may assume that $|\phi'(\dot{x})| \neq 1$ for *all* $\dot{x} \in G/H$, since equality can occur only in a locally negligible set (cf. § 3.9). We can now put $\Phi' = \phi'/(1-|\phi'|)$, and $\Phi = \phi/(1-|\phi|)$ coincides l.a.e. on G with $\Phi' \circ \pi_H$.

For a generalization of the results in 6.4 and 6.5, and references, see Chapter 8, §§ 2.5 and 2.6.

6.6. We insert here a remark on certain functions in $L^1_B(G)$, where B is a Banach space, G a l.c. group (cf. § 2.3; μ is the Haar measure). Let $\mathfrak{g}$ be a bounded continuous function on G with values in B. Then for any f in the ordinary space $L^1(G)$ the function $y \to f(y)\mathbf{g}(y)$, $y \in G$, is in $L^1_B(G)$, so that $\int f(y)\mathbf{g}(y)\,dy$ is defined; we have

$$\left\| \int f(y)\mathbf{g}(y)\,dy \right\| \leqslant \int \|f(y)\mathbf{g}(y)\|\,dy = \int |f(y)|\,.\,\|\mathbf{g}(y)\|\,dy. \tag{8}$$

Now let $B = L^1(G)$ and consider the vector-valued function $y \to L_y g$, $y \in G$, for $g \in L^1(G)$. Then *the integral* $\int f(y) L_y g\,dy$ *in the 'abstract' sense above coincides with the convolution* $f \star g$ defined in 5.1. Thus we have another interpretation of the formulae 5.5 (λ_5, ρ_5).

The integral is an element of $B = L^1(G)$ satisfying (cf. § 2.3 (viii))

$$\left\langle \int f(y)\, L_y g\,dy, \phi \right\rangle = \int \langle f(y)\, L_y g, \phi\rangle\,dy \quad \text{for all } \phi \in L^\infty(G), \tag{*}$$

where $\langle\ ,\ \rangle$ has the usual meaning. But the right-hand side is simply $\langle f \star g, \phi\rangle$, by an interchange of integrations; thus (*) implies $\int f(y)\, L_y g\,dy = f \star g$. We also observe that, in the present case, (8) reduces to a familiar inequality, since $\|L_y g\|_1$ is independent of $y \in G$.

6.7. A function $\phi \in L^\infty(G)$ coincides l.a.e. with a right uniformly continuous function if (and only if) $\|L_y\phi - \phi\|_\infty \to 0$ when $y \to e$ (cf. D. A. Edwards [42]; the proof is based, essentially, on 3.10 and 4.8). This contrasts with 5.5 (λ_4).

7. Beurling algebras

7.1. Let G be a locally compact group. A real-valued function w on G is said to be a *weight function* if it has the following properties:

(1 a) $\qquad w(x) \geqslant 1 \qquad x \in G,$

(1 b) $\qquad w(xy) \leqslant w(x)w(y) \qquad x, y \in G,$

(1 c) $\qquad w$ is measurable and locally bounded.

The functions f such that $fw \in L^1(G)$ form a subalgebra of $L^1(G)$ which is a Banach algebra under the norm

$$\|f\|_{1,w} = \int |f(x)|\,w(x)\,dx.$$

We denote it by $L^1_w(G)$ and call it a *Beurling algebra* (cf. Chap. 1, § 6).

$L^1_w(G)$ contains $\mathscr{K}(G)$, since w is locally bounded; moreover, $\mathscr{K}(G)$ *is dense in* $L^1_w(G)$ (the proof is the same as in Chap. 1, § 6.1).

Let w_1 be a weight function on G and put† $w(x) = \limsup_{y\to e} w_1(xy)$. Then w is upper semi-continuous and, as is readily seen, is also a weight function. Moreover, $w_1(x) \leqslant w(x) \leqslant w(e)w_1(x)$ for all $x \in G$, so that w and w_1 define the same subalgebra of $L^1(G)$, with equivalent norms.

† The right-hand side is defined as $\inf_U \sup_{y\in U} w_1(xy)$, where U ranges over the compact nds. of e. The introduction of this function was suggested by H. Freudenthal.

Hence *we shall assume from now on, without loss of generality, that the weight function w defining a Beurling algebra $L^1_w(G)$ is upper semi-continuous*. Examples were given in Chapter 1, § 6.1.

7.2. An important difference between $L^1_w(G)$ and $L^1(G)$ is this: the isometric involution 5.2 does not, in general, apply to $L^1_w(G)$, even when G is abelian. In fact, f^* need not even be in $L^1_w(G)$ for *all* $f \in L^1_w(G)$ if w is not symmetric; this is readily shown by examples (cf. Chap. 1, § 6.1).

The operators L_a, R_a (§ 5.5 (λ_1, ρ_1)) are defined in $L^1_w(G)$, but are not isometric (unless $w(x) =$ const. l.a.e.). We have for $f \in L^1_w(G)$, $a \in G$,

$$(2\lambda) \qquad \|L_a f\|_{1,w} \leqslant w(a)\|f\|_{1,w}, \qquad (2\rho) \qquad \|R_a f\|_{1,w} \leqslant w(a)\|f\|_{1,w}.$$

The analogue of 5.5 (λ_4, ρ_4) holds: given $f \in L^1_w(G)$ and $\epsilon > 0$, there are neighbourhoods U, V of e such that

$$(3\lambda) \quad \|L_y f - f\|_{1,w} < \epsilon \ (y \in U), \qquad (3\rho) \quad \|R_y f - f\|_{1,w} < \epsilon \ (y \in V).$$

See Chapter 1, § 6.3: the proof there carries over to groups, for L_y *and* R_y.

Thus the functions $y \to L_y f$ and $y \to R_y f$, $y \in G$, are continuous.

$L^1_w(G)$ possesses approximate left units and approximate right units. This follows from (3 λ, ρ) in the same way as for $L^1(G)$ in 5.6.

The closed left [right] ideals of $L^1_w(G)$ coincide with the closed left [right] invariant subspaces. The proofs are as for $L^1(G)$.

See 5.7. Even Proposition 5.8 carries over to $L^1_w(G)$ *for the case* $H = G$, with almost the same proof, and we can interpret it just as in 5.9.

7.3. The dual space of $L^1_w(G)$ is $L^\infty_w(G)$, formed by all complex-valued measurable functions ϕ on G such that†

$$(4) \qquad \|\phi\|_{\infty,w} = \operatorname*{ess.sup}_{x \in G} \frac{|\phi(x)|}{w(x)} < \infty.$$

That is, the continuous linear functionals on $L^1_w(G)$ can be put in the form

$$(5) \qquad f \to \langle f, \phi \rangle \equiv \int f(x)\overline{\phi(x)}\,dx \qquad \phi \in L^\infty_w(G),$$

and the norm of such a functional is precisely $\|\phi\|_{\infty,w}$. This follows from the corresponding fact for $L^1(G)$ (§ 6.1).

It is of importance in the applications that, *if $\phi \in L^\infty_w(G)$ is continuous, then $|\phi(x)| \leqslant \|\phi\|_{\infty,w} \cdot w(x)$ for* ALL $x \in G$. This follows from the upper semi-continuity of w.

† Two such functions coinciding l.a.e. represent the same element of $L^\infty_w(G)$.

If $f \in L_w^1(G)$, $\phi \in L_w^\infty(G)$, then† $f^* \star \phi \in L_w^\infty(G)$. Also

(6) $$\|f^* \star \phi\|_{\infty,w} \leqslant \|f\|_{1,w} \cdot \|\phi\|_{\infty,w}$$

and $f^* \star \phi$ is continuous.

We have $$|f^* \star \phi(x)| \leqslant \|f\|_{1,w} \cdot \|\phi\|_{\infty,w} \cdot w(x)$$

and $$|f^* \star \phi(yx) - f^* \star \phi(x)| \leqslant \|R_y f - f\|_{1,w} \cdot \|\phi\|_{\infty,w} \cdot w(x).$$

Remark 1. The relations (4)–(7) in 6.1 also hold for $L_w^1(G)$ and $L_w^\infty(G)$; likewise the notion of orthogonality and the analogues of 6.2 and 6.3.

Remark 2. Let μ be a measure on G such that $w\mu \in M^1(G)$. Let $\phi \in L_w^\infty(G)$ be continuous. Then the function $\mu^* \star \phi$ is continuous and is in $L_w^\infty(G)$. This is analogous to 6.1, Remark 2.

7.4. Let H be a closed normal subgroup of G. The algebra $L^1(G/H)$ is the image of $L^1(G)$ under the morphism T_H (§ 5.3), thus T_H maps a subalgebra of $L^1(G)$ onto a subalgebra of $L^1(G/H)$. We shall prove: *the image of a Beurling algebra on G under the map T_H is a Beurling algebra on G/H* [117]. More precisely: define for a weight function w on G

(7) $$\dot{w}(\dot{x}) = \inf_{\xi \in H} w(x\xi) \qquad \dot{x} = \pi_H(x),$$

then *$\dot{w}$ is a weight function on G/H and T_H maps $L_w^1(G)$ onto $L_{\dot{w}}^1(G/H)$. Let $J_w^1(G,H)$ be the kernel of the restriction of T_H to $L_w^1(G)$. Then the isomorphism*

(8) $$L_{\dot{w}}^1(G/H) \cong L_w^1(G)/J_w^1(G,H)$$

is not only algebraic, but also isometric (the right-hand side being provided with the ordinary quotient norm).

We recall that w may be assumed to be upper semi-continuous (§ 7.1). Then $\dot{w}$ in (7) is also upper semi-continuous and hence satisfies, in particular, (1 c); obviously (1 a) holds for $\dot{w}$ and (1 b) is readily verified.

We have $\dot{w}(\pi_H(xy)) = \inf_{\eta \in H} w(xy\eta) \leqslant w(x)\dot{w}(\pi_H(y))$. Replacing here x by $x\xi$, $\xi \in H$, we obtain $\dot{w}(\dot{x}\dot{y}) \leqslant \dot{w}(\dot{x})\dot{w}(\dot{y})$, since $\pi_H(x\xi y) = \pi_H(xy)$ if $\xi \in H$.

It is clear that T_H maps $L_w^1(G)$ into $L_{\dot{w}}^1(G/H)$ and that

(9) $$\|T_H f\|_{1,\dot{w}} \leqslant \|f\|_{1,w}.$$

This follows from the extended Weil formula:

$$\int |f(x)|\, w(x)\, dx \geqslant \int |f(x)|\, \dot{w} \circ \pi_H(x)\, dx \geqslant \int_{G/H} |T_H f(\dot{x})|\, \dot{w}(\dot{x})\, d\dot{x}.$$

The kernel $J_w^1(G,H)$ is a closed subspace of $L_w^1(G)$. It is of course

† This may not be true of $f \star \phi$, if w is not symmetric.

simply $L^1_w(G) \cap J^1(G,H)$, where $J^1(G,H)$ is the kernel of T_H in $L^1(G)$ (cf. § 4.6).

7.5. The two following properties of $J^1_w(G,H)$ will be useful:

(i) A function $\phi \in L^\infty_w(G)$ is orthogonal to $J^1_w(G,H)$ if and only if ϕ is (l.a.e. equal to) a function of the form $\phi' \circ \pi_H$, where ϕ' is a (measurable) function on G/H.

(ii) $J^1_w(G,H)$ is the closure of $\mathscr{J}(G,H)$ (§ 4.2 (6)) in $L^1_w(G)$. This is analogous to 4.6.

(i) If $\phi \in L^\infty_w(G)$ is of the form $\phi' \circ \pi_H$, then $\phi \perp J^1_w(G,H)$ by the extended Weil formula. Conversely, if $\phi \perp J^1_w(G,H)$, then in particular $\langle L_\eta k - k, \phi\rangle = 0$ $(\eta \in H)$ for every $k \in \mathscr{K}(G)$, or $\langle k, L_{\eta^{-1}}\phi - \phi\rangle = 0$. Hence for each fixed $\eta \in H$ we have: $\phi(\eta x) = \phi(x)$ l.a.e. on G (§ 2.7 (vii)) and then we can apply 6.5.

(ii) is obtained by the same method.

7.6. We give first an outline of the proof of the isomorphism (8).† We shall show that for each $f \in L^1_w(G)$ the relation

$$\inf_{f_0 \in J^1_w(G,H)} \|f - f_0\|_{1,w} = \|T_H f\|_{1,\dot{w}} \tag{10}$$

holds. That is, $L^1_w(G)/J^1_w(G,H)$ is *isometrically* injected into $L^1_{\dot{w}}(G/H)$ and since it is a complete space, it must be a closed subset of $L^1_{\dot{w}}(G/H)$. But it is also dense in $L^1_{\dot{w}}(G/H)$, since it contains $\mathscr{K}(G/H)$. Hence it is the whole of $L^1_{\dot{w}}(G/H)$. Thus (8) will be proved once (10) is established.

Denote the left-hand side of (10) by d_f:

$$d_f = \inf_{f_0 \in J^1_w(G,H)} \|f - f_0\|_{1,w}. \tag{11}$$

It is obvious that

$$d_f \geqslant \|T_H f\|_{1,\dot{w}}, \tag{12}$$

since by (9) $\|f - f_0\|_{1,w} \geqslant \|T_H(f - f_0)\|_{1,\dot{w}} = \|T_H f\|_{1,\dot{w}}$ for all $f_0 \in J^1_w(G,H)$. Thus, to prove (10) it is enough to show that

$$d_f \leqslant \|T_H f\|_{1,\dot{w}}, \tag{13}$$

and here we may assume $d_f > 0$, that is: $f \notin J^1_w(G,H)$. But in this case there is a $\phi \in L^\infty_w(G)$ such that

(14 a) $\phi \perp J^1_w(G,H)$, (14 b) $\langle f, \phi\rangle = 1$, (14 c) $\|\phi\|_{\infty,w} = \dfrac{1}{d_f}$.

Now (14 a) implies that ϕ is of the form $\phi' \circ \pi_H$, where ϕ' is a (measurable) function on G/H (§ 7.5 (i)). We shall prove later that $\phi' \in L^\infty_{\dot{w}}(G/H)$ and

$$\|\phi'\|_{\infty,\dot{w}} = \|\phi' \circ \pi_H\|_{\infty,w}. \tag{15}$$

† The reader may find it of interest to consider the reason why the proof for $L^1(G)$ (§ 4) does not apply to $L^1_w(G)$.

Assume (15) *for the moment.* Then the proof of (10) can be completed very simply: from (14 b) we get by the extended Weil formula

$$1 = \langle f, \phi \rangle = \langle f, \phi' \circ \pi_H \rangle = \langle T_H f, \phi' \rangle,$$

and hence by (15) and (14 c)

$$1 \leqslant \|T_H f\|_{1,\dot{w}} \cdot \|\phi'\|_{\infty,\dot{w}} = \|T_H f\|_{1,\dot{w}} \frac{1}{d_f},$$

which proves (13) and hence (10).

7.7. If the group G is countable at infinity, then (15) results from Proposition 4.9 which yields (cf. also (7))

$$\|\phi' \circ \pi_H\|_{\infty,w} = \operatorname{ess.\,sup}_{x \in G} \frac{|\phi'(\pi_H(x))|}{w(x)} = \operatorname{ess.\,sup}_{\dot{x} \in G/H} \frac{|\phi'(\dot{x})|}{\dot{w}(\dot{x})} = \|\phi'\|_{\infty,\dot{w}}.$$

Thus (10) *holds for l.c. groups countable at infinity.* In order to extend (10) to general l.c. groups, we first note some facts concerning open subgroups.

7.8. Let G_* be any *open* subgroup of G. If $f \in \mathscr{K}(G)$ has its support in G_*, then for any closed subgroup H of G

(i) $$\int_H f(x\xi)\, d\xi = \int_{H \cap G_*} f(x\xi)\, d\xi \qquad \text{for all } x \in G_*$$

and, if f_* is the restriction of f to G_*,

(ii) $\int_H f(x\xi)\, d\xi = 0$ for all $x \in G \Leftrightarrow \int_{H \cap G_*} f_*(x\xi)\, d\xi = 0$ for all $x \in G_*$;

(iii) if H is normal, then

(16 a) $$\pi_H(G_*) \cong G_*/H_*,$$

where $H_* = H \cap G_*$ and '$\cong$' is an algebraic and topological isomorphism. Moreover, if f and f_* are as before, then (i) says

(16 b) $$T_H f(\dot{x}) = T_{H_*} f_*(\dot{x}) \qquad \text{for all } \dot{x} \in \pi_H(G_*),$$

$T_{H_*} f_*$ being considered as a function on $\pi_H(G_*)$ via (16 a).

(i) The function $y \to f(xy)$, $y \in G$, has support $x^{-1}\operatorname{Supp} f$ which still lies in G_* if $x \in G_*$. Note also that the Haar measure of $H \cap G_*$ is simply the restriction of that of H.

(ii) First we observe that $\int_{H \cap G_*} f_*(x\xi)\, d\xi = \int_{H \cap G_*} f(x\xi)\, d\xi$ for all $x \in G_*$. The implication '$\Rightarrow$' now follows from (i). To verify '$\Leftarrow$' observe that, if the right-hand side of (i) is zero for all $x \in G_*$, then the left-hand side of (i) is zero for all $x \in G_* H$ by the left invariance of $d\xi$; moreover, for $x \in G - G_* H$ the left-hand side of (i) is zero in any case, since $\operatorname{Supp} f \subset G_*$.

(iii) Since π_H is an open map and G_* is an open subgroup, the restriction of π_H to G_* is also open; its kernel is $H \cap G_*$. Hence $\pi_H(G_*)$ and G_*/H_* are isomorphic as *topological* groups.

REMARK. From (16 a, b) it follows that, if the Haar measures on $G, H, G/H$ are canonically related, then so are their respective restrictions to G_*, H_*, G_*/H_*, for any open subgroup $G_* \subset G$ and $H_* = H \cap G_*$.

7.9. We now prove (10) for general l.c. groups G. First observe that in (10) and (11) we can replace $J_w^1(G,H)$ by $\mathscr{J}(G,H)$ (cf. § 7.5 (ii)):

$$d_f = \inf_{k \in \mathscr{J}(G,H)} \|f-k\|_{1,w}. \tag{17}$$

Now let f be in $\mathscr{K}(G)$. Let G_1 be the subgroup generated by a compact symmetric neighbourhood of e containing Supp f. Consider the family of all subgroups G_* of G generated by compact symmetric neighbourhoods of e and containing G_1; we note that G_1, G_* are open subgroups of G, countable at infinity (cf. § 1.8 (iv)). Also to any two subgroups G'_*, G''_* there is a third one, G_*, such that $G'_* \subset G_*$, $G''_* \subset G_*$. We can now write instead of (17)

$$d_f = \inf_{G_*} \inf_{\substack{k \in \mathscr{J}(G,H) \\ \operatorname{Supp} k \subset G_*}} \int_{G_*} |f(x)-k(x)|\, dx. \tag{18}$$

Let k_* be the restriction of k to G_*; put $H_* = H \cap G_*$. From 7.8 (ii) we get at once, on considering $\mathscr{J}(G_*, H_*)$,

$$k \in \mathscr{J}(G,H) \text{ and } \operatorname{Supp} k \subset G_* \quad \Leftrightarrow \quad k_* \in \mathscr{J}(G_*, H_*). \tag{19}$$

We can apply (10) to G_*, H_* (cf. § 7.7): thus we obtain from (18) and (19) (using also 7.5 (ii) for G_*, H_*)

$$d_f = \inf_{G_*} \int_{\pi_H(G_*)} |T_{H_*} f_*(\dot{x})|\, \dot{w}_*(\dot{x})\, d\dot{x},$$

where f_* is the restriction of f to G_*, $T_{H_*} f_*$ is considered as a function on $\pi_H(G_*)$ (cf. (16 a)), and

$$\dot{w}_*(\dot{x}) = \inf_{\xi \in H_*} w(x\xi) \qquad \dot{x} = \pi_H(x), x \in G_*.$$

By (16 b) we can write

$$d_f = \inf_{G_*} \int_{\pi_H(G_*)} |T_H f(\dot{x})|\, \dot{w}_*(\dot{x})\, d\dot{x}. \tag{20}$$

In (20) we may replace $\pi_H(G_*)$ by $\pi_H(G_1)$, since $T_H f$ has support in $\pi_H(G_1) \subset \pi_H(G_*)$. Moreover, we can then interchange 'inf' and '$\int$' (§ 2.4 (v)). Thus, since

$$\inf_{G_*} \inf_{\xi \in H_*} w(x\xi) = \inf_{\xi \in H} w(x\xi) \qquad \text{for all } x \in G,$$

we obtain

$$d_f = \int_{\pi_H(G_1)} |T_H f(\dot{x})|\, \dot{w}(\dot{x})\, d\dot{x} = \int_{G/H} |T_H f(\dot{x})|\, \dot{w}(\dot{x})\, d\dot{x}.$$

This completes the proof of (10) for general G, if $f \in \mathscr{K}(G)$. Then (10) follows for all $f \in L^1_w(G)$ by approximation.

For any $f \in L^1_w(G)$ and any $\epsilon > 0$ there is a $g \in \mathscr{K}(G)$ such that $\|f-g\|_{1,w} < \epsilon$. Then

$$d_f < d_g+\epsilon = \|T_H g\|_{1,\dot{w}}+\epsilon < \|T_H f\|_{1,\dot{w}}+2\epsilon.$$

Since $\epsilon > 0$ was arbitrary, this yields (13) and hence (10).

Thus the isomorphism (8) has been established for all locally compact groups.

7.10. We can now prove the following result, for general l.c. groups G. Let ϕ' be a (complex-valued) function on G/H, H being a closed normal subgroup of G. Then ϕ' is in $L^\infty_{\dot{w}}(G/H)$ if and only if $\phi' \circ \pi_H$ is in $L^\infty_w(G)$, where $\dot{w}$ is defined by (7); moreover, relation (15) holds.

The 'only if'-part is obvious (cf. § 3.9), since clearly $\|\phi' \circ \pi_H\|_{\infty,w} \leqslant \|\phi'\|_{\infty,\dot{w}}$.

Now suppose $\phi' \circ \pi_H \in L^\infty_w(G)$. Then $\phi' \circ \pi_H \perp J^1_w(G,H)$ (§ 7.5 (i)). Thus by (8) $\phi' \circ \pi_H$ induces a continuous linear functional on $L^1_{\dot{w}}(G/H)$, *of norm equal to* $\|\phi' \circ \pi_H\|_{\infty,w}$: this functional is defined by $f' \to \langle f, \phi' \circ \pi_H\rangle$, $f' \in L^1_{\dot{w}}(G/H)$, where f is *any* function in $L^1_w(G)$ for which $T_H f = f'$. Thus there is a $\phi'' \in L^\infty_{\dot{w}}(G/H)$ such that (i) $\langle f, \phi' \circ \pi_H\rangle = \langle T_H f, \phi''\rangle$ for all $f \in L^1_w(G)$, (ii) $\|\phi''\|_{\infty,\dot{w}} = \|\phi' \circ \pi_H\|_{\infty,w}$. We have $\langle f, \phi' \circ \pi_H\rangle = \int_{G/H} T_H f(\dot{x})\overline{\phi'(\dot{x})}\,d\dot{x}$ by the extended Weil formula and in particular ϕ' is locally integrable; hence (i) implies: $\phi'(\dot{x}) = \phi''(\dot{x})$ l.a.e. on G/H (cf. § 2.7 (vii)). Thus $\|\phi'\|_{\infty,\dot{w}} = \|\phi''\|_{\infty,\dot{w}}$ and (ii) yields the desired result.

We remark that (15) is, of course, equivalent to (8), but a *direct* proof of (15) for *general* l.c. groups seems more difficult. For continuous functions (15) is almost obvious.

4

LOCALLY COMPACT ABELIAN GROUPS AND THE FOUNDATIONS OF HARMONIC ANALYSIS

1. Locally compact abelian groups

1.1. THE most fundamental locally compact abelian (l.c.a.) groups are the following:

$\mathbf{R}$, the additive group of real numbers, and $\mathbf{R}^\nu$, the ν-dimensional real vector space.

$\mathbf{Z}$, the additive group of integers, and $\mathbf{Z}^\nu$, the 'lattice' of points in $\mathbf{R}^\nu$ with coordinates in $\mathbf{Z}$. For any $a \in \mathbf{R}$ we denote the group $\{an \mid n \in \mathbf{Z}\}$ by $a\mathbf{Z}$; in practice we take $a > 0$.

$\mathbf{T}$, the multiplicative group of complex numbers of absolute value 1. It is isomorphic, as a topological group, to $\mathbf{R}/\mathbf{Z}$; this is the significance of† $e^{2\pi ix}$. The group $\mathbf{T}^\nu$ is the ν-dimensional torus (whence the notation).

F, a finite abelian group; we denote the finite cyclic group of order n by Z_n.

REMARKS

1. The *closed* (proper) subgroups of $\mathbf{R}$ are all of the form $a\mathbf{Z}$ ($a \in \mathbf{R}$). The subgroups of $\mathbf{Z}$ are all of the form $n\mathbf{Z}$ ($n \in \mathbf{Z}$); correspondingly, the closed subgroups of $\mathbf{T}$ are of the form $n^{-1}\mathbf{Z}/\mathbf{Z}$ ($n \geqslant 1$).

2. The nd. $V_1 = \{z \mid z = e^{2\pi ix}, |x| < \frac{1}{3}\}$ of 1 in $\mathbf{T}$ does not contain any subgroup (closed or not) of $\mathbf{T}$ except $\{1\}$. (If $z \in V_1$ and if for some $n \geqslant 2$ the powers $z^2,\ldots, z^n$ are all in V_1, then $z = e^{2\pi ix}$ with $|x| < 1/3n$, by induction.)

3. $\mathbf{R}/a\mathbf{Z} \cong \mathbf{R}/\mathbf{Z}$ for all $a \neq 0$ in $\mathbf{R}$.

A group of the form $\mathbf{R}^{\nu_1}\times\mathbf{Z}^{\nu_2}\times\mathbf{T}^{\nu_3}\times F$ ($\nu_1, \nu_2, \nu_3 \geqslant 0$) is said to be an *elementary group*. It can be shown that quotient groups and closed subgroups of elementary groups are again elementary.

References. [13 *e*, Chap. VII, § 1], [77, Chap. III, Theorem 1.4].

1.2. A basic method for forming new groups from given ones, or for resolving a given group into simpler groups, is as follows. Let $(G_n)_{n\geqslant 0}$ be a sequence of locally compact groups (not necessarily abelian) such that $G_n = \pi_n(G_{n+1})$, π_n being a strict morphism with *compact* kernel

† A classical treatment of this topic is given in most books on function theory. For a more abstract version see H. Cartan [22] and also [13 *e*, Chap. VIII, § 2, n° 1]; there the group $\mathbf{T}$ is denoted by $\mathbf{U}$ and the letter $\mathbf{T}$ is used for $\mathbf{R}/\mathbf{Z}$.

for all n. Let G be the subgroup of $\prod_{n\geqslant 0} G_n$ consisting of the sequences $(x_n)_{n\geqslant 0}$ such that $x_n = \pi_n(x_{n+1})$ for all n. G is locally compact (in the induced topology) and is called the *projective limit* of the 'projective sequence' $(G_n)_{n\geqslant 0}$; we write

$$G = \varprojlim_{n\geqslant 0} G_n \tag{1}$$

or simply $\varprojlim G_n$. This notation does not indicate explicitly the mappings π_n (on which G also depends), but it will be sufficient for our purposes.

To show that G is l.c., choose an integer $N \geqslant 0$ and let U_N be a compact nd. of e_N in G_N. The sequences $(x_n)_{n\geqslant 0}$ in G such that $x_N \in U_N$ form a *compact* nd. $V(U_N)$ of e in G. We also observe that, if U_N ranges over a basis at e_N in G_N, for each $N \geqslant 0$, then the corresponding nds. $V(U_N)$ form a basis at e in G.

REMARK. If $G = \varprojlim_{n\geqslant 0} G_n$, then also $G \cong \varprojlim_{n\geqslant 0} G_{n+j}$ for $j = 1, 2,\ldots$, and more generally $G \cong \varprojlim_{j\geqslant 0} G_{n_j}$ for any strictly increasing subsequence $(n_j)_{j\geqslant 0}$.

Examples of projective limits are given below (§ 1.3).

The definition of projective limit can be extended to non-countable 'projective families' of topological groups, but then it is less simple: see [138, § 5], where historical references are given, [106, Chap. VIII], [100, § 2.7], [13 *e*, Chap. III, § 7].

1.3. Let G be a locally compact group, not necessarily abelian, and $(H_n)_{n\geqslant 0}$ a sequence of closed normal subgroups such that

(i) $H_n \supset H_{n+1}$ and H_n/H_{n+1} is compact ($n \geqslant 0$).

Then $(G/H_n)_{n\geqslant 0}$ is a projective sequence; put

$$G_* = \varprojlim_{n\geqslant 0} G/H_n. \tag{2}$$

The elements of G_* can be represented in a concrete way as 'nested sequences of cosets' $(x_n H_n)_{n\geqslant 0}$ with $x_n \in G$ and $x_n H_n \supset x_{n+1} H_{n+1}$ (or, equivalently, $x_{n+1} = x_n h_n$, $h_n \in H_n$) for all n.†

Examples

$$(a)\quad \varprojlim_{n\geqslant 0} \mathbf{R}/\lambda_n \mathbf{Z}, \qquad (b)\quad \varprojlim_{n\geqslant 0} \mathbf{Z}/\lambda_n \mathbf{Z},$$

where $(\lambda_n)_{..\geqslant 0}$ is a sequence of integers such that $0 < \lambda_n < \lambda_{n+1}$ and λ_{n+1}/λ_n is also an integer. These groups are compact. If in (*b*) we put $\lambda_n = l^n$, with $l > 1$ an integer, we obtain the *group of l-adic integers*, denoted by $\mathbf{Z}_l$ and written additively. $\mathbf{Z}_l$ is most familiar for $l = p$, a prime.

† This construction goes back to van Dantzig and van der Waerden: see [27, especially p. 113, footnote 27].

Example (c). Let $(\lambda_n)_{n\geqslant 0}$ be a sequence of integers as in (a) and (b). Let $M(2,\mathbf{Z})$ be the ring of all 2×2 matrices with entries from $\mathbf{Z}$ and let $G = SL(2,\mathbf{Z})$, the multiplicative group of all $x\in M(2,\mathbf{Z})$ with $\det x = 1$. Let H_n consist of all $x \in G$ such that $x \equiv e \,(\mathrm{mod}\,\lambda_n)$, where e is the 2×2 unit matrix (for any $A, B \in M(2,\mathbf{Z})$ we write $A \equiv B\,(\mathrm{mod}\,\lambda)$, for an integer λ, if $A-B = \lambda C$, with $C \in M(2,\mathbf{Z})$). H_n is a normal subgroup. G/H_n is finite; it is not abelian (unless $n = 0$ and $\lambda_0 = 1$).

Now suppose the sequence $(H_n)_{n\geqslant 0}$ considered above also satisfies

(ii) $\bigcap_{n\geqslant 0} H_n = \{e\}$.

Then there exists an injective morphism of G into G_*, the image of G being dense in G_*. Moreover, if H_n is compact for (some and hence for) all $n \geqslant 0$, then this morphism is an isomorphism of G onto G_*.

Proof. Let U be a nd. of e in G. For a fixed integer $N \geqslant 0$ let $V_N(U)$ consist of all $x = (x_n H_n)_{n\geqslant 0}$ in G_* such that $x_N \in U$; then $V_N(U)$ is a nd. of e_* in G_*. The family $(V_N(U))$ forms a basis at e_*; here U ranges over a basis at e in G and N over the positive integers. To $a \in G$ we make correspond $a_* = (aH_n)_{n\geqslant 0}$ in G_*. This is an injection of G into G_* by (ii), and also a morphism (for the continuity observe that each map $G \to G/H_n$ is continuous). Moreover, the image of G is dense in G_*: given $x_* = (x_n H_n)_{n\geqslant 0}$ in G_*, choose $a = x_N$, then a_* is close to x_* for large N.

If H_n is compact for $n \geqslant 0$, then each $x_* \in G_*$ is a nested sequence of *compact* cosets $(x_n H_n)_{n\geqslant 0}$ in G, hence $\bigcap_{n\geqslant 0} x_n H_n \neq \emptyset$ and by (ii) it contains only one point of G, say a. Then $x_* = a_* = (aH_n)_{n\geqslant 0}$. Thus the morphism $G \to G_*$ is bijective; moreover, it is readily seen that it is strict, hence $G \cong G_*$.

1.4. Let $G = \varprojlim G_n$ be a projective limit as in (1). If the kernel of the map $\pi_n\colon G_{n+1} \to G_n$, reduces to $\{e_{n+1}\}$ for all sufficiently large n, then $G \cong G_n$ for large n. If the kernel of π_n is non-trivial for infinitely many n, then G contains a sequence $(K_n)_{n\geqslant 0}$ of distinct compact normal subgroups such that $K_n \supset K_{n+1}$ for all $n \geqslant 0$ and $\bigcap_{n\geqslant 0} K_n = \{e\}$.

For the first case see 1.2, Remark. In the second case let, for each $n \geqslant 0$, K_n be the compact normal subgroup of all $x = (x_j)_{j\geqslant 0}$ in G such that $x_j = e_j$ for $0 \leqslant j \leqslant n$. The sequence $(K_n)_{n\geqslant 0}$ or a subsequence has the required properties. We note that $G/K_n \cong G_n$ and $G = \varprojlim G/K_n$.

Combining this fact with 1.3, we see that a *locally compact group G can be represented as a projective limit in the sense of* 1.2, *in a non-trivial way, if and only if G contains a sequence $(K_n)_{n\geqslant 0}$ of distinct compact normal subgroups such that $K_n \supset K_{n+1}$ for all $n \geqslant 0$ and $\bigcap_{n\geqslant 0} K_n = \{e\}$.*

Remark. It is readily seen that, for *any* such sequence $(K_n)_{n\geqslant 0}$, K_n is arbitrarily small for large n (i.e. $K_n \subset U$, if $n \geqslant N_U$, for any nd. U of e).

1.5. Let G be a locally compact group containing a sequence of *open* subgroups $(G_n)_{n\geqslant 0}$ such that $G_n \subset G_{n+1}$ and $G = \bigcup_{n\geqslant 0} G_n$. Then we call G the *inductive limit* of the 'inductive sequence' $(G_n)_{n\geqslant 0}$ and write

$$G = \varinjlim_{n\geqslant 0} G_n \tag{3}$$

or simply $\varinjlim G_n$.

More generally, let G be a locally compact group containing a sequence of closed subgroups $(G_n)_{n\geqslant 0}$ such that each G_n is open in G_{n+1} (not necessarily open in G). We consider the abstract group $\bigcup_{n\geqslant 0} G_n$ and provide it with a topology by taking as neighbourhoods of e the neighbourhoods of e in G_0. In this way we get a locally compact group, denoted by $\varinjlim_{n\geqslant 0} G_n$. It contains G_0, and thus each G_n, as *open* subgroup and the topology of G_n as a subgroup of $\varinjlim_{n\geqslant 0} G_n$ is the same as originally in G.

Examples (a) $\varinjlim_{n\geqslant 0} \lambda_n^{-1}\mathbf{Z}$, (b) $\varinjlim_{n\geqslant 0} \lambda_n^{-1}\mathbf{Z}/\mathbf{Z}$,

where $(\lambda_n)_{n\geqslant 0}$ is a sequence of integers as in 1.3, Examples. We remark that there is no ambiguity of notation in (b). If we choose $\lambda_n = (n+1)!$, then (a) is the discrete additive group of rationals and (b) is (isomorphic to) the discrete multiplicative group of all complex roots of unity.

The definition of inductive limit can be given in a more general way (cf. [138, p. 109] and also expositions of the theory of categories).

2. Duality theory and structure theory

This section is a brief survey of the relevant concepts and results, with references to the literature. Compare also § 4.

2.1. A *character* of a locally compact abelian group is a *continuous* function χ on G such that $\chi(x) \in \mathbf{C}$, $|\chi(x)| = 1$ and $\chi(xy) = \chi(x)\chi(y)$ for all $x, y \in G$—in short, a morphism of G into the group $\mathbf{T}$. The character identically equal to 1 is called *unit character*.

The first fundamental fact is that *every l.c.a. group has 'sufficiently many' characters*: given any $a \in G$, $a \neq e$, there is a character χ such that $\chi(a) \neq 1$.

REMARK 1. For *compact* abelian groups this is a special case of the famous *theorem of Peter–Weyl* [138, § 21] which is proved by analytical methods. The result can then be extended to general l.c.a. groups by means of the structure theory of these groups [138, p. 99]. This will be discussed again in § 4 (cf. § 4.7, Remark).

REMARK 2. The following fact is of importance in the applications. Let V_1 be the nd. $\{z \mid z = e^{2\pi i x}, |x| < \frac{1}{3}\}$ of 1 in **T**. If χ is such that $\chi(x) \in V_1$ for all $x \in G$, then $\chi(x) = 1$ for all $x \in G$. This follows from 1.1, Remark 2.

2.2. The characters of G form themselves an abelian group under pointwise multiplication, with the unit character as neutral element. We introduce a topology in this group as follows. For a compact set $C \subset G$ and $\epsilon > 0$ let $U(C, \epsilon)$ be the set of all characters χ such that $|\chi(x) - 1| < \epsilon$ for all $x \in C$; the sets $U(C, \epsilon)$, for all compact $C \subset G$ and all $\epsilon > 0$, are taken as a basis of neighbourhoods of the unit character. It can be shown that in this topology the characters of G form a *locally compact* topological group, denoted by $\hat{G}$ and called the *dual group* of G ([138, § 27], [106, § 34]).

2.3. (i) If G is discrete, then $\hat{G}$ is compact; if G is compact, then $\hat{G}$ is discrete.

(ii) If G is countable at infinity, then $\hat{G}$ possesses a countable basis at $\hat{e}$ (and hence is metrizable, cf. e.g. [100, Theorem 1.22]), and conversely.

(i) Cf. the references in 2.2. The case of compact G is very easy: we consider $U(C, \epsilon)$ (cf. § 2.2) with $C = G$; for small $\epsilon > 0$ we can apply 2.1, Remark 2.

(ii) The direct part is obvious from the definitions. For the converse, consider a sequence $U(C_n, \epsilon_n)$ forming a basis at $\hat{e}$ in $\hat{G}$. Thus, if a character χ is 1 on each compact set $C_n \subset G$, then $\chi = 1$; this implies (§ 2.1) that the closed subgroup generated by $\bigcup_n C_n$ is G. It follows that G is countable at infinity.

2.4. *The dual of an elementary group* (§ 1.1) *is again an elementary group.* The proof is in four parts:

(i) The dual of **R** is (isomorphic to) **R**. Every character of **R** is of the form $x \to e^{2\pi i x y}$ for some $y \in \mathbf{R}$.

(ii) The dual of **Z** is (isomorphic to) **T**, and conversely.

(iii) The dual of a finite cyclic group Z_n is (isomorphic to) Z_n.

(iv) The dual of a product $G_1 \times G_2$ is (isomorphic to) $\hat{G}_1 \times \hat{G}_2$, where G_1, G_2 are any l.c.a. groups and $\hat{G}_1$, $\hat{G}_2$ their respective duals.

A detailed account of (i)–(iv) is given in [106, § 36]. Concerning (iv) we observe that every character χ of $G = G_1 \times G_2$ can be written uniquely in the form $\chi = \chi_1 \chi_2$, where χ_1, χ_2 are characters of G having the value 1 on G_2, G_1 respectively. The set of all such characters χ_j is a closed subgroup of $\hat{G}$ isomorphic to $\hat{G}_j$ $(j = 1, 2)$, thus $\hat{G} = \hat{G}_1 \times \hat{G}_2$ algebraically and it can be verified that the topology of $\hat{G}$ is the product topology.

REMARK. The correspondence by means of which, in a concrete case, a l.c.a. group G' may be exhibited as the dual of G is, of course, only determined up to an automorphism of G'. For example, in (i) above we may choose any $c \neq 0$ in $\mathbf{R}$ and make correspond to $y \in \mathbf{R}$ the character χ: $\chi(x) = e^{cixy}$ of $\mathbf{R}$. The choice $c = 2\pi$ has certain advantages.

2.5. The celebrated *duality theorem* of Pontryagin and van Kampen states that, if $\hat{G}$ is the dual of a l.c.a. group G, then the dual of $\hat{G}$ is isomorphic, as a topological group, to G. More precisely: write the characters of G in the form†

$$(1) \qquad x \to \langle x, \hat{x} \rangle \qquad x \in G,$$

with $\hat{x} \in \hat{G}$. Fix $x \in G$; then

$$(2) \qquad \hat{x} \to \langle x, \hat{x} \rangle \qquad \hat{x} \in \hat{G}$$

is clearly a character of $\hat{G}$. The duality theorem asserts that (i) all characters of $\hat{G}$ are obtained in this way and (2) yields an algebraic isomorphism of $(\hat{G})^\wedge$ with G; (ii) this isomorphism is a topological.

REMARK. We can also consider the function $(x, \hat{x}) \to \langle x, \hat{x} \rangle$ on $G \times \hat{G}$: we say that $\langle x, \hat{x} \rangle$ *puts* G and $\hat{G}$ *in duality*; cf. also 2.4, Remark. For an interesting analogy see [13 *b*, Chap. II, § 6, n° 1].

The theorem was first proved by Pontryagin [105] for abelian groups that are compact or discrete and satisfy the 'second axiom of countability'. The proof was based on the existence of sufficiently many characters (for compact groups see 2.1, Remark 1; for discrete (countable) abelian groups it is much easier to show). Then van Kampen [81] extended the theorem to general l.c.a. groups, using the structure theory of such groups. A. Weil gave the proof in a novel way, likewise based on structure theory [138, §§ 27, 28 and p. 159]. The second edition of Pontryagin's book [106, Chap. VI] takes Weil's methods into account, while the first edition gives the proofs only for l.c.a. groups satisfying the 'second axiom of countability'.

For the special role which the group $\mathbf{T}$ plays in connexion with the duality theorem see [106, Example 72].

2.6. There is a rather interesting analogy between l.c.a. groups and Banach spaces; this was stressed by Bourbaki [13 *a*, Chap. III, Note hist., p. 153].

The fundamental result on the existence of sufficiently many characters (§ 2.1) is, of course, the analogue of a familiar corollary of the Hahn–Banach theorem. The analogue of the Hahn–Banach theorem itself—or rather of that part of it due to Hahn—is the following *extension theorem for characters*. Let H be a closed subgroup of a l.c.a. group G and χ_H a character of H. Then there is a character χ of G coinciding with χ_H on H.

† This notation is analogous to that for linear functionals (Chap. 3, § 2.8 (13)); there will be no danger of confusion, as the letters are different. See also 2.6 below.

This can be proved by transfinite induction if G is discrete or, more generally, if G/H is discrete (cf. [138, § 26, Lemma 1], but in the general case the proof is connected with the duality theorem; see [106, Theorem 55] and 2.8 (ii) below. Also the relationship between the extension theorem above and the result in 2.1 is not quite as simple as that of their analogues for Banach spaces: this is due to the fact that the cyclic subgroup generated by an element of G need not be closed in G.

2.7. We shall now follow up the analogy noted in 2.6. Let H be a closed subgroup of G. Then we define the *orthogonal subgroup* $H^\perp \subset \hat{G}$ by

$$H^\perp = \{\hat{x} \mid \hat{x} \in \hat{G}, \langle x, \hat{x}\rangle = 1 \text{ for all } x \in H\}.$$

Likewise, if Γ is a closed subgroup of $\hat{G}$, we put

$$\Gamma^\perp = \{x \mid x \in G, \langle x, \hat{x}\rangle = 1 \text{ for all } \hat{x} \in \Gamma\}.$$

Example. For $G = \mathbf{R}$, $H = a\mathbf{Z}$ ($a \in \mathbf{R}$, $a \neq 0$) we have $(a\mathbf{Z})^\perp = a^{-1}\mathbf{Z}$ (cf. § 2.4, (i) and § 2.5, Remark); in particular, $\mathbf{Z}^\perp = \mathbf{Z}$.

The relation

$$(H^\perp)^\perp = H \tag{3}$$

holds for all closed subgroups of G.

This is shown by means of the following proposition, the analogue of a familiar fact for Banach spaces. *Let H be a closed subgroup of G. Then for any $x_0 \in G - H$ there is an $\hat{x}_0 \in \hat{G}$ such that $\langle x, \hat{x}_0\rangle = 1$ for all $x \in H$, but $\langle x_0, \hat{x}_0\rangle \neq 1$.* This follows from 2.1, applied to G/H: every character of G/H defines a character of G which is 1 on H. We note that in the proof of (3) the duality theorem 2.5 is not used; the proof is precisely like that for Banach spaces.

2.8. Let $H \subset G$ and $H^\perp \subset \hat{G}$ be as in 2.7. Then the following holds (cf. [138, pp. 108–9]).

(i) *The dual group of G/H is $H^\perp$.* This can be verified directly.

Every character of G/H defines a character of G that is 1 on H, and conversely; moreover, the algebraic isomorphism $(G/H)^\wedge \cong H^\perp$ is also topological (this is shown by means of Chap. 3, § 1.8 (ii)).

(ii) *The dual group of H is $\hat{G}/H^\perp$.* This follows from (i) by duality, in view of (3). More precisely: by restricting the characters of G to H we obtain all characters of H and the algebraic isomorphism $\hat{G}/H^\perp \cong \hat{H}$ is also topological. Thus (ii) contains the extension theorem 2.6 (and is, in fact, equivalent to it).

Remark 1. From (i) and (ii) it follows that, if one of the groups H, $H^\perp$ is open [compact], then the other is compact [open] (cf. § 2.3 (i)).

Remark 2. If H is a closed subgroup of G, then by (ii) there is a canonical strict morphism of $\hat{G}$ onto $\hat{H}$. Now let H be any subgroup of G, closed or not. Let G_d be the group with the same algebraic structure as G, but with the discrete topology, and $(G_d)^\wedge$ the (compact) dual. There is a canonical injection of $\hat{G}$ into $(G_d)^\wedge$ which is a morphism. Let H_d be the subgroup of G_d corresponding to H; using (ii) for $H_d \subset G_d$ and $(G_d)^\wedge$, we obtain a canonical (strict) morphism of $(G_d)^\wedge$ onto $(H_d)^\wedge$. By composition we get a *canonical morphism of* $\hat{G}$ *into* $(H_d)^\wedge$.

(iii) More generally, *if* H_1, H_2 *are closed subgroups of* G *such that* $H_1 \subset H_2$, *then* $H_1^\perp \supset H_2^\perp$ *and* $(H_2/H_1)^\wedge \cong H_1^\perp/H_2^\perp$.

This follows from (i) and (ii) if H_2/H_1 is considered as subgroup of G/H_1: then the orthogonal group of H_2/H_1 in $(G/H_1)^\wedge = H_1^\perp$ is $H_2^\perp$.

(iv) *Let* H_1, H_2 *be any two closed subgroups of* G. *If the product* $H_1 H_2$ *is closed in* G, *then* $H_1^\perp H_2^\perp$ *is closed in* $\hat{G}$, *and conversely.* This also has its counterpart for Banach spaces [111, VI, Lemma 1.3].

We shall show that, *if* $H_1 H_2$ *is closed, then*

(*) $$(H_1 \cap H_2)^\perp = H_1^\perp H_2^\perp.$$

This will imply in particular that $H_1^\perp H_2^\perp$ is closed. The converse part then follows by duality.

Proof of (*) (cf. [111, VI, Lemma 1.4]). The inclusion $(H_1 \cap H_2)^\perp \supset H_1^\perp H_2^\perp$ is trivial. Also, if $H_1 H_2 = G$ and $H_1 \cap H_2 = \{e\}$, then (*) is clearly true (cf. § 2.4 (iv)). We then obtain (*) whenever $H_1 H_2 = G$, by considering $G/(H_1 \cap H_2)$. Now we prove the general case as follows. Given a character χ of G that is 1 on $H_1 \cap H_2$, we first consider the restriction χ_H of χ to the locally compact group $H = H_1 H_2$: by the result already obtained, we have $\chi_H = \chi_{H,1}\chi_{H,2}$, where $\chi_{H,j}$ is a character of H equal to 1 on H_j $(j = 1, 2)$. We can extend $\chi_{H,1}$ to a character χ_1 of G (§ 2.6), and we define χ_2 on G by $\chi = \chi_1\chi_2$; then χ_j is a character of G which is 1 on H_j, hence $(H_1 \cap H_2)^\perp \subset H_1^\perp H_2^\perp$. Thus (*) is proved.

(v) *Let* H_1, H_2 *be closed subgroups of* G. *Then* $\overline{H_1 H_2}^\perp = H_1^\perp \cap H_2^\perp$ *and dually* $(H_1 \cap H_2)^\perp = (H_1^\perp H_2^\perp)^-$. Likewise for the closure of the product of any family of closed subgroups of G, and their intersection. These relations follow directly from the definitions.

2.9. We state here, for later reference, the main *structure theorems for l.c.a. groups.* Proofs will be found in [138, §§ 26, 29] and [106, § 39]; they are both difficult and very enlightening (cf. also [13 *d*, Chap. II, § 2]).

(i) *Every compactly generated l.c.a. group is a product* $\mathbf{R}^{\nu_1}\times\mathbf{Z}^{\nu_2}\times K$, *where* K *is a compact abelian group* $(\nu_1, \nu_2 \geqslant 0)$.

This is a generalization, due in effect to A. Weil [138, pp. 98–99 and 110], of a famous theorem of Pontryagin [105, p. 377] stating that every connected l.c.a. group is of the form $\mathbf{R}^{\nu}\times K$. See also the historical discussion by Freudenthal [46, p. 313].

REMARK 1. The proof of (i) makes use of the following result, due to A. Weil ([138, pp. 96–97], cf. also [100, § 2.21, Lemma]) which is of independent interest and not difficult to prove. Let x be an element of an arbitrary l.c. group; then the set $(x^n)_{n \in \mathbf{Z}}$ either forms a discrete subgroup or else its closure is a compact subgroup (necessarily abelian).

(ii) Every l.c.a. group is a product $\mathbf{R}^\nu \times G_1$ ($\nu \geqslant 0$), where G_1 is a l.c.a. group containing a compact subgroup K such that G_1/K is discrete ([81, Theorem 2]; see also [138, p. 110]).†

REMARK 2. The properties of groups of the type G_1 above are, as it were, opposite to those of $\mathbf{R}^\nu$ (which has discrete subgroups with compact quotient groups). The family of all groups G_1 contains with a group G_1 also the quotient groups, the closed subgroups, and the dual of G_1. A well-known example is the additive group of p-adic numbers (cf. § 3.2).

3. Some examples

The examples below occur frequently.

3.1. (i) Let $G = \prod_{\alpha \in A} K_\alpha$, where each K_α is a compact abelian group. Then $\hat{G}$ is algebraically isomorphic to the subgroup of $\prod_{\alpha \in A} \hat{K}_\alpha$ consisting of the elements $(\hat{x}_\alpha)_{\alpha \in A}$ such that $\hat{x}_\alpha = \hat{e}_\alpha$ for almost all $\alpha \in A$. The topology of $\hat{G}$ is discrete (and thus does not coincide with the induced topology if A is infinite and K_α non-trivial).‡

(ii) Let $G = \varprojlim_{n \geqslant 0} G_n$ (cf. § 1.2), with G_n abelian. Then the dual $\hat{G}$ contains a sequence of open subgroups (isomorphic to) $\hat{G}_n$ such that $\hat{G} = \varinjlim_{n \geqslant 0} \hat{G}_n$ (cf. § 1.5). That is, *projective and inductive limits are dual to one another*.

(iii) Let $G_* = \varprojlim_{n \geqslant 0} G/H_n$ (cf. § 1.3), where G is abelian. Then

$$(G_*)^\wedge = \varinjlim_{n \geqslant 0} H_n^\perp$$

(cf. § 1.5). We note that $(G_*)^\wedge$ lies in $\hat{G}$, but the topology of $(G_*)^\wedge$ is not necessarily that induced by $\hat{G}$: it is such that $H_0^\perp$ (and hence each $H_n^\perp$) is open in $(G_*)^\wedge$.

(i) If A is countable, the proof is precisely like that of (ii) (cf. below; § 2.4 (iv) is also used); in the general case it is entirely analogous.

(ii) There is a nested sequence of compact subgroups $(K_n)_{n \geqslant 0}$ such that $G/K_n \cong G_n$ and K_n is 'arbitrarily small' for large n (§ 1.4). Hence every character χ of G is 1 on some K_n (§ 2.1, Remark 2), or $\chi \in K_n^\perp$ for some n; thus $\hat{G} = \bigcup_{n \geqslant 0} K_n^\perp$.

† An extension of (ii) to a larger class of groups has very recently been given in [57, Theorem 4.4].

‡ A generalization of (i), due to Braconnier [14, pp. 7–8], has applications in algebraic number theory [137, § I].

Moreover, $K_n^\perp$ is open in $\hat{G}$ (§ 2.8, Remark 1) and $K_n^\perp \subset K_{n+1}^\perp$, so $\hat{G} = \underrightarrow{\lim}_{n\geqslant 0} K_n^\perp$; also $K_n^\perp = (G/K_n)^\wedge \cong \hat{G}_n$.

(iii) follows from (ii) in view of 2.8 (i).

Examples. The examples (*a*), (*b*) in 1.3 and 1.5 are dual groups, respectively. In particular we obtain the dual of the discrete additive group of rationals, the dual of the discrete multiplicative group of all complex roots of unity, and the fact that $\mathbf{Z}_l$ is the dual of a very simple group.

3.2. Let G be a l.c.a. group containing a sequence $\mathfrak{H} = (H_n)_{n\in\mathbf{Z}}$ of distinct discrete subgroups such that G/H_n is compact and $H_n \supset H_{n+1}$ for all n. Let $\mathfrak{H}^*$ be the sequence $(H_n^*)_{n\in\mathbf{Z}}$, where $H_n^* = H_{-n}^\perp \subset \hat{G}$; then $\mathfrak{H}^*$ has the same properties: $\hat{G}/H_n^*$ is compact and $H_n^* \supset H_{n+1}^*$ for all n.

We consider $\bigcup_{n'\geqslant 0} H_{-n'}$ with the discrete topology, that is, $\underrightarrow{\lim}_{n'\geqslant 0} H_{-n'}$ (cf. § 1.5). Then we form $\underrightarrow{\lim}_{n'\geqslant 0} H_{-n'}/H_n$ for each $n \geqslant 0$ (note that there is, in fact, no ambiguity in this notation). We now have a projective sequence of discrete abelian groups and put

(1) $$Q_{\mathfrak{H}} = \underleftarrow{\lim}_{n\geqslant 0}\underrightarrow{\lim}_{n'\geqslant 0} H_{-n'}/H_n.$$

$Q_{\mathfrak{H}}$ contains the subgroups

(2) $$\Gamma_{n'} = \underleftarrow{\lim}_{n\geqslant 0} H_{-n'}/H_n \qquad n' \geqslant 0.$$

$\Gamma_{n'}$ consists of all nested sequences $(x_n H_n)_{n\geqslant 0}$ such that $x_0 \in H_{-n'}$ (and thus $x_n \in H_{-n'}$ also for $n > 0$). By the definition of the topology of projective limits, $\Gamma_{n'}$ is clearly an *open* subgroup of $Q_{\mathfrak{H}}$. Hence we can write

(3 a) $$Q_{\mathfrak{H}} = \underrightarrow{\lim}_{n'\geqslant 0}\Gamma_{n'}$$

or

(3 b) $$Q_{\mathfrak{H}} = \underrightarrow{\lim}_{n'\geqslant 0}\underleftarrow{\lim}_{n\geqslant 0} H_{-n'}/H_n,$$

that is, we can 'interchange limits' in (1). Then we have (cf. §§ 3.1 (ii) and 2.8 (iii)):

(4) $$(Q_{\mathfrak{H}})^\wedge = \underleftarrow{\lim}_{n'\geqslant 0}\underrightarrow{\lim}_{n\geqslant 0} H_{-n}^*/H_{n'}^* = Q_{\mathfrak{H}^*}.$$

In particular, *if $\hat{G} \cong G$ and if under this isomorphism H_n^* coincides with H_n for all*† $n \in \mathbf{Z}$, *then* $(Q_{\mathfrak{H}})^\wedge \cong Q_{\mathfrak{H}}$.

Example 1. Let $G = \mathbf{R}$ and put $H_n = \lambda_n \mathbf{Z}$, where $\lambda = (\lambda_n)_{n\in\mathbf{Z}}$ is a sequence in $\mathbf{R}$ such that $\lambda_0 = 1$, $0 < \lambda_n < \lambda_{n+1}$, and $\lambda_{n+1}/\lambda_n \in \mathbf{Z}$ for all

† Actually, it is enough to consider $n \geqslant 0$ (cf. § 2.7 (3)).

$n \in \mathbf{Z}$. Then $(H_n)_{n\in\mathbf{Z}}$ satisfies the conditions above. Denote the corresponding group (1) by Q_λ. Now consider $\lambda^* = (\lambda_n^*)_{n\in\mathbf{Z}}$, where $\lambda_n^* = \lambda_{-n}^{-1}$. Then, in the notation above, $H_n^* = \lambda_n^* \mathbf{Z}$ and (4) becomes: $(Q_\lambda)^\wedge = Q_{\lambda^*}$; *in particular, if* $\lambda_{-n} = \lambda_n^{-1}$ *for* $n \geqslant 1$, *then* $(Q_\lambda)^\wedge = Q_\lambda$. Usually Q_λ is written additively. The elements $x \in Q_\lambda$ have a *canonical representation*

$$x = \sum_{n\in\mathbf{Z}} a_n \lambda_n, \tag{5}$$

where the coefficients a_n are integers such that $0 \leqslant a_n < \lambda_{n+1}/\lambda_n$ for all n and $a_n = 0$ for almost all $n < 0$. *The 'infinite series'* (5) *is to be interpreted as the nested sequence of cosets* $(x_n + \lambda_n \mathbf{Z})_{n\geqslant 0}$, *where* $x_n = \sum_{j<n} a_j \lambda_j$, *the ordinary sum.* With the canonical representation (5) we can effect the addition $x+y$ 'term-by-term', with appropriate reduction and carry-over at each step.

Reference. [74, § 10, Notes.]

Example 2. Let $l > 1$ be an integer and put in Example 1 $\lambda_n = l^n$, $n \in \mathbf{Z}$. Then we obtain a self-dual additive group: this is the *group* $\mathbf{Q}_l$ *of l-adic numbers*, which is most familiar when $l = p$, a prime. $\mathbf{Q}_l$ contains $\mathbf{Z}_l$ (§ 1.3, Example (*b*)) as *open* subgroup; we recall that $\mathbf{Z}_l$ is compact.

Remark 1. Let $H_l = \bigcup_{n'\geqslant 0} l^{-n'}\mathbf{Z}$. Then H_l is a ring and, if we put for $x \in H_l$, $x \neq 0$, $|x|_l = l^{-k}$, where k is the largest integer such that $x \in l^k\mathbf{Z}$, and $|0|_l = 0$, then $|x+y|_l \leqslant \max(|x|_l, |y|_l)$, $|xy|_l \leqslant |x|_l . |y|_l$ for all $x, y \in H_l$, with equality in the second relation if $l = p$, a prime. It is readily seen that $\mathbf{Q}_l$ *is simply the completion of* H_l *with respect to* $|\,.\,|_l$, and $|\,.\,|_l$ can be extended to $\mathbf{Q}_l$. Usually, $\mathbf{Q}_l$ is defined in a slightly different way, as a completion of the *rationals*.

Let Q be the *discrete* group of rationals and Q_l the subgroup of Q consisting of all $r \in Q$ such that $r = a/b$ $(a, b \in \mathbf{Z})$ with b relatively prime to l. The usual definition of $\mathbf{Q}_l$ amounts to letting $\mathbf{Q}_l = \varprojlim_{n\geqslant 0} Q/l^n Q_l$. To see the equivalence with the definition above, we note that $Q = \varinjlim_{n'\geqslant 0} l^{-n'}Q_l$ and that for $H_l = \varinjlim_{n'\geqslant 0} l^{-n'}\mathbf{Z}$ we have $H_l + Q_l = Q$ and $H_l \cap Q_l = \mathbf{Z}$, whence $Q/Q_l \cong H_l/\mathbf{Z}$ and therefore also $Q/l^n Q_l \cong H_l/l^n\mathbf{Z}$ for $n > 0$.

Remark 2. *In* $\mathbf{Q}_l$ *division by l is defined.* This follows at once from the corresponding property of H_l (cf. Remark 1). This division is essential for the application of certain methods of harmonic analysis to $L^1(\mathbf{Q}_l)$ (cf. Chap. 6, § 3.4). More generally, $\mathbf{Q}_l$ *is a ring*, but it is a field only if $l = p$, a prime.†

† A detailed exposition is given in [96].

We have explicitly: if $x = (x_n + l^n\mathbf{Z})_{n \geq 0} \in \mathbf{Q}_l$, then $y = x/l$ is given by

$$y = (y_n + l^n\mathbf{Z})_{n \geq 0},$$

where $y_n = x_{n+1}/l$ (note that $x_{n+1} \equiv x_n \pmod{l^n}$). Also, for arbitrary $x, y \in \mathbf{Q}_l$, the *product* $z = xy$ is $(z_n + l^n\mathbf{Z})_{n \geq 0}$, where $z_n = x_{n+r}\, y_{n+r}$, r being any positive integer such that $x_0 l^r$, $y_0 l^r$ (and hence also $x_n l^r$, $y_n l^r$, $n > 0$) are in $\mathbf{Z}$. We observe that $z_n + l^n\mathbf{Z}$ is the (uniquely determined) coset of $l^n\mathbf{Z}$ in H_l that contains

$$(x_m + l^m\mathbf{Z})(y_m + l^m\mathbf{Z})$$

for all sufficiently large m.

REMARK 3. We obtain a topological injection of $\mathbf{Q}_l$ into $\mathbf{R}_+$ by letting correspond to $x = \sum_{n \in \mathbf{Z}} a_n l^n$ in $\mathbf{Q}_l$ (cf. (5)) the real number

$$x' = \sum_{n \in \mathbf{Z}} 2a_n (2l-1)^{-n}.$$

The Haar measure of $\mathbf{Q}_l$ is then represented by the function $x' \to \sum_{n \in \mathbf{Z}} a_n l^{-n}$, or by a Cantor function defined on $\mathbf{R}_+$ (cf. Chap. 3, § 3.8, Example (vi)).

3.3. We discuss here the explicit form of the characters of projective limits.

Let $G = \varprojlim_{n \geq 0} G_n$ (§ 1.2). Then for every character χ of G there is some value of n, say N, and a character χ_N of G_N such that for all $x = (x_n)_{n \geq 0}$ in G we have $\chi(x) = \chi_N(x_N)$ (cf. § 3.1 (ii)).

In the case $G_* = \varprojlim_{n \geq 0} G/H_n$ (§ 1.3) we have, correspondingly,

$$x = (x_n H_n)_{n \geq 0} \quad \text{and} \quad \chi(x) = \chi_N(x_N H_N)$$

for some χ_N in $(G/H_N)^\wedge = H_N^\perp$, or $\chi(x) = \langle x_N, \hat{x} \rangle$ for some $\hat{x}$ $(= \hat{x}_\chi)$ in $H_N^\perp$ (cf. § 3.1 (iii)). We can express this more conveniently as follows:

$$\chi\big((x_n H_n)_{n \geq 0}\big) = \lim_{n \to \infty} \langle x_n, \hat{x} \rangle, \qquad \hat{x}\ (= \hat{x}_\chi) \in \bigcup_{n \geq 0} H_n^\perp. \tag{6}$$

The notation is justified by the fact that, if $\hat{x} \in \bigcup_{n \geq 0} H_n^\perp$, then $\langle x_n, \hat{x} \rangle$ is constant for large n (viz. for $n \geq N$, if $\hat{x} \in H_N^\perp$), since $(x_n H_n)_{n \geq 0}$ is a nested sequence. (6) gives the general expression of the characters of G_* in terms of those of G.

Now let $Q_{\mathfrak{H}}$ be the group (1) above and let $\Gamma_{n'}$ be the subgroup (2). Given a character χ of $Q_{\mathfrak{H}}$, consider its restriction to $\Gamma_{n'}$. Since $\Gamma_{n'}$ is also a (closed) subgroup of $G_* = \varprojlim_{n \geq 0} G/H_n$, we have by (6): there is some $\hat{x}_{n'} \in \bigcup_{n \geq 0} H_{-n}^*$ (cf. § 3.2 for the notation) such that for every $(x_n H_n)_{n \geq 0}$ in $\Gamma_{n'}$

$$\chi\big((x_n H_n)_{n \geq 0}\big) = \lim_{n \to \infty} \langle x_n, \hat{x}_{n'} \rangle.$$

The *coset* $\hat{x}_{n'} H_{n'}^*$ of $H_{n'}^*$ in $\bigcup_{n \geq 0} H_{-n}^*$ is uniquely determined by χ. Now consider χ on $Q_{\mathfrak{H}}$ itself. Every $x \in Q_{\mathfrak{H}}$ lies in some $\Gamma_{n'}$, $n' \geqslant 0$ (cf. (3a)). Hence to every character χ of $Q_{\mathfrak{H}}$ there is a nested sequence of cosets $(\hat{x}_{n'} H^*_{n'})_{n' \geqslant 0}$ in $\bigcup_{n \geqslant 0} H^*_{-n}$, uniquely determined by χ, such that for all $x = (x_n H_n)_{n \geqslant 0}$ in $Q_{\mathfrak{H}}$

$$\chi\big((x_n H_n)_{n \geqslant 0}\big) = \lim_{n' \to \infty} \lim_{n \to \infty} \langle x_n, \hat{x}_{n'} \rangle. \tag{7}$$

Actually, if $x_0 \in H_{-N'}$ and $\hat{x}_0 \in H^*_{-N}$, say, then $\langle x_n, \hat{x}_{n'} \rangle$ is constant for $n \geqslant N$, $n' \geqslant N'$.

Example. The characters of Q_λ (§ 3.2, Example 1) are of the form

$$\chi_y(x) = \lim_{n' \to \infty} \lim_{n \to \infty} e^{2\pi i x_n y_{n'}},$$

where $x = (x_n + \lambda_n \mathbf{Z})_{n \geqslant 0} \in Q_\lambda$, $y = (y_n + \lambda^*_n \mathbf{Z})_{n \geqslant 0} \in Q_{\lambda^*}$. In particular we can write the characters of $\mathbf{Q}_l$ (§ 3.2, Example 2 and Remark 2) in the form

$$\chi_y(x) = \chi_1(xy) \qquad x, y \in \mathbf{Q}_l, \tag{8}$$

where

$$\chi_1(x) = \lim_{n \to \infty} e^{2\pi i x_n} \qquad x = (x_n + l^n \mathbf{Z})_{n \geqslant 0} \in \mathbf{Q}_l$$

is a particular character of $\mathbf{Q}_l$. The kernel of χ_1 is $\mathbf{Z}_l$ and in the 'canonical duality' (8) of $\mathbf{Q}_l$ with itself we have $\mathbf{Z}_l^\perp = \mathbf{Z}_l$; there is, in general, a perfect analogy with $\mathbf{R}$ (cf. § 2.7, Example).

Reference. Cf. also [74, § 25].

4. The foundations of harmonic analysis

Here a summary of the principal definitions, methods and results is given, with references to the literature.

4.1. Let G be a locally compact abelian group with Haar measure dx. The *Fourier transform* (F.t.) $\hat{f}$ of a function $f \in L^1(G)$ is defined by†

$$\hat{f}(\hat{x}) = \int_G f(x) \overline{\langle x, \hat{x} \rangle}\, dx \qquad \hat{x} \in \hat{G}. \tag{1}$$

Thus

$$|\hat{f}(\hat{x})| \leqslant \|f\|_1 \qquad \hat{x} \in \hat{G}. \tag{2}$$

Also $\hat{f}$ is a *continuous* function on $\hat{G}$: for $f \in \mathscr{K}(G)$ this is clear from the definition of the topology in $\hat{G}$ (§ 2.2) and for general $f \in L^1(G)$ it follows by approximation.

Fourier transforms have three fundamental *analytical properties* (see also Chap. 1, § 1.2):

† In [138, p. 112] and also in classical analysis on $\mathbf{R}$ or $\mathbf{R}^\nu$ slightly different normalizations are used.

(i) RIEMANN–LEBESGUE LEMMA. The F.t. of a function $f \in L^1(G)$ vanishes at infinity.

(ii) UNIQUENESS THEOREM. A function $f \in L^1(G)$ is uniquely determined a.e. by its F.t.: if $\hat{f}(\hat{x}) = 0$ for all $\hat{x} \in \hat{G}$, then $f(x) = 0$ a.e. on G.

(iii) INVERSION THEOREM. Let the Haar measure dx on G be given. Then there is a normalization of the Haar measure on $\hat{G}$, say $d\hat{x}$, such that for all functions $f \in L^1(G)$ which are continuous and have a F.t. $\hat{f}$ belonging to $L^1(\hat{G})$ the following relation holds:

$$f(x) = \int_{\hat{G}} \langle x, \hat{x} \rangle \hat{f}(\hat{x})\, d\hat{x} \qquad \text{for all } x \in G. \tag{3}$$

We call $d\hat{x}$ the *dual Haar measure* (relative to dx); the right-hand side of (3) is termed *inverse Fourier transform* (of $\hat{f}$). The formulae (1) and (3) are quite similar.†

REMARK. There is a more general formulation of (iii) that includes (ii) and which we call the *general inversion theorem*: if $f \in L^1(G)$ and $\hat{f} \in L^1(\hat{G})$, then (3) holds a.e. on G ($d\hat{x}$ being dual to dx). It should be observed, though, that, conversely, this theorem can readily be deduced from (iii).

Suppose (iii) holds and let f be *any* function in $L^1(G)$ such that $\hat{f} \in L^1(\hat{G})$. Take a sequence $(u_n)_{n \geq 1}$ in $\mathscr{K}_+(G)$ satisfying $\int u_n = 1$ and $\|u_n \star f - f\|_1 < 2^{-n}$ (Chap. 3, § 5.6); then $u_n \star f(x) \to f(x)$ a.e. on G (Chap. 3, § 2.4 (i)). Now $u_n \star f$ is continuous (Chap. 3, § 6.1, Remark 1) and $(u_n \star f)^\wedge = \hat{u}_n \hat{f}$ (cf. § 4.2 below). Thus by (iii) we have

$$u_n \star f(x) = \int_{\hat{G}} \langle x, \hat{x} \rangle \hat{u}_n(\hat{x}) \hat{f}(\hat{x})\, d\hat{x} \to \int_{\hat{G}} \langle x, \hat{x} \rangle \hat{f}(\hat{x})\, d\hat{x} \quad (n \to \infty)$$

and the general inversion theorem follows.

4.2. The *algebraic properties* of Fourier transforms are as follows (cf. Chap. 1, § 1.2). Let f, g be in $L^1(G)$.

(i) $(\alpha f + \beta g)^\wedge = \alpha \hat{f} + \beta \hat{g}$, $\quad \alpha, \beta \in \mathbf{C}$;

(ii) $(f \star g)^\wedge = \hat{f}\hat{g}$;

(iii *a*) $[L_a f]^\wedge = \bar{\chi}_a \hat{f}$, $\quad a \in G$ $[\chi_a(\hat{x}) = \langle a, \hat{x} \rangle]$;

(iii *b*) $[\chi_{\hat{a}} f]^\wedge = L_{\hat{a}} \hat{f}$, $\quad \hat{a} \in \hat{G}$ $[\chi_{\hat{a}}(x) = \langle x, \hat{a} \rangle]$;

(iv) $[f^*]^\wedge = \bar{\hat{f}}$ $\quad [f^*(x) = \overline{f(x^{-1})}]$.

The proof of (ii) for $f, g \in \mathscr{K}(G)$ is quite elementary; then (ii) extends to $L^1(G)$ by continuity. Alternatively, (ii) can be proved directly for $f, g \in L^1(G)$ by means of the Lebesgue–Fubini theorem; compare Chapter 3, § 5.1.

† In practice, if dx is given, we obtain $d\hat{x}$ by considering one (continuous) function $f \neq 0$ in $L^1(G)$ such that $\hat{f} \in L^1(\hat{G})$ and determining the arbitrary factor of the Haar measure on $\hat{G}$ so that (3) holds for this f.

The Fourier transforms of the functions in $L^1(G)$ form a function algebra $\mathscr{F}^1(\hat{G})$ on $\hat{G}$, the *Fourier algebra* of $\hat{G}$, with the ordinary point-wise algebraic operations (cf. (i) and (ii) above). $\mathscr{F}^1(\hat{G})$ is a subalgebra of $\mathscr{C}^0(\hat{G})$ (cf. § 4.1 (i)) and is algebraically isomorphic to $L^1(G)$ (cf. § 4.1 (ii)). We define a norm in $\mathscr{F}^1(\hat{G})$ by putting

$$\|\hat{f}\| = \|f\|_1 \qquad f \in L^1(G). \tag{4}$$

Then $\mathscr{F}^1(\hat{G})$ *is isomorphic to* $L^1(G)$ *as a Banach algebra.* (2) can now be written

$$\|\hat{f}\|_\infty \leqslant \|\hat{f}\|. \tag{5}$$

4.3. We discuss here another property of Fourier transforms. Let H be a closed subgroup of G and consider the canonical mapping T_H of $L^1(G)$ onto $L^1(G/H)$ (Chap. 3, § 4). Let f be in $L^1(G)$. *The F.t. of* $T_H f \in L^1(G/H)$ *is the restriction of the F.t.* $\hat{f}$ *of* f *to the orthogonal subgroup* $H^\perp \subset \hat{G}$ (§ 2.8 (i)), if the Haar measures on G, H, G/H are canonically related (Chap. 3, § 3.3 (i)) [109, Theorem 1.3].

If $\hat{x} \in H^\perp$, then $\langle x\xi, \hat{x}\rangle = \langle x, \hat{x}\rangle$ for all $\xi \in H$, $x \in G$, and by the extended Weil formula

$$[T_H f]^\wedge(\hat{x}) = \int_{G/H} T_H f(\dot{x})\overline{\langle \dot{x}, \hat{x}\rangle}\, d\dot{x} = \int_G f(x)\overline{\langle x, \hat{x}\rangle}\, dx = \hat{f}(\hat{x}).$$

4.4. The *Fourier transform of a bounded measure* μ on G is defined by

$$\hat{\mu}(\hat{x}) = \int_G \overline{\langle x, \hat{x}\rangle}\, d\mu(x) \qquad \hat{x} \in \hat{G}.$$

If $d\mu(x) = f(x)\, dx, f \in L^1(G)$, then $\hat{\mu} = \hat{f}$. We note that $\hat{\mu}$ is a continuous function† on $\hat{G}$ and $\|\hat{\mu}\|_\infty \leqslant \|\mu\|$. As to analytical properties of $\hat{\mu}$, we merely state: 4.1 (i) need not hold for $\hat{\mu}$, but the analogue of 4.1 (ii) remains valid and there is also an analogue of 4.1 (iii).

The bounded complex measures on G form a Banach algebra (Chap. 3, § 5.4, Remark) and the algebraic properties 4.2 (i–iv) simply carry over (with an obvious definition of L_a in 4.2 (iii a)). There is also an analogue of 4.3; we shall consider this in Chapter 8, § 2.7.

Now let H be a closed subgroup of G, with Haar measure $d\xi$. Every function $h \in L^1(H)$ defines a bounded measure μ_h on G (Chap. 3, § 5.4 (9)). We have $[\mu_h]^\wedge = \hat{h} \circ \pi_{H^\perp}$, where $\hat{h}$, the F.t. of h, is a function on $\hat{H} = \hat{G}/H^\perp$ and $\pi_{H^\perp}$ is the canonical map $\hat{G} \to \hat{G}/H^\perp$. The consideration of the measure $\mu_h \in M^1(G)$ corresponding to a function $h \in L^1(H)$ is often useful in harmonic analysis.

† The proof of the continuity is quite analogous to that in 4.1 for $\hat{f}$, since the measures with compact support are dense in $M^1(G)$.

4.5. We now state, for later reference, a basic result about the spaces $L^2(G)$ and $L^2(\hat{G})$.

PLANCHEREL'S THEOREM. Let the Haar measure dx on G be given. Then there is a normalization of the Haar measure on $\hat{G}$, say $d\hat{x}$, and a bijective linear map $F \leftrightarrow \hat{F}$ of $L^2(G)$ onto $L^2(\hat{G})$ with the following properties:

(i) $\|F\|_2 = \|\hat{F}\|_2$;

(ii) if $F \in L^1 \cap L^2(G)$, then $\hat{F}(\hat{x}) = \int_G F(x)\overline{\langle x, \hat{x}\rangle}\, dx$ a.e. on $\hat{G}$ and if $\hat{F} \in L^1 \cap L^2(\hat{G})$, then $F(x) = \int_{\hat{G}} \langle x, \hat{x}\rangle \hat{F}(\hat{x})\, d\hat{x}$ a.e. on G.

REMARK 1. The normalization of the Haar measure on $\hat{G}$ is the same as in the inversion theorem (§ 4.1 (iii)): $d\hat{x}$ is the 'dual Haar measure'. The duality between dx and $d\hat{x}$ is a symmetrical relationship.

REMARK 2. From (i) and (ii) above it follows that for $F_1, F_2 \in L^2(G)$ the relations

$$\int_G F_1(x)\overline{F_2(x)}\, dx = \int_{\hat{G}} \hat{F}_1(\hat{x})\overline{\hat{F}_2(\hat{x})}\, d\hat{x}$$

and

$$[F_1 . F_2]^\wedge = \hat{F}_1 \star \hat{F}_2 \tag{6}$$

hold. We note that $F_1 . F_2$ is in $L^1(G)$ and that the convolution $\hat{F}_1 \star \hat{F}_2$ exists, since $\hat{F}_1$, $\hat{F}_2$ are in $L^2(\hat{G})$. These formulae are very useful in practice.

For the classical background of Plancherel's theorem see the references in [138, p. 122]. For l.c.a. groups the theorem was first proved by A. Weil [138, pp. 113–15] by a novel method, significant far beyond the immediate context (cf. § 5 below); the proof also made use of the structure theory of l.c.a. groups.

Weil first proves the following *Proposition*: if Plancherel's theorem holds for H and G/H, where H is a closed subgroup of G, then it holds for G. The proof is entirely analytical: it is based on the introduction of the '*relativized Fourier transform*' [138, p. 115]

$$\int_H f(x\xi)\overline{\langle \xi, \hat{\xi}\rangle}\, d\xi \qquad f \in \mathscr{K}(G), \tag{*}$$

that is, the F.t. of the function $\xi \to f(x\xi)$, $\xi \in H$ ($x \in G$ being fixed). For the general principle underlying (*) see § 5.

By means of the proposition above, Plancherel's theorem can be deduced from the theorem of Peter–Weyl, via the structure theory of l.c.a. groups; this is described in detail in [138, p. 113].

Theorems 4.1 (i) and (iii) follow from Plancherel's theorem: this was shown by A. Weil [138. pp. 116–18] who also extended some other classical results from the line or the circle to l.c.a. groups [138, pp. 117–19]. Weil's approach showed very clearly the fundamental role of Plancherel's theorem in harmonic analysis on groups—a role that will appear again in 4.6 and in the applications to be given later.

4.6. A complex-valued function ϕ on a group G (not necessarily locally compact or abelian) is said to be *positive-definite* if

$$\sum_{1\leqslant m\leqslant N}\sum_{1\leqslant n\leqslant N}\phi(x_m^{-1}x_n)\,c_m\,\overline{c_n}\geqslant 0 \tag{7}$$

for all finite sequences $(x_n)_{1\leqslant n\leqslant N}$ in G and $(c_n)_{1\leqslant n\leqslant N}$ in $\mathbf{C}$. This property implies: $|\phi(x)|\leqslant\phi(e)$ for all $x\in G$.

If G is locally compact, then for a *continuous* function ϕ the condition (7) is equivalent to

$$\iint\phi(x^{-1}y)f(x)\overline{f(y)}\,dxdy\geqslant 0 \qquad f\in\mathscr{K}(G),$$

or

$$\langle f^{*}\star f,\phi\rangle\geqslant 0 \qquad f\in\mathscr{K}(G),$$

as can readily be shown.

We now state another fundamental result.

BOCHNER'S THEOREM. If G is a locally compact abelian group, then the continuous positive-definite functions on G are precisely those of the form

$$\phi(x)=\int_{\hat G}\langle x,\hat x\rangle\,d\mu(\hat x), \tag{8}$$

where μ is a positive bounded measure on the dual group $\hat G$.

References to the classical cases of Bochner's theorem are given by A. Weil [138, pp. 122–3] who proved it for general l.c.a. groups by deducing it from Plancherel's theorem (§ 4.5).

Later, but independently, Raikov [107] proved Bochner's theorem for l.c.a. groups by a method based on Gelfand's theory of commutative Banach algebras, but requiring neither the structure theory of l.c.a. groups nor the Pontryagin–van Kampen duality theorem (§ 2.5).† In Raikov's approach the Riemann–Lebesgue lemma 4.1 (i) appears as an immediate consequence of Gelfand's theory and the uniqueness and inversion theorems 4.1 (ii) and (iii) as well as Plancherel's theorem are deduced from Bochner's theorem. Moreover, Raikov showed that the Pontryagin–van Kampen duality theorem can also be deduced from Bochner's theorem, without using the structure theory of l.c.a. groups. Thus an entirely analytical approach to the duality theorem became possible. A very enlightening discussion of all these matters, including

† Raikov's work was done in 1940–1, but [107] was published only in 1945.

references to the work of others, is given by Raikov in the Introduction to his monograph [107].

4.7. In 1947 H. Cartan and R. Godement [23] published a new proof of Bochner's theorem, based on the Krein–Milman theorem.† This proof does not require the theory of commutative Banach algebras; the possibility of such a proof is also hinted at by Raikov at the very end of the Introduction in [107].

The Krein–Milman theorem provides, in fact, the basis for an extension of harmonic analysis to general locally compact groups.

By using this theorem, Gelfand and Raikov [50] succeeded in proving the existence of 'sufficiently many' (topologically) irreducible unitary representations (in general infinite-dimensional) for arbitrary l.c. groups, thus generalizing profoundly the theorem of Peter–Weyl (cf. § 2.1, Remark 1).

REMARK. For *abelian* l.c. groups this yields a new proof, independent of structure theory, for the existence of sufficiently many characters (cf. § 2.1). It will be observed that this result follows, of course, also from Bochner's theorem, as was pointed out by Raikov in [107, § 9]. Indeed, given any $x_0 \in G$, $x_0 \neq e$, there is a continuous positive-definite function ϕ on G such that $\phi(e) > 0, \phi(x_0) = 0$ (e.g. $\phi(x) = \int f(xy)\overline{f(y)}\,dy$, with $f \in \mathscr{K}(G), f(e) \neq 0$ and Supp$f$ sufficiently small) and by Bochner's theorem $\phi(x) = \int_G \langle x, \hat{x}\rangle\, d\mu(\hat{x})$ for some positive $\mu \in M^1(\hat{G})$; hence $\langle x_0, \hat{x}\rangle = 1$ for *all* $\hat{x} \in \hat{G}$ would imply $\phi(x_0) = \phi(e)$, a contradiction.

The memoir of Cartan and Godement [23] gives a systematic construction, by purely analytical methods, of the foundations of harmonic analysis on locally compact abelian groups and provides a clear insight into the theory.‡ In this context the book of Dixmier [38] will be found of considerable interest; see also the survey articles of Mackey [93] and Braconnier [15] and the exposition by Bourbaki [13 *d* Chap. II, § 1].

Reference should also be made to a remarkable result of Eymard [44] yielding an extension of the Fourier algebras $\mathscr{F}^1(\hat{G})$ to arbitrary l.c. groups.

5. The principle of relativization

5.1. Let G be an arbitrary locally compact group, H a closed subgroup, and f a (complex-valued) function on G. Put for $x \in G$

$$f_{x,H}(\xi) = f(x\xi) \qquad \xi \in H. \tag{1}$$

It is frequently of interest to investigate the relations that may exist

† For simple proofs of the Krein–Milman theorem see E. Artin [3] and H. Bauer [5], where an extension and references are given (the reference to R. Arens on p. 391 of [5] is meant to be to E. Artin, as Professor Bauer has kindly informed the author).

‡ Small changes of a technical or expository nature will, of course, suggest themselves to a reader of [23]; see also [18].

between the properties of the functions $f_{x,H}$ and those of the function f. For $x = e$ we obtain the restriction of f to H, denoted simply by f_H.

Suppose next that in a formula, or in a theorem, there occurs explicitly the group G. If we replace G by H, it may happen that the formula, or the theorem, remains meaningful and, perhaps, still true, possibly with some slight additional modifications.

This general, systematic passage from a group to a subgroup may be called the *principle of relativization*. The principle is, of course, familiar in topology, but in the case of functions on locally compact groups it acquires much wider scope and significance, owing to the group structure, and often leads to new problems and new results. This already appears, implicitly, in the book of A. Weil [138].

5.2. Let us illustrate the remarks above by some examples.

(i) Let G be a locally compact *abelian* group. For $f \in \mathscr{K}(G)$ let f_H be the restriction of f to the closed subgroup H, as above, and consider the *relativized Fourier transform*

$$[f_H]\hat{}(\hat{\xi}) = \int_H f(\xi)\overline{\langle \xi, \hat{\xi} \rangle}\, d\xi \qquad \hat{\xi} \in \hat{H}. \tag{2}$$

(2) was introduced by A. Weil in his proof of Plancherel's theorem, in even a more general form (cf. § 4.5 (*)); he discussed it again in [139, §§ 16, 17].

What are the relations, if any, between $\hat{f}$, defined on $\hat{G}$, and $[f_H]\hat{}$, which is defined on $\hat{G}/H^{\perp} = \hat{H}$? This will be considered in Chapter 5, § 5.5 and Chapter 7, § 2.2.

(ii) Let again G be a l.c.a. group. We can write for $f \in L^1(G)$

$$\int_G f(x)\, dx = \hat{f}(\hat{e}). \tag{3}$$

What becomes of this if we replace integration over G by integration over a closed subgroup H? This question leads directly to Poisson's formula (Chap. 5, § 5).

(iii) It can be shown that for a certain class of locally compact groups G, including all abelian ones, the following relation holds:

$$\inf \left\| \sum_n c_n L_{y_n} f \right\|_1 = \left| \int_G f(x)\, dx \right| \qquad f \in L^1(G), \tag{4}$$

where the infimum is taken over all finite sums with $c_n > 0$, $\sum_n c_n = 1$, $y_n \in G$. What is the corresponding formula when the elements y_n are restricted to lie in some closed subgroup H of G? This will be discussed in Chapter 8, § 4.

(iv) The lemma in Chapter 3, § 3.2 is a relativization of the particular case stated there.

(v) The lemma in Chapter 3, § 5.8 is a relativization of a familiar result.

In the following chapters the principle of relativization will usually be mentioned explicitly when an application occurs. The part it plays in analysis on groups will appear very clearly.

5.3. Let G be a l.c. group and H a closed normal subgroup. It is often an interesting problem to investigate the relations that may exist between functions, or spaces of functions, on G and those on G/H (compare Chap. 3, § 5.3). This approach may be viewed as being 'dual' to the principle of relativization, especially in the case of abelian groups: here we refer to 4.3 for an example.

5

FUNCTIONS ON LOCALLY COMPACT ABELIAN GROUPS

1. The functions σ and τ

We define here certain auxiliary functions, on arbitrary locally compact abelian groups; they are basic tools in later developments.

1.1. Let G be a l.c.a. group with Haar measure dx; let $\hat{G}$ be the dual group and $d\hat{x}$ the dual Haar measure (Chap. 4, § 4.5, Remark 1). Consider functions $\hat{g}, \hat{h} \in L^2(\hat{G})$ with supports contained in a compact set $\hat{E} \subset \hat{G}$ and let $g, h \in L^2(G)$ be their inverse Fourier transforms, respectively; put $\kappa = g.h$, so that κ is in $L^1(G)$. Then

$$\|L_y\kappa - \kappa\|_1 \leqslant 2\,\|\hat{g}\|_2\,\|\hat{h}\|_2 \max_{\hat{x}\in\hat{E}} |\langle y, \hat{x}\rangle - 1| \qquad y \in G \tag{1}$$

and the F.t. of κ is

$$\hat{\kappa} = \hat{g} \star \hat{h}. \tag{2}$$

Write $L_y\kappa - \kappa = (L_y g - g)L_y h + g(L_y h - h)$. By Schwarz's inequality and Plancherel's theorem

$$\|(L_y g - g)L_y h\|_1 \leqslant \|L_y g - g\|_2 \|L_y h\|_2$$
$$= \left\{\int_{\hat{G}} |\hat{g}(\hat{x})|^2 |\overline{\langle y, \hat{x}\rangle} - 1|^2 \, d\hat{x}\right\}^{\frac{1}{2}} \|\hat{h}\|_2 \leqslant \|\hat{g}\|_2 \|\hat{h}\|_2 \max_{\hat{x}\in\hat{E}} |\langle y, \hat{x}\rangle - 1|.$$

The same estimate holds for $\|g(L_y h - h)\|_1$; thus (1) follows. (2) is also an application of Plancherel's theorem (cf. Chap. 4, § 4.5 (6)).

Remark. When $G = \mathbf{R}^\nu$, we can express (1) simply in the form $\|L_y\kappa - \kappa\|_1 \leqslant C|y|$, where $C = C_\kappa$ is independent of $y \in \mathbf{R}^\nu$; compare also Chapter 1, § 2.1.

1.2. Let $\hat{S}$ be a compact, symmetric neighbourhood of $\hat{e}$ in $\hat{G}$. Let $\phi_{\hat{S}}$ be the characteristic function of $\hat{S}$ and $F_{\hat{S}}$ the inverse Fourier transform of $\phi_{\hat{S}}$,

$$F_{\hat{S}}(x) = \int_{\hat{G}} \langle x, \hat{x}\rangle \phi_{\hat{S}}(\hat{x}) \, d\hat{x} \qquad x \in G,$$

with $d\hat{x}$ as in 1.1. Since $\hat{S}^{-1} = \hat{S}$, $F_{\hat{S}}(x)$ is real and $F_{\hat{S}}(x^{-1}) = F_{\hat{S}}(x)$. Put

$$\sigma(x) = \sigma_{\hat{S}}(x) = \frac{1}{m(\hat{S})}\{F_{\hat{S}}(x)\}^2 \qquad x \in G, \tag{3}$$

where m denotes Haar measure in $\hat{G}$ (normalized as stated). Then $\sigma \in L^1(G)$ and

(i) $\sigma(x) \geqslant 0 \quad (x \in G), \qquad \int \sigma(x)\,dx = 1;$

(ii) $\|L_y\sigma - \sigma\|_1 \leqslant 2\max_{\hat{x}\in\hat{S}} |\langle y, \hat{x}\rangle - 1| \quad (y \in G).$

(i) By Plancherel's theorem $F_{\hat{S}} \in L^2(G)$, thus $\sigma \in L^1(G)$ and since $\sigma \geqslant 0$ we have $\int \sigma = \|F_{\hat{S}}\|_2^2/m(\hat{S}) = \|\phi_{\hat{S}}\|_2^2/m(\hat{S}) = 1$.

(ii) follows from 1.1, with $\hat{g} = \hat{h} = \phi_{\hat{S}}$, $\hat{E} = \hat{S}$.

The F.t. of $\sigma_{\hat{S}}$ is given by (cf. (2))

$$\hat{\sigma}(\hat{x}) = \hat{\sigma}_{\hat{S}}(\hat{x}) = \frac{1}{m(\hat{S})}\phi_{\hat{S}} \star \phi_{\hat{S}}(\hat{x}) = \frac{m(\hat{x}\hat{S} \cap \hat{S})}{m(\hat{S})}. \tag{4}$$

Thus $0 \leqslant \hat{\sigma}_{\hat{S}} \leqslant 1$, $\hat{\sigma}_{\hat{S}}(\hat{e}) = 1$ and $\operatorname{Supp}\hat{\sigma}_{\hat{S}} \subset \hat{S}^2$.

REMARK. Since we can take $\hat{S}$ arbitrarily small, the functions $\hat{\sigma}_{\hat{S}}$ and their translates show that the Fourier algebra $\mathscr{F}^1(\hat{G})$ has the second 'standard property' (Chap. 2, § 1.1 (ii)).

The Fourier transforms $\hat{\sigma}_{\hat{S}}$ are generalized 'triangle functions': if $\hat{G} = \mathbf{R}$ and $\hat{S} = [-\frac{1}{2}\epsilon, \frac{1}{2}\epsilon]$ $(\epsilon > 0)$, then $\hat{\sigma}_{\hat{S}}(\hat{x})$ reduces to $\Delta_1(t/\epsilon)$ (cf. Chap. 1, § 1.3, Examples (ii) and (iv)), and $\sigma_{\hat{S}}(x) = \epsilon^{-1}(\sin \pi\epsilon x/\pi x)^2$; similarly for $\mathbf{R}^\nu$.

As we shall see, the functions $\sigma_{\hat{S}}$ play the same role on a general l.c.a. group as the familiar 'averaging functions' f_T on $\mathbf{R}$, where $f_T(x) = 1/2T$ if $|x| \leqslant T$, $f_T(x) = 0$ if $|x| > T$ $(T > 0)$.

1.3. Consider now two compact symmetric neighbourhoods $\hat{S}$ and $\hat{T}$ of $\hat{e}$ in $\hat{G}$. Then the product $\hat{S}\hat{T}$ is a neighbourhood of the same kind. Let $\phi_{\hat{S}}$, $\phi_{\hat{S}\hat{T}}$ be the characteristic functions of $\hat{S}$, $\hat{S}\hat{T}$ respectively; put

$$F_{\hat{S}}(x) = \int_{\hat{G}} \langle x, \hat{x}\rangle \phi_{\hat{S}}(\hat{x})\,d\hat{x}, \qquad F_{\hat{S}\hat{T}}(x) = \int_{\hat{G}} \langle x, \hat{x}\rangle \phi_{\hat{S}\hat{T}}(\hat{x})\,d\hat{x}$$

and define, with $m(\hat{S})$ as in 1.2,

$$\tau(x) = \tau_{\hat{S},\hat{T}}(x) = \frac{1}{m(\hat{S})} F_{\hat{S}}(x) F_{\hat{S}\hat{T}}(x) \qquad x \in G. \tag{5}$$

Then $\tau \in L^1(G)$ and

$$\|\tau_{\hat{S},\hat{T}}\|_1 \leqslant \left\{\frac{m(\hat{S}\hat{T})}{m(\hat{S})}\right\}^{\frac{1}{2}}, \tag{6}$$

$$\|L_y\tau_{\hat{S},\hat{T}} - \tau_{\hat{S},\hat{T}}\|_1 \leqslant 2\left\{\frac{m(\hat{S}\hat{T})}{m(\hat{S})}\right\}^{\frac{1}{2}} \max_{\hat{x}\in\hat{S}\hat{T}} |\langle y, \hat{x}\rangle - 1| \qquad y \in G. \tag{7}$$

(6) is obtained by Schwarz's inequality and Plancherel's theorem:

$$\|\tau_{\hat{S},\hat{T}}\|_1 \leqslant \frac{1}{m(\hat{S})}\|F_{\hat{S}}\|_2\|F_{\hat{S}\hat{T}}\|_2 = \frac{1}{m(\hat{S})}\|\phi_{\hat{S}}\|_2\|\phi_{\hat{S}\hat{T}}\|_2 = \left\{\frac{m(\hat{S}\hat{T})}{m(\hat{S})}\right\}^{\frac{1}{2}}.$$

(7) follows from 1.1, with $\hat{g} = \phi_{\hat{S}}$, $\hat{h} = \phi_{\hat{S}\hat{T}}$, $\hat{E} = \hat{S}\hat{T}$.

The F.t. of $\tau_{\hat{S},\hat{T}}$ is given by (cf. (2))

$$\hat{\tau}(\hat{x}) = \hat{\tau}_{\hat{S},\hat{T}}(\hat{x}) = \frac{1}{m(\hat{S})}\phi_{\hat{S}} \star \phi_{\hat{S}\hat{T}}(\hat{x}) = \frac{m(\hat{x}\hat{S} \cap \hat{S}\hat{T})}{m(\hat{S})}. \tag{8}$$

Hence $0 \leqslant \hat{\tau}_{\hat{S},\hat{T}} \leqslant 1$, $\hat{\tau}_{\hat{S},\hat{T}}(\hat{x}) = 1$ for $\hat{x} \in \hat{T}$, and $\operatorname{Supp} \hat{\tau}_{\hat{S},\hat{T}} \subset \hat{S}^2\hat{T}$.

The functions $\hat{\tau}$ may be called 'generalized trapezium functions': when $\hat{G} = \mathbf{R}$, they reduce to the functions introduced in Chapter 1, § 1.3, Examples (iii) and (iv).

REMARK 1. The support of $\hat{\tau}$ can be made arbitrarily small: given any neighbourhood $\hat{U}$ of $\hat{e}$, we can take $\hat{S}$ such that $\hat{S}^3 \subset \hat{U}$ and then $\operatorname{Supp} \hat{\tau}_{\hat{S},\hat{T}} \subset \hat{U}$ if $\hat{T} \subset \hat{S}$.

REMARK 2. We have not used the fact that $\hat{T}$ is a *neighbourhood* of $\hat{e}$: $\hat{T}$ may be any compact symmetric set containing $\hat{e}$; if $\hat{T} = \{\hat{e}\}$, then $\tau_{\hat{S},\hat{T}} = \sigma_{\hat{S}}$.

REMARK 3. For the groups **R** and **Z** the functions σ and τ defined by (3) and (5) are familiar. On compact groups we can simply put $\sigma = \tau = 1$. Hence we can obtain σ and τ explicitly for all groups $\mathbf{R}^{\nu_1} \times \mathbf{Z}^{\nu_2} \times K$, with K compact, that is, for all compactly generated l.c.a. groups (Chap. 4, § 2.9 (i)). The case of a general l.c.a. group G reduces to this as follows: G is the union of compactly generated open subgroups G_V (Chap. 3, § 1.8 (iv)); on each G_V the functions σ, τ are already defined and we extend them to G by putting $\sigma(x) = \tau(x) = 0$ if $x \notin G_V$. Thus σ and τ can be obtained, essentially, by means of the corresponding 'classical' functions on **R** and **Z**. But the significance of 1.2 and 1.3 is precisely that *the definition and fundamental properties of the functions σ, τ, on a general l.c.a. group, are independent of structure theory and of the particular functions on* **R**, **Z**.

References. [143, § 12, Lemma 6_{12}], [138, p. 50], [109, pp. 404–6].

2. Lemmas on functions in $L^1(G)$

We establish here some results of a technical nature, for later applications.

2.1. A locally compact abelian group has the following properties:

(i) If $K \subset G$ is compact, then for every $\epsilon > 0$ there is a function $s \in L^1(G)$ which is positive (Chap. 3, § 1.9) and such that $\int s(x)\,dx = 1$ and $\|L_y s - s\|_1 < \epsilon$ for all $y \in K$.

(ii) If $\alpha > 1$, then, given any compact set $K \subset G$, there is for every $\epsilon > 0$ a function $\tau \in L^1(G)$ such that $\|\tau\|_1 < \alpha$, the F.t. $\hat{\tau}$ is 1 near $\hat{e}$, and $\|L_y \tau - \tau\|_1 < \epsilon$ for all $y \in K$.

(i) Given K and ϵ, let $\hat{S}$ be any compact, symmetric nd. of $\hat{e}$ contained in the open nd.

$$\{\hat{x} \mid \hat{x} \in \hat{G},\ |\langle y, \hat{x}\rangle - 1| < \tfrac{1}{2}\epsilon \text{ for all } y \in K\}.$$

Then we can put $s = \sigma_{\hat{S}}$ (cf. § 1.2).

(ii) Given α, K, and ϵ, put

$$\hat{U} = \{\hat{x} \mid \hat{x} \in \hat{G},\ |\langle y, \hat{x}\rangle - 1| < \tfrac{1}{2}(\epsilon/\alpha) \text{ for all } y \in K\}.$$

Let $\hat{S}$ be a compact, symmetric nd. of $\hat{e}$ contained in $\hat{U}$; then we can choose a second such nd. $\hat{T}$ satisfying

$$\hat{S}\hat{T} \subset \hat{U} \quad \text{and} \quad m(\hat{S}\hat{T}) < \alpha m(\hat{S}),$$

since $\hat{U}$ is open and $\alpha > 1$ (cf. Chap. 3, §§ 1.8 (i) and 2.5 (11 a)). Now define $\tau = \tau_{\hat{S},\hat{T}}$ by 1.3 (5): then (cf. § 1.3 (6–8)) $\|\tau\|_1 < \alpha^{\frac{1}{2}} < \alpha$ and for $y \in K$ we have

$$\|L_y \tau - \tau\|_1 < 2\alpha^{\frac{1}{2}}(\epsilon/2\alpha) < \epsilon;$$

also $\hat{\tau}(\hat{x}) = 1$ if $\hat{x} \in \hat{T}$.

The significance of (i) and (ii) for analysis on groups will appear in the chapters that follow. One fact should be mentioned here: (i) can be extended, as it stands, to certain non-abelian locally compact groups (Chap. 8, §§ 3.1 and 3.3); but a generalization of (ii), if possible at all, would involve infinite-dimensional unitary group representations which are outside the scope of this monograph.

2.2. Now let G be any locally compact group, not necessarily abelian, and consider a family $\mathfrak{F}$ in $L^1(G)$ with the following properties:

(i) $\mathfrak{F}$ is bounded in $L^1(G)$, that is, there is a $C > 0$ such that $\|\kappa\|_1 \leqslant C$ for all $\kappa \in \mathfrak{F}$.

(ii) If K is a compact set in G, then for every $\epsilon > 0$ there is a function κ $(= \kappa_{K,\epsilon})$ in $\mathfrak{F}$ such that $\|L_y \kappa - \kappa\|_1 < \epsilon$ for all $y \in K$ (i.e. $\mathfrak{F}$ contains functions 'approximately invariant' under all L_y, $y \in K$, if K is compact).

Remark 1. On a locally compact abelian group there exist non-trivial families with the properties (i) and (ii), for example

$$\mathfrak{F}_1 = \{s \mid s \in L^1(G),\ s \geqslant 0,\ \int s(x)\, dx = 1\}$$

and, for any $\alpha > 1$,

$$\mathfrak{F}_\alpha = \{\tau \mid \tau \in L^1(G),\ \|\tau\|_1 < \alpha,\ \hat{\tau}(\hat{x}) = 1 \text{ near } \hat{e}\}.$$

This is, in fact, the significance of 2.1 (i) and (ii). We shall see later that the family $\mathfrak{F}_1$ possesses the required properties also in the case of certain non-abelian groups; this is trivial for all compact groups, of course.

Lemma. Let G be any locally compact group and $\mathfrak{F}$ a family in $L^1(G)$ with the properties (i) and (ii) above. Then, given any $f \in L^1(G)$, there exists, for every $\epsilon > 0$, a κ $(= \kappa_{f,\epsilon})$ in $\mathfrak{F}$ such that

$$\left\| f \star \kappa - \left[\int f(x)\, dx\right]\kappa \right\|_1 < \epsilon \tag{1}$$

and hence also

$$\|f \star \kappa\|_1 < C\left|\int f(x)\, dx\right| + \epsilon, \tag{2}$$

C being the constant in (i).

First let $\kappa \in \mathfrak{F}$ be arbitrary. Writing

$$\int f(y)\kappa(y^{-1}x)\,dy - \Big[\int f(y)\,dy\Big]\kappa(x) = \int f(y)\{\kappa(y^{-1}x)-\kappa(x)\}\,dy,$$

we obtain

$$(*) \qquad \Big\|f \star \kappa - \Big[\int f(y)\,dy\Big]\kappa\Big\|_1 \leqslant \int |f(y)|\,.\,\|L_y\kappa-\kappa\|_1\,dy.$$

There is a compact set $K \subset G$ such that $\int_{G-K} |f(y)|\,dy < \epsilon/3C$ (Chap. 3, § 2.3 (vi)). Since by (i) $\|L_y\kappa-\kappa\|_1 \leqslant 2C$ for any $\kappa \in \mathfrak{F}$, we obtain

$$\Big\|f \star \kappa - \Big[\int f(y)\,dy\Big]\kappa\Big\|_1 < \int_K |f(y)|\,dy\,.\max_{y\in K}\|L_y\kappa-\kappa\|_1 + \tfrac{2}{3}\epsilon.$$

By (ii) there is a $\kappa \in \mathfrak{F}$ such that $\|f\|_1 . \max_{y\in K}\|L_y\kappa-\kappa\|_1 < \frac{1}{3}\epsilon$ and (1) then holds for this κ. Moreover, (1) implies $\|f \star \kappa\|_1 < |\int f(x)\,dx|\,.\,\|\kappa\|_1+\epsilon$ and (2) follows.

Remark 2. For later applications the case of functions with compact supports is of especial interest. In this case the relation (1) holds *uniformly*, in the following sense: let a compact set $K \subset G$ and a number $M_1 > 0$ be given; then for every $\epsilon > 0$ there is a $\kappa \in \mathfrak{F}$ satisfying (1) and (2) for *all* functions $f \in L^1(G)$ vanishing (a.e.) outside K and such that $\|f\|_1 \leqslant M_1$.

If f vanishes a.e. outside K and $\|f\|_1 \leqslant M_1$, then (*) yields

$$\Big\|f \star \kappa - \Big[\int f(y)\,dy\Big]\kappa\Big\|_1 \leqslant M_1 . \max_{y\in K}\|L_y\kappa-\kappa\|_1,$$

thus the choice of $\kappa \in \mathfrak{F}$ in (1) depends only on K and M_1 (and ϵ).

2.3. We state here two applications of the results in 2.2 for later reference. Every locally compact *abelian* group G has the following properties.

(i) Given any $f \in L^1(G)$ and any $\epsilon > 0$, there is a positive function $s \in L^1(G)$ such that $\int s(x)\,dx = 1$ and

$$(3) \qquad \Big\|f \star s - \Big[\int f(x)\,dx\Big]s\Big\|_1 < \epsilon,$$

$$(4) \qquad \|f \star s\|_1 < \Big|\int f(x)\,dx\Big| + \epsilon.$$

(ii) Let $\alpha > 1$ be fixed. Then, given any $f \in L^1(G)$ and any $\epsilon > 0$, there is a function $\tau \in L^1(G)$ such that $\|\tau\|_1 < \alpha$, the F.t. $\hat{\tau}$ is 1 near $\hat{e} \in \hat{G}$, and

$$(5) \qquad \Big\|f \star \tau - \Big[\int f(x)\,dx\Big]\tau\Big\|_1 < \epsilon,$$

$$(6) \qquad \|f \star \tau\|_1 < \alpha\Big|\int f(x)\,dx\Big| + \epsilon.$$

(i) and (ii) result at once from 2.2 if the lemma and Remark 1 there are combined.

Remark. (ii) extends Chapter 1, § 2.4 from $\mathbf{R}^\nu$ to general l.c.a.

groups. An analysis of the various steps leading to (ii) shows that *the basic tool for this generalization is Plancherel's theorem*—the proof for $\mathbf{R}^\nu$ is much easier; the same applies to (i).

References. [143, § 12, Lemma 6_{13}], [109, p. 407].

3. Lemmas on functions in $L^1(G)$ (continued)†

3.1. We shall prove here a result which is a relativization (Chap. 4, § 5) of a part of Lemma 2.2; the rather general formulation is required for the applications.

Let G be an arbitrary locally compact group, H a closed normal subgroup of G. Let $\mathfrak{F}_H$ be a family of functions in $L^1(H)$ with the following properties:

(i) $\mathfrak{F}_H$ is bounded in $L^1(H)$: there is a $C > 0$ such that $\|\kappa\|_{L^1(H)} \leqslant C$ for all $\kappa \in \mathfrak{F}_H$.

(ii) If K' is a compact set in H, then for every $\epsilon > 0$ there is a $\kappa \in \mathfrak{F}_H$ such that $\|L_y \kappa - \kappa\|_{L^1(H)} < \epsilon$ for all $y \in K'$.

Let k be any function in $\mathscr{K}(G)$.

Then for every $\epsilon > 0$ there is a $\kappa \in \mathfrak{F}_H$ such that

$$(1) \qquad \int_G \left| \int_H k(x\xi)\kappa(\xi^{-1})\, d\xi \right| dx < C \int_{G/H} \left| \int_H k(x\xi)\, d\xi \right| d\dot{x} + \epsilon,$$

where C is the constant in (i) and the Haar measure $d\dot{x}$ on G/H is normalized so that $d\xi\, d\dot{x} = dx$ (Chap. 3, § 3.3 (i)). Moreover, if K is any compact set in G and M a positive number, then $\kappa \in \mathfrak{F}_H$ can be so chosen that (1) holds for all $k \in \mathscr{K}(G)$ with $\operatorname{Supp} k \subset K$ and $\|k\|_\infty \leqslant M$.

We note that, if H coincides with G, then (1) reduces to the relation (2) in Lemma 2.2, so (1) is indeed a relativization.

The proof of 3.1 is in two steps. First we establish the following proposition.

3.2. Let H be a closed subgroup‡ of a locally compact group G. Suppose $\mathfrak{F}_H$ is a family in $L^1(H)$ such that (i) $\|\kappa\|_{L^1(H)} \leqslant C$ for all $\kappa \in \mathfrak{F}_H$; (ii) if $K' \subset H$ is compact, then for every $\epsilon > 0$ there is a $\kappa \in \mathfrak{F}_H$ satisfying $\|L_\eta \kappa - \kappa\|_{L^1(H)} < \epsilon$ for all $\eta \in K'$.

Let k be a function in $\mathscr{K}(G)$, vanishing outside a compact set K, say. Then, given any $\epsilon > 0$, there is a $\kappa \in \mathfrak{F}_H$ such that for all $x \in G$

$$(2) \qquad \int_H \left| \int_H k(x\eta\xi)\kappa(\xi^{-1})\, d\xi \right| d\eta \leqslant C \left| \int_H k(x\xi)\, d\xi \right| + \epsilon . \phi_{KH}(x),$$

† The contents of this section will only be required later (Chaps. 7 and 8).

‡ Here H is not assumed to be normal; this more general hypothesis will be useful in Chapter 8, § 4.

where C is the constant in (i) and ϕ_{KH} is the characteristic function of the product KH.

Moreover, if K is a given compact set in G and M a positive number, the function $\kappa \in \mathfrak{F}_H$ can be so chosen that (2) holds for all $k \in \mathscr{K}(G)$ with $\operatorname{Supp} k \subset K$ and $\|k\|_\infty \leqslant M$.

Suppose $k \in \mathscr{K}(G)$ has support in K and $\|k\|_\infty \leqslant M$. Put, for any $x \in G$, $k_{x,H}(\xi) = k(x\xi)$, $\xi \in H$. The function $k_{x,H}$ is in $\mathscr{K}(H)$; moreover, *if* $x \in K$, then $\operatorname{Supp} k_{x,H}$ lies in the *fixed* compact set $K' = (K^{-1}K) \cap H$ and $\|k_{x,H}\|_{L^1(H)} \leqslant M_1$, where $M_1 = M . m_H(K')$ *is independent of* $x \in K$. Hence we can apply the lemma and Remark 2 in 2.2 (with $G = H$): there is a $\kappa \in \mathfrak{F}_H$ depending only on K' and M_1 (and ϵ) such that

$$\|k_{x,H} \star \kappa\|_{L^1(H)} < C \left| \int_H k_{x,H}(\xi)\, d\xi \right| + \epsilon \qquad x \in K,$$

the convolution being that in $L^1(H)$. This shows that (2) holds for all $x \in K$.

Now observe that both sides of (2) are invariant under translations $x \to x\eta'$, $\eta' \in H$: thus (2) holds for all $x \in KH$. For $x \in G - KH$ the left-hand side of (2) is 0, since $k_{x,H} = 0$ when $x \notin KH$. Hence (2) holds for all $x \in G$; moreover, the choice of $\kappa \in \mathfrak{F}_H$ depends only on K, M, and ϵ.

Remark. In (2) the left-hand side is a continuous function of x; this can be shown by means of the lemma in Chapter 3, § 3.2. The first term on the right in (2) is also continuous.

3.3. We can now obtain the result in 3.1 by applying Weil's formula to the inequality (2); the details follow.

Let $K \subset G$ be compact and M positive. By 3.2 there is a $\kappa \in \mathfrak{F}_H$ such that (2) holds for all $k \in \mathscr{K}(G)$ with $\operatorname{Supp} k \subset K$, $\|k\|_\infty \leqslant M$.

Since the three terms in (2) are invariant under translations $x \to x\eta'$, $\eta' \in H$, they are of the form $F' \circ \pi_H$, where F' is a function on G/H (cf. Chap. 3, § 1.8 (vii)); e.g. $\phi_{KH} = \phi_{\dot K} \circ \pi_H$, where $\dot K = \pi_H(K)$. We also note the Remark at the end of 3.2.

Thus we can integrate (2) over G/H, obtaining

$$\int_{G/H} \left\{ \int_H \left| \int_H k(x\eta\xi)\kappa(\xi^{-1})\, d\xi \right| d\eta \right\} d\dot x \leqslant C \int_{G/H} \left| \int_H k(x\xi)\, d\xi \right| d\dot x + \epsilon . m_{G/H}(\dot K),$$

where $m_{G/H}(\dot K) < \infty$, since $\dot K$ is compact. By the extended Weil formula (or, if κ has compact support, by Weil's formula itself)

$$\int_G \left| \int_H k(x\xi)\kappa(\xi^{-1})\, d\xi \right| dx \leqslant C \int_{G/H} \left| \int_H k(x\xi)\, d\xi \right| d\dot x + \epsilon . m_{G/H}(\dot K).$$

Here $m_{G/H}(\dot K)$ depends only on K, but not on ϵ. Since $\epsilon > 0$ was arbitrary, (1) is proved.

Proposition 3.1 is the basis for further developments: we shall see its applications in Chapter 7, § 4.4 and Chapter 8, §§ 3.3 and 4.2.

4. Some properties of $L^1(G)$ and $\mathscr{F}^1(\hat{G})$

Let G be a l.c.a. group, $\hat{G}$ its dual.

4.1. (i) *The functions in $L^1(G)$ whose Fourier transforms have compact support are dense in $L^1(G)$.* The proof below is due to Segal [123, Theorem 2.6] (cf. also [54, p. 126]).

Given $f \in L^1(G)$, we can put $f = g.h$ with $g, h \in L^2(G)$ (e.g. $g(x) = |f(x)|^{\frac{1}{2}}$, $h(x) = f(x)/g(x)$ if $f(x) \neq 0$, and $h(x) = 0$, if $f(x) = 0$). Let $\hat{g}, \hat{h} \in L^2(\hat{G})$ be the 'Plancherel transforms' so that $\hat{f} = \hat{g} \star \hat{h}$ (Chap. 4, § 4.5 (6)). Take $\hat{g}_n, \hat{h}_n \in \mathscr{K}(\hat{G})$ such that $\|\hat{g}-\hat{g}_n\|_2 < 1/n$, $\|\hat{h}-\hat{h}_n\|_2 < 1/n$, $n \geqslant 1$; let g_n, h_n be the inverse Fourier transforms and put $f_n = g_n . h_n$, so that $f_n \in L^1(G)$. Writing

$$f - f_n = g(h-h_n) + h_n(g-g_n),$$

we have by Schwarz's inequality and Plancherel's theorem

$$\|f-f_n\|_1 < (1/n)\{\|g\|_2 + \|h\|_2 + 1/n\} \qquad n \geqslant 1$$

and $\hat{f}_n = \hat{g}_n \star \hat{h}_n$ has compact support.

(ii) Equivalently we can say that *in the Fourier algebra $\mathscr{F}^1(\hat{G})$* (Chap. 4, § 4.2) *the functions with compact support are dense.* We note that $\mathscr{F}^1(\hat{G})$ *also has approximate units* (cf. Chap. 3, § 5.6).

(iii) Given any $f \in L^1(G)$ and any $\epsilon > 0$, there is a $v \in L^1(G)$ such that the F.t. $\hat{v}$ has compact support and $\|v \star f - f\|_1 < \epsilon$. Thus $\mathscr{F}^1(\hat{G})$ *has approximate units with compact support.*

(iii) follows at once from (ii); compare Chapter 2, § 4.1.

REMARK. For $G = \mathbf{R}^\nu$ (iii) can be proved directly, in a very elementary way (Chap. 1, § 2.3).

4.2. *$\mathscr{F}^1(\hat{G})$ is dense in $\mathscr{C}^0(\hat{G})$*, in the topology of $\mathscr{C}^0(\hat{G})$.

This follows from the (complex form of the) Weierstrass–Stone theorem, since $\mathscr{F}^1(\hat{G})$ separates points (§ 1.2, Remark) and is closed under conjugation (Chap. 4, § 4.2 (iv)). For another method see [124, Lemma 2].

4.3. *$\mathscr{F}^1(\hat{G})$ is a* PROPER *subalgebra of $\mathscr{C}^0(\hat{G})$, whenever $\hat{G}$ is infinite.* A proof was given by Segal [124]; we present it below, in a slightly modified form, since some of the steps are of independent interest and related to other results.

4.4. Let H be any closed subgroup of G and $H^\perp \subset \hat{G}$ the orthogonal subgroup.

(i) Let u, v be in $\mathscr{K}(G)$ and put $k = u \star v$. Then the F.t. $\hat{k}$ is in $L^1(\hat{G})$. Moreover, the restriction of k to H is the inverse F.t. of $T_{H^\perp} \hat{k} \in L^1(\hat{G}/H^\perp)$ (cf. Chap. 3, § 4.4 (10)), provided the Haar measure on $\hat{G}$ is dual to that of G and the Haar measures on $\hat{G}/H^\perp$, $H^\perp$ are canonically related to that on $\hat{G}$ (Chap. 3, § 3.3 (i)).

We have: $k \in \mathscr{K}(G)$ and $\hat{k} = \hat{u}\hat{v} \in L^1(\hat{G})$, since $\hat{u}, \hat{v} \in L^2(\hat{G})$. Hence (Chap. 4, § 4.1 (iii)) k is the inverse F.t. of $\hat{k}$. It now follows exactly as in Chapter 4, § 4.3 that the inverse F.t. of $T_{H^\perp}\hat{k}$ is the restriction of k to H.

Some extensions of (i) will be given in 5.5.

By means of (i) we can now establish the following general result.

(ii) Let μ be in $M^1(G)$. Then $\operatorname{Supp}\mu \subset H$ if and only if the F.t. $\hat{\mu}$ is $H^\perp$-periodic (cf. Chap. 3, § 1.8 (vii)). In particular, let H be an open subgroup† of G; if $f \in L^1(G)$, then f vanishes a.e. outside H if and only if the F.t. $\hat{f}$ is $H^\perp$-periodic.‡

If $\operatorname{Supp}\mu \subset H$, then $\hat{\mu}$ is clearly $H^\perp$-periodic (cf. Chap. 4, § 4.4). To prove the converse, take any function $k = u \star v$ with $u, v \in \mathscr{K}(G)$, so that k is the inverse F.t. of $\hat{k} \in L^1(\hat{G})$ (cf. (i)). Then we have, by an interchange of integrals,

$$\int_G k(x)\,d\mu(x) = \int_{\hat{G}} \hat{k}(\hat{x})\hat{\mu}(\hat{x}^{-1})\,d\hat{x}$$

or, since $\hat{\mu}$ is $H^\perp$-periodic, say $\hat{\mu} = \hat{\mu}' \circ \pi_{H^\perp}$,

$$\int_G k(x)\,d\mu(x) = \int_{\hat{G}/H^\perp} \hat{\mu}'(\hat{x}'^{-1})\,T_{H^\perp}\hat{k}(\hat{x}')\,d\hat{x}'.$$

Now *consider any* $u \in \mathscr{K}(G)$ *such that* $H \cap \operatorname{Supp} u = \varnothing$. Then, if $v \in \mathscr{K}(G)$ has support in a sufficiently small nd. V of e and $k = u \star v$, we have $H \cap \operatorname{Supp} k = \varnothing$ (cf. Chap. 3, § 1.8 (i)). But if k vanishes on H, then $T_{H^\perp}\hat{k}(\hat{x}') = 0$ a.e. on $\hat{G}/H^\perp$, by (i) (cf. Chap. 4, § 4.1 (ii)); hence $\int u \star v\,d\mu = \int k\,d\mu = 0$. Taking a sequence $(v_n)_{n\geqslant 1}$ in $\mathscr{K}(G)$ such that $\operatorname{Supp} v_n \subset V$ and $\|v_n \star u - u\|_\infty < 1/n$, we get $\int u\,d\mu = 0$. Thus $\operatorname{Supp}\mu \subset H$.

From the particular case stated in (ii) we deduce:

(iii) Let $H \subset G$ be an open subgroup (then $H^\perp \subset \hat{G}$ is compact). Let $\phi \in \mathscr{C}^0(\hat{G})$ be $H^\perp$-periodic: $\phi = \phi' \circ \pi_{H^\perp}$ with $\phi' \in \mathscr{C}^0(\hat{G}/H^\perp)$. Then $\phi \in \mathscr{F}^1(\hat{G})$ if and only if $\phi' \in \mathscr{F}^1(\hat{G}/H^\perp)$.

Suppose $\phi' \in \mathscr{F}^1(\hat{G}/H^\perp)$, say ϕ' is the F.t. of $f_1 \in L^1(H)$. Put $f(x) = f_1(x)$ if $x \in H$, $f(x) = 0$ if $x \in G-H$: then $f \in L^1(G)$ and $\hat{f} = \phi' \circ \pi_{H^\perp}$, so $\phi' \circ \pi_{H^\perp} \in \mathscr{F}^1(\hat{G})$. Conversely, let $\phi = \phi' \circ \pi_{H^\perp}$ be in $\mathscr{F}^1(\hat{G})$, say $\phi = \hat{f}$: then f vanishes a.e. on $G-H$ (cf. (ii)). Thus, if f_H is the restriction of f to H, we have $\hat{f} = (f_H)^\wedge \circ \pi_{H^\perp}$—that is, $\phi' = (f_H)^\wedge \in \mathscr{F}^1(\hat{G}/H^\perp)$.

4.5. We now proceed to the proof of 4.3.

(i) Suppose that the mapping $f \to \hat{f}$ of $L^1(G)$ into $\mathscr{C}^0(\hat{G})$ is surjective. This implies that $L^1(G)$ and $\mathscr{C}^0(\hat{G})$ are isomorphic as Banach spaces, with equivalent norms. Hence for any $\phi \in L^\infty(G)$ the linear functional $f \to \int f(x)\phi(x)\,dx$ defines a *continuous* linear functional on $\mathscr{C}^0(\hat{G})$, or

† For *open* subgroups H the Haar measure is the restriction of that of G to H.

‡ The particular case was proved in [124, Lemma 3]; the general case is in [120, § 2.7.1], but the argument offered is not conclusive.

$$\int_G f(x)\phi(x)\,dx = \int_{\hat G} \hat f(\hat x)\,d\mu_\phi(\hat x), \tag{1}$$

for some $\mu_\phi \in M^1(\hat G)$. Thus

$$\int_G f(x)\phi(x)\,dx = \int_G f(x)\left\{\int_{\hat G} \overline{\langle x, \hat x\rangle}\,d\mu_\phi(\hat x)\right\} dx$$

for all $f \in L^1(G)$, whence

$$\phi(x) = \int_{\hat G} \overline{\langle x, \hat x\rangle}\,d\mu_\phi(\hat x) \qquad \text{l.a.e. on } G. \tag{2}$$

That is, *if* $\mathscr{F}^1(\hat G) = \mathscr{C}^0(\hat G)$, *then every* $\phi \in L^\infty(G)$ *is of the form* (2), *where* $\mu_\phi \in M^1(\hat G)$; we will show that this is impossible if $\hat G$ is infinite.

(ii) *Let G be countable at infinity*, so that $\hat G$ has a countable basis at $\hat e$ (Chap. 4, § 2.3 (ii)). Consider a sequence $(\hat S_n)_{n\geqslant 1}$ of compact, symmetric neighbourhoods shrinking to $\{\hat e\}$; let $\sigma_n = \sigma_{\hat S_n}$ be the corresponding functions in $L^1(G)$ defined by 1.2 (3). *If* $\phi \in L^\infty(G)$ *has the form* (2), *then*

$$\lim_{n\to\infty} \int_G \sigma_n(x)\phi(x)\,dx = \int_{\hat G} \phi_{\hat e}(\hat x)\,d\mu_\phi(\hat x), \tag{3}$$

where $\phi_{\hat e}(\hat x) = 0$ for $\hat x \neq \hat e$, $\phi_{\hat e}(\hat e) = 1$.

The first integral in (3) is equal to $\int_{\hat G} \hat\sigma_n(\hat x)\,d\mu_\phi(\hat x)$. We have $0 \leqslant \hat\sigma_n \leqslant 1$ for all $n \geqslant 1$, and $\hat\sigma_n(\hat x) \to \phi_{\hat e}(\hat x)$ when $n \to \infty$, for each $\hat x \in \hat G$ (cf. § 1.2 (4)); thus (3) follows from Chapter 3, § 2.4 (iv).

(iii) From (i) and (ii) we can deduce that if the l.c.a. group G is countable at infinity, but not compact, then $\mathscr{F}^1(\hat G) \neq \mathscr{C}^0(\hat G)$.

If $\mathscr{F}^1(\hat G) = \mathscr{C}^0(\hat G)$, then all $\phi \in L^\infty(G)$ are of the form (2); thus the limit formula (3) applies to *all* $\phi \in L^\infty(G)$. Hence (Chap. 3, § 2.8 (v)) there is an $f \in L^1(G)$ such that $\int_G f(x)\phi(x)\,dx = \int_{\hat G} \phi_{\hat e}\,d\mu_\phi$ or

$$\int_{\hat G} \hat f(\hat x)\,d\mu_\phi(\hat x) = \int_{\hat G} \phi_{\hat e}(\hat x)\,d\mu_\phi(\hat x) \qquad \text{for all } \phi \in L^\infty(G). \tag{*}$$

But here μ_ϕ ranges over all of $M^1(\hat G)$. Putting $d\mu_\phi(\hat x) = \overline{\hat f(\hat x)}k(\hat x)\,d\hat x$, where k ranges over all functions in $\mathscr{K}_+(\hat G)$ such that $k(\hat e) = 0$, we obtain from (*): $\hat f(\hat x) = 0$ for all $\hat x \neq \hat e$, and hence $\hat f = 0$, since $\hat G$ is not discrete. On the other hand, if we let μ_ϕ be the Dirac measure at $\hat e$, the right-hand side of (*) is 1, a contradiction.

(iv) From (iii) and § 4.4 (iii) we obtain at once that $\mathscr{F}^1(\hat G) \neq \mathscr{C}^0(\hat G)$ whenever G is a l.c.a. group which is not compact: in fact, every such group G contains an *open* subgroup H countable at infinity, but not compact.

Let C be a compact nd. of e and let $(a_n)_{n\geqslant 1}$ be a sequence in G such that $a_1 = e$ and all the sets $a_n C$ are mutually disjoint (cf. the proof in Chap. 3, § 3.1 (iv)).

The subgroup H generated by $\bigcup_{n\geqslant 1} a_n C$ has the desired properties: we can write $H = \bigcup_{n\geqslant 1} (K_n \cup K_n^{-1})^n$, where $K_n = \bigcup_{1\leqslant j\leqslant n} a_j C$.

(v) Finally, let G be compact. *If* $\mathscr{F}^1(\hat{G}) = \mathscr{C}^0(\hat{G})$, then (2) shows, in the present case, that $\mathscr{K}(G)$ $(= \mathscr{C}^0(G))$ coincides with $\mathscr{F}^1(G)$ (observe that $M^1(\hat{G}) = L^1(\hat{G})$, for discrete $\hat{G}$). But this is impossible if $\hat{G}$ is infinite, on account of (iv), with the roles of G and $\hat{G}$ reversed.

Thus the proof of 4.3 is complete.

4.6. For generalizations of Segal's result 4.3 see R. E. Edwards [43] and M. Rajagopalan [108].

Segal's result is equivalent to the following statement (cf. Chap. 2, § 3.6): *whenever G is an infinite l.c.a. group, there are functions $f \in L^1(G)$ such that $\|f\|_1$ is arbitrarily large, but $\|\hat{f}\|_\infty \leqslant 1$.* It would be of interest to construct such functions; if $G = \mathbf{R}$, $\mathbf{Z}$, or $\mathbf{T}$, this can readily be done, but a general construction for l.c.a. G seems more difficult.

5. Poisson's formula

5.1. We prove here the following result for l.c.a. groups G. Let $f \in \mathscr{K}(G)$ be such that $\hat{f} \in L^1(\hat{G})$.† Let H be a closed subgroup of G and $H^\perp$ the orthogonal subgroup of $\hat{G}$. Then the restriction of $\hat{f}$ to $H^\perp$ is in $L^1(H^\perp)$. Moreover, given Haar measures dx, $d\xi$ on G, H respectively, there is a normalization of the Haar measure on $H^\perp$, say $d\xi^\perp$, such that *Poisson's formula*

$$\text{(1)} \qquad \int_H f(\xi)\, d\xi = \int_{H^\perp} \hat{f}(\xi^\perp)\, d\xi^\perp$$

holds for ALL $f \in \mathscr{K}(G)$ *with* $\hat{f} \in L^1(\hat{G})$. We call $d\xi^\perp$ the *orthogonal Haar measure* (of $d\xi$ relative to dx); it is determined as follows: we choose $d\dot{x}$ on G/H so that $d\xi\, d\dot{x} = dx$ (Chap. 3, § 3.3 (i)) and then take $d\xi^\perp$ on $H^\perp = (G/H)^\wedge$ as the dual measure of $d\dot{x}$.

5.2. First we establish some preliminary lemmas. Let the Haar measures on G and H be given and let those on G/H and $H^\perp$ be determined as in 5.1.

(i) Let g be in $\mathscr{K}(G)$. Put $\hat{g}_{\hat{x},H^\perp}(\xi^\perp) = \hat{g}(\hat{x}\xi^\perp)$, $\xi^\perp \in H^\perp$, for fixed $\hat{x} \in \hat{G}$. Then $\hat{g}_{\hat{x},H^\perp} \in L^2(H^\perp)$ and $\|\hat{g}_{\hat{x},H^\perp}\|_2 \leqslant \alpha$, a number independent of $\hat{x} \in \hat{G}$.

The function $\hat{y} \to \hat{g}(\hat{x}\hat{y})$, $\hat{y} \in \hat{G}$, is the F.t. of $g \cdot \bar{\chi}_{\hat{x}}$, hence $\hat{g}_{\hat{x},H^\perp}$ is the F.t. of

$$T_H(g \cdot \bar{\chi}_{\hat{x}}) \in \mathscr{K}(G/H)$$

(Chap. 4, § 4.3). Thus $\hat{g}_{\hat{x},H^\perp} \in L^2(H^\perp)$ and

$$\|\hat{g}_{\hat{x},H^\perp}\|_2 = \|T_H(g \cdot \bar{\chi}_{\hat{x}})\|_2 \leqslant \|T_H(|g|)\|_2 = \alpha,$$

which is independent of $\hat{x}$.

† For example, if $f = g \star h$, with $g, h \in \mathscr{K}(G)$, then $\hat{g}, \hat{h} \in L^2(\hat{G})$ and $\hat{f} = \hat{g} \cdot \hat{h} \in L^1(\hat{G})$.

(ii) Let g, h be in $\mathscr{K}(G)$ and put $\kappa = g \star h$, so that $\kappa \in \mathscr{K}(G)$. Then, in the notation of (i), we have: $\hat{\kappa}_{\hat{x},H^\perp}$ is in $L^1(H^\perp)$ and there is a constant C such that $\|\hat{\kappa}_{\hat{x},H^\perp}\|_1 \leqslant C$, or

$$\int_{H^\perp} |\hat{\kappa}(\hat{x}\xi^\perp)|\, d\xi^\perp \leqslant C \qquad \text{for all } \hat{x} \in \hat{G}. \tag{2}$$

Since $\hat{\kappa}_{\hat{x},H^\perp} = \hat{g}_{\hat{x},H^\perp} . \hat{h}_{\hat{x},H^\perp}$, we have

$$\|\hat{\kappa}_{\hat{x},H^\perp}\|_1 \leqslant \|\hat{g}_{\hat{x},H^\perp}\|_2 . \|\hat{h}_{\hat{x},H^\perp}\|_2 \leqslant \alpha . \beta,$$

with α, β independent of $\hat{x} \in \hat{G}$ by (i).

(iii) If $f \in \mathscr{K}(G)$ and $\hat{f} \in L^1(\hat{G})$, then the restriction of $\hat{f}$ to $H^\perp$ is in $L^1(H^\perp)$. More precisely, given any compact set $K \subset G$, there is a constant C $(= C_K)$ such that, whenever $f \in \mathscr{K}(G)$, $\operatorname{Supp} f \subset K$ and $\hat{f} \in L^1(\hat{G})$, then

$$\int_{H^\perp} |\hat{f}(\xi^\perp)|\, d\xi^\perp \leqslant C \int_{\hat{G}} |\hat{f}(\hat{x})|\, d\hat{x}. \tag{3}$$

Take $g \in \mathscr{K}(G)$ with $\int g = 1$ and choose an $h \in \mathscr{K}(G)$ which is 1 on $(\operatorname{Supp} g)^{-1}K$; put $\kappa = g \star h$. Since $\kappa \in \mathscr{K}(G)$ is 1 on K, we have $f = f.\kappa$ and hence $\hat{f} = \hat{f} \star \hat{\kappa}$ (Chap. 4, § 4.5 (6)); substituting this into the left-hand-side of (3), we obtain (3) by using (2).

Remark. From (3) it clearly follows that $\int_{H^\perp} \hat{f}(\hat{x}\xi^\perp)\, d\xi^\perp$ is a *continuous* function of $\hat{x} \in \hat{G}$.

5.3. We can now prove (1). Let f be as in 5.1. By Chapter 4, § 4.3 the F.t. of $T_H f \in \mathscr{K}(G/H)$ is $\hat{f}_{H^\perp}$, the restriction of $\hat{f}$ to $H^\perp$ (the Haar measure on G/H being normalized as in 5.1). By 5.2 (iii) $\hat{f}_{H^\perp}$ is in $L^1(H^\perp)$, hence we may apply the inversion theorem (Chap. 4, § 4.1 (iii)): $T_H f$ is the inverse F.t. of $\hat{f}_{H^\perp}$ (the Haar measure on $H^\perp$ being chosen as stated in 5.1); in particular this yields (1).

Remark 1. We can state 5.1 more generally. Let f be as in 5.1; put $f_{x,H}(\xi) = f(x\xi)$, $\xi \in H$, for any $x \in G$. Then $f_{x,H} \in \mathscr{K}(H)$ has the F.t. $T_{H^\perp}(\hat{f} \cdot \chi_x)$ $[\chi_x(\hat{x}) = \langle x, \hat{x}\rangle]$, provided the Haar measure of $H^\perp$ is orthogonal to that of H.

The proof consists simply in applying (1) to the function

$$y \to f(xy) . \overline{\langle y, \hat{x}\rangle}, \quad y \in G,$$

which satisfies the conditions in 5.1 if f does.

Remark 2. Let $f \in L^1(G)$ be of the form $f = f_0 \star f_1$ with $f_0 \in L^1(G)$ arbitrary, but f_1 in $\mathscr{K}(G)$ and such that $\hat{f}_1 \in L^1(\hat{G})$. Then the restriction f_H of f to H is in $L^1(H)$ and Poisson's formula (1) holds for f; more generally, $f_{x,H}$ is in $L^1(H)$ for every $x \in G$ and its F.t. is as in Remark 1.

The function f is continuous; also $\int_H |f(x\xi)|\, d\xi \leqslant \|f_0\|_1 \cdot \|T_H(|f_1|)\|_\infty$. Moreover, $T_H f$ is continuous (cf. Chap. 3, § 3.2, Lemma) and the restriction of $\hat{f} = \hat{f}_0 \cdot \hat{f}_1$ to $H^\perp$ is in $L^1(H^\perp)$, since that of $\hat{f}_1$ is (cf. § 5.1). Poisson's formula readily follows, in the same way as above. For $f_{x,H}$ cf. Remark 1 (note that for $a \in G$ and any character $\chi_{\hat{x}}$ of G we have $\bar{\chi}_{\hat{x}} \cdot L_a f = (\bar{\chi}_{\hat{x}} \cdot f_0) \star (\bar{\chi}_{\hat{x}} \cdot L_a f_1)$).

REMARK 3. Poisson's formula also holds for continuous functions in $L^1(G)$ whose Fourier transforms have compact support. This is merely a dual formulation of the conditions in 5.1.

References. [11, § 46], [143, § 11, Lemma 6_7], [111, II, Hilfssatz 1]; cf. also [91, 37 E], where the argument is not valid, and [13 *d*, Chap. II, § 1, n° 8]. For applications and generalizations cf. [24] and [139, pp. 168–9].

5.4. As an application of Poisson's formula we now prove that, *if the Haar measures of G, H, G/H are canonically related, then so are the dual Haar measures of $\hat{G}$, $H^\perp = (G/H)^\wedge$ and $\hat{G}/H^\perp = \hat{H}$.*

Let dx, $d\xi$, $d\dot{x}$ on G, H, G/H be given and such that $d\xi\, d\dot{x} = dx$. Choose $d\hat{x}$, $d\xi^\perp$ on $\hat{G}$, $H^\perp$ dual to dx, $d\dot{x}$, respectively, and *define* $d\hat{x}'$ on $\hat{G}/H^\perp$ by $d\xi^\perp d\hat{x}' = d\hat{x}$. We will show that then $d\hat{x}'$ is dual to $d\xi$, which gives the desired result.

Choosing $f \in \mathscr{K}(G)$ such that $\hat{f} \in L^1(\hat{G})$, we have, as a special case of 5.3, Remark 1: the F.t. of the restriction f_H of f to H is $T_{H^\perp}\hat{f}$ ($d\xi^\perp$ being orthogonal to $d\xi$). On the other hand, f is the inverse F.t. of $\hat{f}$ ($d\hat{x}$ being dual to dx) and from this we obtain, exactly as in Chapter 4, § 4.3, that the inverse F.t. of $T_{H^\perp}\hat{f} \in L^1(\hat{G}/H^\perp)$ is f_H (the Haar measure $d\hat{x}'$ on $\hat{G}/H^\perp$ being determined as stated above). Choosing f also such that $f_H \neq 0$, we obtain: $d\hat{x}'$ is dual to $d\xi$.

This was proved by A. Weil, first in implicit form [138, pp. 114–5], then explicitly in [139, §§ 16 and 17]; a comparison of the two expositions is enlightening (cf. also [13 *d*, Chap. II, § 1, n° 8, Prop. 9]).

Example. Let H be discrete, G/H compact. Normalize the Haar measure of G/H by $m(G/H) = 1$, and take ordinary summation on H; likewise for $\hat{G}/H^\perp$ and $H^\perp$. Then the Haar measures on G and $\hat{G}$ given by Weil's formula are dual to one another. Analogously for H compact, G/H discrete.

5.5. Let H be a closed subgroup of G, $H^\perp$ the orthogonal subgroup of $\hat{G}$. Let the Haar measure on $H^\perp$ be orthogonal to that on H (§ 5.1).

(i) Let f be a continuous function in $L^1(G)$ such that the restriction f_H of f to H is in $L^1(H)$. If $\hat{f} \in L^1(\hat{G})$, then $T_{H^\perp}\hat{f}$ coincides a.e. on $\hat{G}/H^\perp$ with the F.t. of f_H.

(ii) If $f \in \mathscr{K}(G)$ and $\hat{f} \in L^1(\hat{G})$ or, dually, if $f \in L^1(G)$ is continuous and $\operatorname{Supp}\hat{f}$ compact, then $(f_H)^\wedge = T_{H^\perp}\hat{f}$. This is a first answer to a question raised in Chapter 4, § 5.2 (i) (cf. also Chap. 7, § 2.2).

(i) The inverse F.t. of $T_{H^\perp}\hat{f} \in L^1(\hat{G}/H^\perp)$ is f_H: this is shown as in 4.4 (i). The assertion then follows from the general inversion theorem (Chap. 4, § 4.1, Remark).

(ii) This is contained in 5.3, Remarks 1 and 3.

6

WIENER'S THEOREM AND LOCALLY COMPACT ABELIAN GROUPS

1. Wiener's theorem for groups

1.1. FIRST we establish the *theorem of Wiener–Lévy for locally compact abelian groups.* Let $\hat{f}$ be in $\mathscr{F}^1(\hat{G})$, the algebra of Fourier transforms of the functions in $L^1(G)$. Let $\hat{K}$ be a compact set in $\hat{G}$; denote by $\hat{f}(\hat{K})$ the image of $\hat{K}$ in $\mathbf{C}$ under the map $z = \hat{f}(\hat{x})$, $\hat{x} \in \hat{K}$. If $z \to \mathrm{A}(z)$ is an analytic function defined (and single-valued) on an open neighbourhood of $\hat{f}(\hat{K})$, then there is a function $\hat{g} \in \mathscr{F}^1(\hat{G})$ such that $\hat{g}(\hat{x}) = \mathrm{A}(\hat{f}(\hat{x}))$ for all $\hat{x} \in \hat{K}$.

This is the analogue of Chapter 1, § 3.1 for l.c.a. groups; the modified formulation in Chapter 1, § 3.5 also holds for l.c.a. groups.

The analytical tools needed for the proof were already prepared in Chapter 5; the remaining part of the proof is quite easy.

Chapter 1, § 3.2 extends to all l.c.a. groups, on account of Chapter 5, § 2.3 (ii), relation (5). The proofs in Chapter 1, §§ 3.3–3.5 then carry over to $\mathscr{F}^1(\hat{G})$, merely with a change in notation. The auxiliary 'trapezium functions' in $\mathscr{F}^1(\hat{G})$ are now those defined for l.c.a. groups in Chapter 5, § 1.3 and their translates.

The proof of the theorem of Wiener–Lévy given here for l.c.a. groups is entirely analogous to the classical proofs, where $\hat{G}$ is the circle or the line. The main problem, in the general case, consisted in extending Chapter 1, § 2.4 to l.c.a. groups: it is Plancherel's theorem which plays the leading part in this extension (cf. Chap. 5, § 2.3, Remark).

The theorem above is contained, essentially, in a result of Segal [123, Theorem 3.8]; see also the references in 1.2 below. For the proof given here see [109, p. 406]; this method readily extends to analytic functions of several complex variables (cf. also [120, § 6.2] for 'real-analytic' functions).

A converse to the theorem of Wiener–Lévy—showing that it is in a certain sense a 'best possible' result—has been proved for $\mathscr{F}^1(\mathbf{T})$ by Y. Katznelson [84] and for $\mathscr{F}^1(\hat{G})$ in general by H. Helson and J.-P. Kahane [62]; for a detailed exposition of this and related results we refer to *Acta Math.* **102,** 135–57 (1959) or [120, §§ 6.5 and 6.6]. For $\mathscr{F}^1(\mathbf{T})$ it is the solution of a classical problem going back to [89, p. 10].

1.2. There is a well-known and very powerful generalization of the theorem of Wiener–Lévy to commutative Banach algebras with unit element: see I. M. Gelfand [48, Theorem 20] (an English translation is given as Appendix in [95]) and [51, §§ 6 and 13], [101, § 11, nos. 6 and 7]. For the extension of Gelfand's theory to the case of algebras without unit see [123], [95]; cf. also [13 *d*, Chap. I, § 4].

1.3. The theorem of Wiener–Lévy (§ 1.1) implies in particular that the algebra $\mathscr{F}^1(\hat{G})$ satisfies the first condition for standard function algebras (Chap. 2, § 1.1), and it also satisfies the second, as we know (Chap. 5, § 1.2, Remark); thus $\mathscr{F}^1(\hat{G})$ is a standard function algebra. It is also a normed algebra, isomorphic to $L^1(G)$ (Chap. 4, § 4.2). Moreover, the functions with compact support are dense in $\mathscr{F}^1(\hat{G})$ (Chap. 5, § 4.1 (i)). Hence $\mathscr{F}^1(\hat{G})$ *is a Wiener algebra* and the 'abstract' form of *Wiener's theorem* (Chap. 2, § 2.4) *holds for* $\mathscr{F}^1(\hat{G})$; it is customary to state it for $L^1(G)$: *if a closed ideal of* $L^1(G)$ *has empty cospectrum,*† *then it is the whole of* $L^1(G)$. We may replace here 'closed ideal' by 'closed, translation invariant linear subspace' (cf. Chap. 3, § 5.7); in particular, *Wiener's approximation theorem* (Chap. 1, § 4.1) *holds for* $L^1(G)$.

Another, more algebraic, formulation of Wiener's theorem is this: *every closed ideal of* $L^1(G)$ *distinct from* $L^1(G)$ *is contained in a closed maximal ideal,* namely in an ideal $I_\lambda = \{f \mid f \in L^1(G), \hat{f}(\lambda) = 0\}$, λ being any (fixed) element of the cospectrum.

Wiener's general Tauberian theorem (Chap. 1, § 4.5) extends to l.c.a. groups G if we interpret '∞' as the 'point of infinity' in the one-point compactification of G; the proof is essentially the same.

The extension of Wiener's theorem to l.c.a. groups was given by Godement [54, p. 125] and Segal [123, Part II] as an application of Gelfand's theory. Godement calls Wiener's theorem *théorème tauberien*, but this terminology should be avoided: a Tauberian theorem is, traditionally, a theorem on limits. It is also appropriate that this theorem should bear simply (and eloquently) the name of its discoverer.

The reader will find it instructive to prove Wiener's theorem for *finite* abelian groups in a purely algebraic way (cf. [31]).

1.4. *For any l.c.a. group* $\hat{G}$, *a closed ideal* I *of* $\mathscr{F}^1(\hat{G})$ *contains every function* $\hat{f} \in \mathscr{F}^1(\hat{G})$ *such that* $\hat{f}$ *vanishes on* cosp I *and* Bdr cosp $\hat{f} \cap$ Bdr cosp I *is a scattered set* (*generalization of Wiener's theorem*). Indeed, the condition

† The cospectrum of an ideal $I \subset L^1(G)$ is the set of all $\hat{x} \in \hat{G}$ for which the Fourier transforms of all $f \in I$ vanish; this terminology is due to L. Schwartz [121]. The cospectrum of a function $f \in L^1(G)$ is the set of zeros of its F.t.

of Wiener–Ditkin (Chap. 2, § 4.3) holds for $\mathscr{F}^1(\hat{G})$: this follows at once from the relation (5) in Chapter 5, § 2.3, in the same way as for $\mathscr{F}^1(\mathbf{R}^\nu)$ in Chapter 1, § 3.2. $\mathscr{F}^1(\hat{G})$ also has approximate units (Chap. 3, § 5.6). Thus the result above follows from Chapter 2, § 4.6; it can of course be stated, equivalently, for $L^1(G)$.

REMARK. The condition of Wiener–Ditkin already occurs, implicitly, in the proof of the theorem of Wiener–Lévy in 1.1; the role of Plancherel's theorem in this context has been pointed out there. Plancherel's theorem is also used in proving that the functions with compact support are dense in $\mathscr{F}^1(\hat{G})$ (Chap. 5, § 4.1 (i)). Thus *the theorem of Plancherel is a basic tool in the proof that Wiener's theorem and its generalization hold for all l.c.a. groups.* The original proof given by Wiener for the line does not use Plancherel's theorem at all (nor does that in Chapter 1). This is the main difference between the 'classical' and the 'group-theoretic' proofs.

1.5. COROLLARY. *Every closed set in $\hat{G}$ with a scattered (in particular, countable) boundary is a Wiener set for $\mathscr{F}^1(\hat{G})$.* In the algebraic terminology already introduced in 1.3 this is expressed as follows: every closed ideal I of $\mathscr{F}^1(\hat{G})$—or of $L^1(G)$—such that cosp I has a scattered boundary is the intersection of all closed maximal ideals containing I.

REMARK. Wiener sets for $\mathscr{F}^1(\hat{G})$ are also called 'sets of spectral synthesis' in the literature; for this terminology cf. Chapter 7, § 1.1, Remark 4.

1.6. *The theorems on Wiener–Ditkin sets* (Chap. 2, § 5.3) *also apply to* $\mathscr{F}^1(\hat{G})$. This, of course, again entails Corollary 1.5.

1.7. For historical references concerning Theorem 1.4 on the line see Chapter 2, § 4.8. In 1949 Kaplansky [82, Theorem 1] proved for general l.c.a. groups $\hat{G}$ that a single point is a Wiener set for $\mathscr{F}^1(\hat{G})$ or, in algebraic terminology, that every closed primary ideal of $\mathscr{F}^1(\hat{G})$ is maximal; for $\hat{G} = \mathbf{R}$ this was first proved in 1935 by Carleman [20, p. 78] by means of complex function theory. Kaplansky's proof made essential use of the structure theory of l.c.a. groups; for a large class of l.c.a. groups Kaplansky [82, Theorem 2] also obtained the result that a closed set with a countable boundary is a Wiener set for $\mathscr{F}^1(\hat{G})$ (§ 1.5). In 1951 Helson [60] gave a new proof, not based on structure theory, of Kaplansky's theorem on single points and then used this result, together with an unpublished proof of Beurling's for the line, to prove that a closed set with a scattered boundary is a Wiener set for $\mathscr{F}^1(\hat{G})$ for *all* l.c.a. groups $\hat{G}$ (Corollary 1.5); his methods yield Theorem 1.4 as well. Another proof of 1.4, based on the methods of Carleman [20], Mandelbrojt–Agmon [98], and A. Weil [138], was given in [109, Part II and Appendix].

The work of Shilov [125] and Mackey [95] mentioned in Chapter 2, § 4.8 implied, of course, that Theorem 1.4, and 1.5, could be proved if the condition of Wiener–Ditkin could be shown to hold in $\mathscr{F}^1(\hat{G})$, for general l.c.a. $\hat{G}$; Kaplansky made an attempt in this direction [82, p. 134, Remark].

That the condition of Wiener–Ditkin does, in fact, hold for all l.c.a. groups was later proved [109, pp. 404–7] in a somewhat different context (cf. § 1.4, Remark) and without knowledge of the work of Shilov and Mackey. The proof has found its way into the literature, being reproduced in the book of Loomis [91, p. 150], where it is combined with Shilov's theory to establish 1.4.

The 'condition D' employed in [91, p. 86] is a slight modification of the original condition of Wiener–Ditkin and, in fact, equivalent to it (cf. Chap. 2, § 5.1); it was introduced (in somewhat greater generality) by Mackey [95, p. 117].

A remarkable extension of 1.5 to certain *non-abelian* l.c. groups has been given by Eymard.†

The result in 1.6 is implicit in the literature; see the references at the end of Chapter 2, § 5.3.

1.8. After L. Schwartz showed in 1948 that in $\mathbf{R}^\nu$, $\nu \geqslant 3$, there exist closed sets which are not Wiener sets for $\mathscr{F}^1(\mathbf{R}^\nu)$ (cf. Chap. 2, § 7.1), the corresponding problem for $\mathscr{F}^1(\hat{G})$ for general non-discrete l.c.a. $\hat{G}$, especially for $\hat{G} = \mathbf{R}$, remained open for some years. The solution—as difficult as it is profound—was given in 1959 by Malliavin [97]: *on every non-discrete l.c.a. group $\hat{G}$ there exist closed sets which are not Wiener sets for $\mathscr{F}^1(\hat{G})$.* See also [66] and [119].

On the other hand, Herz [63] has proved the remarkable result that *the classical ternary Cantor set and also certain other Cantor sets are Wiener sets for $\mathscr{F}^1(\mathbf{R})$*, and has given various extensions for groups [65, pp. 228–30].

We mention here the following problem. Does every non-discrete l.c.a. group $\hat{G}$ contain a Wiener set which is not a Wiener–Ditkin set for $\mathscr{F}^1(\hat{G})$? In case the answer is affirmative, it would generalize Malliavin's result stated above (cf. Chap. 2, § 5.2 (iii)).

Additional references. [134*a*, *c*]; see also [71], [72], and [70].

2. Segal algebras

2.1. In Chapter 1, § 5 we discussed some normed subalgebras of

† [44, (4.19)]. We observe that the condition (R) stated there on page 188 holds for all l.c. groups which are soluble or a semi-direct product of a soluble l.c. group and a compact group (cf. Chap. 8, §§ 3.3, 3.5, 3.6).

$L^1(\mathbf{R}^\nu)$; here we shall consider the general theory underlying those examples.

Let G be a l.c.a. group and let $S^1(G)$ be a subalgebra of $L^1(G)$ satisfying the following conditions:

(i) $S^1(G)$ is dense in $L^1(G)$ in the (norm) topology of $L^1(G)$ and is invariant under translations: $f \in S^1(G) \Rightarrow L_a f \in S^1(G)$ for all $a \in G$.

(ii) $S^1(G)$ is a Banach algebra under some norm $\|.\|_S$ invariant under translations ($\|L_a f\|_S = \|f\|_S$ for all $f \in S^1(G)$, $a \in G$) and the following condition holds: given any $f \in S^1(G)$, there is for every $\epsilon > 0$ a neighbourhood U ($= U_\epsilon$) of e such that

$$(1) \qquad \|L_y f - f\|_S < \epsilon \qquad \text{for all } y \in U.$$

An algebra of this kind will be called a *Segal algebra*. The examples discussed in Chapter 1, § 5 are Segal algebras and the first four of them carry over to l.c.a. groups; there are many other examples. We shall see that Segal algebras share several properties of $L^1(G)$—which is itself the largest Segal algebra on G—and the axiomatic approach will provide some insight.

The Fourier transforms of the functions in $S^1(G)$ form a subalgebra $\mathscr{S}^1(\hat{G})$ of $\mathscr{F}^1(\hat{G})$, isomorphic to $S^1(G)$, with the norm carried over from $S^1(G)$.

The first example of a Segal algebra was given, essentially, by Wiener (cf. Chap. 1, § 5 (iii)), but it was Segal [123, Theorem 3.1] who noticed an underlying abstract structure. The axioms above differ only slightly from those introduced by Segal.

We can define Wiener's example, Chapter 1, § 5 (iii), for any l.c.a. group G with a discrete subgroup Γ such that G/Γ is compact.

There is a compact set $K \subset G$ such that $G = K\Gamma$. It can be shown that the complex-valued continuous functions f on G for which

$$(*) \qquad \|f\|_S = \sup_{u \in G} \sum_{\gamma \in \Gamma} \max_{x \in K} |f(ux\gamma)|$$

is finite form a Segal algebra with (*) as norm. Since $u \to \sum_{\gamma \in \Gamma} \max_{x \in K} |f(ux\gamma)|$ is Γ-periodic, it is enough to restrict u to K in (*). Next we observe that $\|f\|_S$ is finite for all $f \in \mathscr{K}(G)$, since there are only finitely many $\gamma \in \Gamma$ for which $(K^2.\gamma) \cap \operatorname{Supp} f \neq \emptyset$. It can also readily be shown that this algebra does not depend on the choice of K (subject to the condition $K\Gamma = G$).

It should be observed that a Segal algebra, being a subalgebra of $L^1(G)$, is a convolution algebra of equivalence classes of functions in $\mathfrak{L}^1(G)$; in some cases, of course, it may be possible to use only functions instead of equivalence classes: see Chapter 1, § 5 (i), but also § 5 (ii).

2.2. We discuss now some general properties of Segal algebras $S^1(G)$.

(i) *There is a constant C such that*

$$\|f\|_1 \leqslant C\|f\|_S \qquad \textit{for all } f \in S^1(G), \tag{2}$$

where $\|.\|_1$ is the L^1-norm, $\|.\|_S$ the S^1-norm.

Consider the isomorphic algebra $\mathscr{S}^1(\hat{G}) \subset \mathscr{F}^1(\hat{G})$: then (i) follows at once from Chapter 2, § 3.6.

In the examples mentioned in 2.1 we have $C = 1$.

(ii) *$S^1(G)$ is an ideal in $L^1(G)$ and*

$$\|h \star f\|_S \leqslant \|h\|_1 . \|f\|_S \qquad \textit{for all } f \in S^1(G),\ h \in L^1(G). \tag{3}$$

Consider the space $L^1_B(G)$ with $B = S^1(G)$. For $h \in L^1(G)$ and $f \in S^1(G)$ the function $y \to h(y) L_y f$, $y \in G$, is in $L^1_B(G)$, hence $\int h(y) L_y f\, dy$ exists and is an element f_1 in $S^1(G) \subset L^1(G)$. To prove that $f_1 = h \star f$, we proceed as in Chapter 3, § 6.6 (but note that B is now a different Banach space). If $\phi \in L^\infty(G)$, then the restriction of $f \to \langle f, \phi\rangle$ to $S^1(G)$ is a *continuous* linear functional on $S^1(G)$, by (2). Hence f_1 must satisfy, in particular,

$$\langle f_1, \phi\rangle = \int \langle h(y) L_y f, \phi\rangle\, dy \qquad \text{for all } \phi \in L^\infty(G),$$

whence $f_1 = h \star f$ (cf. Chap. 3, § 6.6). For the S^1-norm we have

$$\left\| \int h(y) L_y f\, dy \right\|_S \leqslant \int \|h(y) L_y f\|_S\, dy = \int |h(y)| . \|L_y f\|_S\, dy$$

and since $\|L_y f\|_S = \|f\|_S$, (3) follows.

In practice the relation (3) can be verified directly, of course.

(iii) *The algebra $\mathscr{S}^1(\hat{G}) \cong S^1(G)$ (§ 2.1) contains all $\hat{f} \in \mathscr{F}^1(\hat{G})$ with compact support* and hence is a standard function algebra (Chap. 2, § 1.1).

$\mathscr{S}^1(\hat{G})$ is an ideal in $\mathscr{F}^1(\hat{G})$ (cf. (ii)) and has empty cospectrum by 2.1 (i). The assertion now follows from Chapter 2, § 1.4 (ii) and the fact that $\mathscr{F}^1(\hat{G})$ is a standard function algebra.

(iv) *To any compact set $\hat{K} \subset \hat{G}$ there is a constant $C_{\hat{K}}$ such that every $f \in S^1(G)$ whose Fourier transform vanishes outside $\hat{K}$ satisfies*

$$\|f\|_S \leqslant C_{\hat{K}} \|f\|_1. \tag{4}$$

Take $\tau \in L^1(G)$ such that $\hat{\tau}$ is 1 on $\hat{K}$ and $\operatorname{Supp} \hat{\tau}$ is compact. Then $\tau \in S^1(G)$, by (iii), and hence $\|f\|_S = \|f \star \tau\|_S \leqslant \|f\|_1 . \|\tau\|_S$, by (3); thus we may put $C_{\hat{K}} = \|\tau\|_S$. Compare [143, p. 82, (11.04)].

Remark. There is also a more abstract proof. Consider the ideal $I(\hat{K})$ of $\mathscr{F}^1(\hat{G})$ consisting of all $\hat{f} \in \mathscr{F}^1(\hat{G})$ with $\operatorname{Supp} \hat{f} \subset \hat{K}$. This is also the ideal of all $\hat{f} \in \mathscr{S}^1(\hat{G})$ with $\operatorname{Supp} \hat{f} \subset \hat{K}$ (cf. (iii)). Moreover, $I(\hat{K})$ is a *complete* normed function algebra in each of the norms inherited from $L^1(G)$ and $S^1(G)$, hence (iv) holds (cf. Chap. 2, § 3.6).

This is a 'local' converse to (2).

2.3. Next we show: given any $f \in S^1(G)$, there is for every $\epsilon > 0$ a $v \in S^1(G)$ such that the F.t. $\hat{v}$ has compact support and $\|v \star f - f\|_S < \epsilon$.

Thus $\mathscr{S}^1(\hat{G}) \cong S^1(G)$ (cf. § 2.1) is a Wiener algebra and has approximate units.

By (1) there is a nd. U of e such that $\|L_y f - f\|_S < \frac{1}{2}\epsilon$ for all $y \in U$. Take $u \in \mathscr{K}_+(G)$ with $\int u = 1$ and $\operatorname{Supp} u \subset U$; then $u \star f - f = \int u(y)\{L_y f - f\}\, dy$ (cf. the proof in § 2.2 (ii)) and

$$(*) \qquad \left\| \int u(y)\{L_y f - f\}\, dy \right\|_S \leqslant \int u(y) \|L_y f - f\|_S\, dy.$$

Hence $\|u \star f - f\|_S < \frac{1}{2}\epsilon$. This first part of the proof is like that in Chapter 3, § 5.6, but the justification of (*) is more abstract.

Next there is a $v \in L^1(G)$ such that the F.t. $\hat{v}$ has compact support and

$$\|u - v\|_1 < \epsilon/2\|f\|_S$$

(Chap. 5, § 4.1 (i)). Now $v \in S^1(G)$ (since $\hat{v} \in \mathscr{S}^1(\hat{G})$, by § 2.2 (iii)); this might not be true of u. Finally (cf. (3))

$$\|v \star f - f\|_S \leqslant \|v - u\|_1 . \|f\|_S + \|u \star f - f\|_S < \epsilon.$$

Thus *Wiener's theorem holds for* $S^1(G)$. We can also prove (in the same way as for $L^1(G)$, cf. Chap. 3, §§ 5.7–5.9) that *the closed ideals of* $S^1(G)$ *coincide with the closed translation invariant linear subspaces of* $S^1(G)$, but this is contained in 2.4 below.

2.4. We now state the main result on Segal algebras. *The ideal theory of a Segal algebra* $S^1(G)$ *is the same as that of* $L^1(G)$. More precisely, there is a bijective correspondence between the family of all closed ideals of $S^1(G)$ and the family of all closed ideals of $L^1(G)$: *every closed ideal* I_S *of* $S^1(G)$ *is of the form* $I \cap S^1(G)$, *where* I *is a (unique) closed ideal of* $L^1(G)$; I is the closure of I_S in $L^1(G)$ so that, in particular, I_S *and* I *have the same cospectrum.*†

To prove this, we go over to Fourier transforms, taking as our starting point the properties 2.2 (iii) and 2.3. We shall prove the general proposition which follows.

2.5. Let $\mathscr{A}(X)$ be a complete normed Wiener algebra with approximate units. Let $\mathscr{A}_1(X)$ be a subalgebra of $\mathscr{A}(X)$ containing $\mathscr{A}_0$, the ideal of all functions in $\mathscr{A}(X)$ with compact support. Suppose that $\mathscr{A}_1(X)$ is a complete normed algebra in some other norm and has approximate units, and that $\mathscr{A}_0$ is dense in $\mathscr{A}_1(X)$ (thus $\mathscr{A}_1(X)$ is also a Wiener algebra). Then every closed ideal I_1 of $\mathscr{A}_1(X)$ is of the form

$$I_1 = I \cap \mathscr{A}_1(X),$$

where I is a unique closed ideal of $\mathscr{A}(X)$; I is the closure of I_1 in $\mathscr{A}(X)$. Conversely, if I is any closed ideal of $\mathscr{A}(X)$, then $I \cap \mathscr{A}_1(X)$ is a closed ideal of $\mathscr{A}_1(X)$, by Chap. 2, § 3.6.

† Cf. [117]; for certain special Segal algebras cf. also the results given in [86, Theorem 5] and [135].

(i) We first consider $\mathscr{A}_0$ as a normed algebra in its own right, with the norm inherited from $\mathscr{A}(X)$, and prove: there is a bijective correspondence between the family of all closed ideals I of $\mathscr{A}(X)$ and the family of all (relatively) closed ideals I_0 of $\mathscr{A}_0$; this correspondence is defined by $I_0 = I \cap \mathscr{A}_0$.

If I is a closed ideal of $\mathscr{A}$, then $I \cap \mathscr{A}_0$ is a relatively closed ideal of $\mathscr{A}_0$. Moreover, I is the closure of $I \cap \mathscr{A}_0$ in $\mathscr{A}$ (Chap. 2, § 4.1) and hence, if I, I' are closed ideals of $\mathscr{A}$ such that $I \cap \mathscr{A}_0 = I' \cap \mathscr{A}_0$, then $I = I'$. Also every relatively closed ideal I_0 of $\mathscr{A}_0$ is of the form $I_0 = I \cap \mathscr{A}_0$, for some closed ideal I of $\mathscr{A}$: we can take for I the closure of I_0 in $\mathscr{A}$, which is an ideal of $\mathscr{A}$, since $\mathscr{A}_0$ is dense in $\mathscr{A}$.

(ii) An ideal of $\mathscr{A}_0$ relatively closed in $\mathscr{A}_0$ in the topology induced by $\mathscr{A}_1(X)$ is also relatively closed in $\mathscr{A}_0$ in the topology induced by $\mathscr{A}(X)$, and conversely.

Suppose the ideal $I_0 \subset \mathscr{A}_0$ is relatively closed ($\mathscr{A}_1$). Let $(f_n)_{n \geqslant 1}$ be a sequence in I_0 tending to $f \in \mathscr{A}_0$ in the $\mathscr{A}$-norm; we will show that $f \in I_0$. As f has compact support, there is a $\tau \in \mathscr{A}$ such that $\operatorname{Supp} \tau$ is compact and τ is 1 on $\operatorname{Supp} f$ (Chap. 2, § 1.3, Remark); thus $\tau \in \mathscr{A}_0 \subset \mathscr{A}_1$ and $f\tau = f$. Now $f_n \tau \to f$ in the $\mathscr{A}$-norm ($n \to \infty$). The functions $f_n \tau$, $n \geqslant 1$, and f vanish outside the fixed compact set $\operatorname{Supp} \tau$; hence $f_n \tau \in I_0$ also tends to f in the $\mathscr{A}_1$-norm (cf. § 2.2 (iv), especially the Remark). Thus $f \in I_0$, i.e. I_0 is relatively closed ($\mathscr{A}$). The converse is clear from Chapter 2, § 3.6.

(iii) We can now prove that, if I_1 is any closed ideal of $\mathscr{A}_1(X)$, then $I_1 = I \cap \mathscr{A}_1(X)$, where I is a unique closed ideal of $\mathscr{A}(X)$ and is the closure of I_1 in $\mathscr{A}(X)$.

Clearly I is unique: if $I' \cap \mathscr{A}_1 = I \cap \mathscr{A}_1$, then $I' \cap \mathscr{A}_0 = I \cap \mathscr{A}_0$, whence $I' = I$ by (i). Now, given any closed ideal I_1 of $\mathscr{A}_1$, put $I_0 = I_1 \cap \mathscr{A}_0$. Then there is a closed ideal I of $\mathscr{A}$ such that also $I_0 = I \cap \mathscr{A}_0$ (cf. (i) and (ii)); moreover, we have $I_1 = I \cap \mathscr{A}_1$: indeed, $I \cap \mathscr{A}_1$ is a *closed* ideal of $\mathscr{A}_1$ (cf. Chap. 2, § 3.6), say I_1', and

$$I_1' \cap \mathscr{A}_0 = I \cap \mathscr{A}_0 = I_1 \cap \mathscr{A}_0,$$

whence $I_1' = I_1$ (cf. the proof of (i), replacing $\mathscr{A}$ by $\mathscr{A}_1$). I is the closure of I_1 in $\mathscr{A}$, since $I_0 \subset I_1 \subset I$, and I is the closure of I_0 in $\mathscr{A}$.

Thus the proof is complete; in particular we have proved Theorem 2.4.

Remark. One can also consider infinite intersections of Segal algebras; here Theorem 2.4 still applies. The examples in Chapter 2, § 3.3 (ii) and (iii) are of this type.

The property expressed by Theorem 2.4 distinguishes Segal algebras rather sharply from Beurling algebras (compare, in this respect, Chap. 7, § 3.10). It will be observed that Beurling algebras have all but one of the defining properties (§ 2.1) of a Segal algebra. $L^1(G)$, of course, is both a Segal and a Beurling algebra.

2.6. The theory of Segal algebras was developed in 2.1–2.4 for locally compact *abelian* groups G. But the axioms in 2.1 make sense, as they stand, for *any* locally compact group G. The property 2.2 (i), however, must now be stated as an additional axiom, since its proof made use of the commutativity of G; it then follows (cf. § 2.2 (ii)) that a Segal algebra is a *left* ideal of $L^1(G)$.

The examples in Chapter 1, § 5 (i) and (ii) carry over to general l.c. groups; another example is given below.

Example. Let G be a unimodular l.c. group containing a discrete subgroup H such that the quotient space G/H is compact (or has finite invariant measure, cf. Chap. 8, § 1 and [136, § 5]). We introduce a new norm in $\mathscr{K}(G)$:

$$\|k\|_S = \sup_{\dot{x}\in G/H} \sum_{\xi\in H} |k(x\xi)| \qquad k \in \mathscr{K}(G).$$

This norm also satisfies (1) (cf. Chap. 3, § 3.2, Lemma) and (2), and by completion we obtain a Segal algebra: since H is a discrete subgroup, this algebra consists entirely of continuous functions.

REMARK. An examination of Wiener's example (cf. § 2.1) shows that it could be extended to l.c. groups G containing a discrete *normal* subgroup Γ such that G/Γ is compact.

2.7. Let $S^1(G)$ be a Segal algebra, G being any l.c. group. Let H be a closed normal subgroup of G. *The image of* $S^1(G)$ *under the map* T_H (Chap. 3, §§ 4.1, 4.4, 5.3) *is a Segal algebra on* G/H. We simply introduce in $T_H[S^1(G)]$ the ordinary quotient norm; then the axioms for a Segal algebra hold, as is readily seen. The analogous result for Beurling algebras rests on a more delicate proof (Chap. 3, § 7).

2.8. It may be of interest to mention here an open question concerning Segal algebras on abelian l.c. groups. In all examples that we have discussed (cf. § 2.1 and Chap. 1, § 5), $S^1(G)$ contains with a function f also the product χf, for any character χ of G. How far is this true in general? The relation $\|\chi f\|_S = \|f\|_S$ certainly need not hold, as the example in Chapter 1, § 5 (v) shows.

Segal algebras on non-abelian l.c. groups have not yet been investigated more closely.

3. Beurling algebras

3.1. We introduced in Chapter 1, § 6 the algebras $L^1_\alpha(\mathbf{R}^\nu)$ and also some other Beurling algebras $L^1_w(\mathbf{R}^\nu)$ with a weight function w satisfying certain special conditions; for these particular Beurling algebras we gave elementary proofs of the analogues of the theorem of Wiener–Lévy and Wiener's theorem. Now let G be a l.c.a. group and let $L^1_w(G)$ be any Beurling algebra on G (cf. Chap. 3, § 7.1); denote by $\mathscr{F}^1_w(\hat{G})$ the corresponding subalgebra of $\mathscr{F}^1(\hat{G})$ on the dual group $\hat{G}$, with the norm carried over from $L^1_w(G)$, so that $\mathscr{F}^1_w(\hat{G}) \cong L^1_w(G)$.

$\mathscr{F}^1_w(\hat{G})$ is a Wiener algebra and Wiener's theorem holds for $\mathscr{F}^1_w(\hat{G})$ if and only if the weight function w satisfies

$$\sum_{n \geqslant 1} \frac{\log w(x^n)}{n^2} < \infty \qquad \textit{for every } x \in G. \tag{1}$$

REMARK. If additive notation is used for G, then x^n should be replaced by nx.

This theorem, due to Domar [39, Theorem 2.11], is a generalization of a theorem of Beurling [7, Theorem V B] for $G = \mathbf{R}$. We call (1) the *condition of Beurling–Domar.*

Domar's theorem is proved by means of a theorem of classical analysis on the line and the theory of commutative Banach algebras; it is a remarkable example of the power of that theory. The proof also uses the structure theory of l.c.a. groups.

3.2. Another known result is the following. *Let w be a weight function on G satisfying Shilov's condition: for every $x \in G$*

(i) $w(x^n) = O(|n|^\alpha)$, $|n| \to \infty$, *for some* $\alpha\ (= \alpha_x) > 0$;

(ii) $\displaystyle \liminf_{|n| \to \infty} \frac{w(x^n)}{|n|} = 0.$

Cf. also § 3.1, Remark.

Then every closed primary ideal of $L^1_w(G)$ is maximal.

For $G = \mathbf{Z}$ this result is due to Shilov, in a somewhat more general form.† For l.c.a. groups it follows from this special case, as will be shown in Chapter 7, § 2.4. The result is contained in a more abstract one of Domar's [39, pp. 40–47, especially p. 47].

Shilov's condition is much more restrictive than that of Beurling–Domar.

The algebra $\mathscr{F}^1_w(\hat{G})$ always possesses approximate units (cf. Chap. 3, § 7.2). Thus, if w satisfies (1) and if the condition of Wiener–Ditkin holds for $\mathscr{F}^1_w(\hat{G})$, then so does the generalization of Wiener's theorem (Chap. 2, § 4.6). However, the question of the validity of the condition of Wiener–Ditkin for $\mathscr{F}^1_w(\hat{G})$ is rather difficult and no general sufficient condition for w is known; we shall consider particular cases in 3.3 and 3.4.

3.3. We prove here that *the algebra $\mathscr{F}^1_\alpha(\mathbf{R}^\nu)$ satisfies the condition of Wiener–Ditkin if $0 \leqslant \alpha < 1$, for all $\nu \geqslant 1$* (cf. Chap. 2, § 7.4 (i)).

† [125, Chap. I, § 3, Theorem 10], [51, § 41]; the reader will see that some simplifications are possible in the proofs given there. We observe that, if (i) and (ii) above hold for *one* $x \in \mathbf{Z}$, $x \neq 0$, then they hold for all $x \in \mathbf{Z}$.

Lemma 1. Let $w_\alpha(x) = (1+|x|)^\alpha$, $x \in \mathbf{R}^\nu$, with $0 \leqslant \alpha < 1$. Let $\hat{g}$, $\hat{h}$ be functions in $L^2(\mathbf{R}^\nu)$ vanishing outside some ball $|t| \leqslant K$ and suppose that their inverse Fourier transforms g, h are such that $g.w_\alpha$ and $h.w_\alpha$ are in $L^2(\mathbf{R}^\nu)$. Let $\kappa = g.h$, so that κ is in $L^1_\alpha(\mathbf{R}^\nu)$, and put

$$M_{1/\rho}\kappa(x) = \rho^{-\nu}\kappa(x/\rho)$$

for $\rho > 0$. Then there is a constant C (depending on $\hat{g}$, $\hat{h}$, and α) such that

$$\|L_y M_{1/\rho}\kappa - M_{1/\rho}\kappa\|_{1,\alpha} \leqslant C\,|y|^\alpha \Big\{\max_{|t|\leqslant K} |\langle y/\rho, t\rangle - 1|\Big\}^{1-\alpha} \tag{2}$$

for all $y \in \mathbf{R}^\nu$ and all $\rho \geqslant 1$, where $\|.\|_{1,\alpha}$ is the norm in $L^1_\alpha(\mathbf{R}^\nu)$.

Lemma 1 is a modification of Chapter 5, § 1.1 and so is its proof.

By Schwarz's inequality $\|\kappa.w_\alpha\|_1 \leqslant \|g\|_2.\|h.w_\alpha\|_2$, thus $\kappa \in L^1_\alpha(\mathbf{R}^\nu)$. Denoting the left-hand side of (2) by $I_{y,\rho}$ we can write

$$I_{y,\rho} = \int |\kappa(x-y/\rho)-\kappa(x)|\, w_\alpha(\rho x)\, dx \leqslant \rho^\alpha \int |\kappa(x-y/\rho)-\kappa(x)|\, w_\alpha(x)\, dx,$$

if $\rho \geqslant 1$. Writing

$$L_{y/\rho}\kappa - \kappa = (L_{y/\rho}g - g).L_{y/\rho}h + g.(L_{y/\rho}h - h)$$

and using Schwarz's inequality again, we obtain

$$(\sharp) \qquad I_{y,\rho} \leqslant \rho^\alpha(A_{y,\rho}+B_{y,\rho}) \qquad \rho \geqslant 1,$$

where

$$A_{y,\rho} = \|L_{y/\rho}g - g\|_2.\Big\{\int |h(x-y/\rho)|^2 w_\alpha^2(x)\, dx\Big\}^{\frac{1}{2}},$$

$$B_{y,\rho} = \|g.w_\alpha\|_2.\|L_{y/\rho}h - h\|_2.$$

By Plancherel's theorem and the relation $w_\alpha(x+y/\rho) \leqslant w_\alpha(x)w_\alpha(y/\rho)$ we have

$$A_{y,\rho} \leqslant \Big\{\int |\hat{g}(t)|^2.|\overline{\langle y/\rho, t\rangle} - 1|^2\, dt\Big\}^{\frac{1}{2}}.\|h.w_\alpha\|_2.w_\alpha(y/\rho)$$

and thus, since $\hat{g}(t) = 0$ for $|t| > K$,

$$(*) \qquad A_{y,\rho} \leqslant \|\hat{g}\|_2.\max_{|t|\leqslant K}|\langle y/\rho, t\rangle - 1|.\|h.w_\alpha\|_2.w_\alpha(y/\rho).$$

Next we observe that for some constant C_0

$$(**) \qquad \Big\{\max_{|t|\leqslant K} |\langle y/\rho, t\rangle - 1|\Big\}^\alpha.w_\alpha(y/\rho) \leqslant C_0\,|y|^\alpha/\rho^\alpha \quad \text{for all } y \in \mathbf{R}^\nu,\ \rho > 0.$$

[For all $x \in \mathbf{R}^\nu$ we have $|\langle x, t\rangle - 1| \leqslant \min\{2\pi|x|.|t|, 2\}$ and $w_\alpha(x) \leqslant 1+|x|^\alpha$, since $\alpha < 1$; thus

$$\Big\{\max_{|t|\leqslant K} |\langle x, t\rangle - 1|\Big\}^\alpha.w_\alpha(x) \leqslant (2\pi K)^\alpha|x|^\alpha + 2^\alpha|x|^\alpha.]$$

From (*) and (**) we now obtain

$$\rho^\alpha A_{y,\rho} \leqslant C_1|y|^\alpha\Big\{\max_{|t|\leqslant K}|\langle y/\rho, t\rangle - 1|\Big\}^{1-\alpha}$$

with C_1 independent of y and ρ; likewise

$$\rho^\alpha B_{y,\rho} \leqslant C_2|y|^\alpha\Big\{\max_{|t|\leqslant K}|\langle y/\rho, t\rangle - 1|\Big\}^{1-\alpha},$$

where C_2 is independent of y and ρ. By ($\sharp$) Lemma 1 follows.

LEMMA 2. Suppose $0 \leqslant \alpha < 1$ and let $\kappa \in L^1_\alpha(\mathbf{R}^\nu)$ and $M_{1/\rho}\kappa$ be as in Lemma 1. Then for every $f \in L^1_\alpha(\mathbf{R}^\nu)$

$$\left\|f \star M_{1/\rho}\kappa - \left[\int f(x)\,dx\right].M_{1/\rho}\kappa\right\|_{1,\alpha} \to 0 \qquad \rho \to \infty.$$

We have as in Chapter 1, § 2.2

$$\left\|f \star M_{1/\rho}\kappa - \left[\int f(x)\,dx\right]M_{1/\rho}\kappa\right\|_{1,\alpha} \leqslant \int |f(x)|.\|L_x M_{1/\rho}\kappa - M_{1/\rho}\kappa\|_{1,\alpha}\,dx$$
$$\leqslant C\int |f(x)|.|x|^\alpha \phi(x/\rho)\,dx,$$

where C is some constant and $\phi(x) = \{\max_{|t|\leqslant K}|\langle x,t\rangle - 1|\}^{1-\alpha}$ (see Lemma 1). Now $\int |f(x)|.|x|^\alpha\,dx < \infty$, ϕ is bounded, continuous and $\phi(0) = 0$, thus Lemma 2 follows.

REMARK. It is easily seen that there are functions κ of the type used in Lemmas 1 and 2 such that the F.t. $\hat{\kappa}$ is 1 near the origin.

Consider first the case of $\mathbf{R}$. Let $t \to \mathrm{T}_1(t)$ be the 'trapezium function' defined in Chapter 1, § 1.3 (iii). Put $\hat{g}_1(t) = \mathrm{T}_1(\frac{1}{2}t)$, $\hat{h}_1(t) = \frac{2}{3}\mathrm{T}_1(2t)$: then $\hat{g}_1$, $\hat{h}_1$ satisfy the conditions of Lemma 1 and $\kappa_1 = g_1.h_1$ is such that $\hat{\kappa}_1 = \hat{g}_1 \star \hat{h}_1$ is 1 for $|t| \leqslant 1$. We can immediately extend this to $\mathbf{R}^\nu$ (cf. Chap. 1, § 1.3 (vi)).

Lemma 2, combined with the Remark above, is the precise analogue of Chapter 1, § 2.4 for $L^1_\alpha(\mathbf{R}^\nu)$. It follows at once, as in Chapter 1, § 3.2, that $\mathscr{F}^1_\alpha(\mathbf{R}^\nu)$ satisfies the condition of Wiener–Ditkin if $0 \leqslant \alpha < 1$.†

3.4. As another example we consider the additive group $\mathbf{Q}_l^\nu$ of all ν-tuples $x = (x_1,\dots, x_\nu)$ of l-adic numbers, l being any integer greater than 1 (Chap. 4, § 3.2, Example 2). The 'points' x whose coordinates are l-adic integers form a compact open subgroup $\mathbf{Z}_l^\nu$. We normalize the Haar measure on $\mathbf{Q}_l$ so that $\mathbf{Z}_l$ has Haar measure 1 and then take on $\mathbf{Q}_l^\nu$ the ν-fold product measure.

$\mathbf{Q}_l$ is self-dual (cf. loc. cit.), hence so is $\mathbf{Q}_l^\nu$ (Chap. 4, § 2.4 (iv)); the characters of $\mathbf{Q}_l^\nu$ can be written (cf. Chap. 4, § 3.3 (8))

(3) $$x \to \langle x,t\rangle = \chi_1(x_1 t_1 + \dots + x_\nu t_\nu) \qquad x, t \in \mathbf{Q}_l^\nu$$

and in this canonical duality of $\mathbf{Q}_l^\nu$ with itself we have $(\mathbf{Z}_l^\nu)^\perp = \mathbf{Z}_l^\nu$. We also note that

(4) $$\int_{\mathbf{Z}_l^\nu} \langle x,t\rangle\,dx = \begin{cases} 1 & \text{if } t \in \mathbf{Z}_l^\nu, \\ 0 & \text{if } t \notin \mathbf{Z}_l^\nu. \end{cases}$$

The restriction of $x \to \langle x,t\rangle$ to $\mathbf{Z}_l^\nu$ is a character of $\mathbf{Z}_l^\nu$. For a character χ of a *compact* abelian group G we have $\chi(a).\int \chi(x)\,dx = \int \chi(ax)\,dx = \int \chi(x)\,dx$ for all $a \in G$.

† [117]; it was also proved, in somewhat greater generality, by C. S. Herz [69].

In the algebra $L^1(\mathbf{Q}_l^\nu)$ we can define a *multiplication operator* M_{l^n} by

(5) $$M_{l^n}f(x) = l^{-n\nu}f(l^n x) \qquad n \in \mathbf{Z},$$

where $l^n x = (l^n x_1,..., l^n x_\nu)$. Then

(6) $$\int M_{l^n}f(x)\,dx = \int f(x)\,dx \qquad f \in L^1(\mathbf{Q}_l^\nu).$$

The map $x \to lx$ is an (algebraic and topological) automorphism of $\mathbf{Q}_l^\nu$ (cf. Chap. 4, § 3.2, Remark 2). Hence $\int k(lx)\,dx = c\int k(x)\,dx$, with c independent of $k \in \mathscr{K}(\mathbf{Q}_l^\nu)$, by the uniqueness of Haar measure. Taking for k the characteristic function of $\mathbf{Z}_l^\nu$, we obtain $c = l^\nu$ and thus (6).

The Fourier transform of $M_{l^n}f$ is (cf. (3))

(7) $$[M_{l^n}f]\hat{}\,(t) = \hat{f}(t/l^n) \qquad t \in \mathbf{Q}_l^\nu.$$

All this is quite analogous to the classical case of $L^1(\mathbf{R}^\nu)$ (Chap. 1, §§ 1.1 and 1.2), with some subtle differences.

We now consider Beurling algebras on $\mathbf{Q}_l^\nu$. First we observe that *on $\mathbf{Q}_l^\nu$ every weight function satisfies the condition of Beurling–Domar and even that of Shilov* (§§ 3.1 and 3.2), since for every $x \in \mathbf{Q}_l^\nu$ the sequence $(nx)_{n\geqslant 1}$ is contained in a compact set. *Thus $\mathscr{F}_w^1(\mathbf{Q}_l^\nu)$ is a Wiener algebra and single points of $\mathbf{Q}_l^\nu$ are Wiener sets for $\mathscr{F}_w^1(\mathbf{Q}_l^\nu)$, for every weight function w on $\mathbf{Q}_l^\nu$.*

Let us now discuss some special Beurling algebras on $\mathbf{Q}_l^\nu$. We define for $\alpha \geqslant 0$ a weight function w_α by

$$w_\alpha(x) = (1+|x|)^\alpha \qquad x \in \mathbf{Q}_l^\nu,$$

where $|x|$ is the norm of $x = (x_1,..., x_\nu)$ given by

$$|x| = \{|x_1|_l^2+...+|x_\nu|_l^2\}^{\frac{1}{2}},$$

with $|\,.\,|_l$ as in Chapter 4, § 3.2, Remark 1. We denote the Beurling algebra corresponding to w_α simply by $L_\alpha^1(\mathbf{Q}_l^\nu)$ and the isomorphic algebra of Fourier transforms by $\mathscr{F}_\alpha^1(\mathbf{Q}_l^\nu)$; these are analogues of $L_\alpha^1(\mathbf{R}^\nu)$ and $\mathscr{F}_\alpha^1(\mathbf{R}^\nu)$ (§ 3.3 and Chap. 1, § 6). The norm of $f \in L_\alpha^1(\mathbf{Q}_l^\nu)$ is again denoted by $\|f\|_{1,\alpha}$.

In $L_\alpha^1(\mathbf{Q}_l^\nu)$ we have not only the translation operator L_y—which is defined for all Beurling algebras—but also the multiplication operator M_{l^n} given by (5); this is analogous to Chapter 1, § 6.4. M_{l^n} corresponds for $n \geqslant 0$ to $M_{1/\rho}$, $\rho \geqslant 1$ (loc. cit.): we have

$$\|M_{l^n}f\|_{1,\alpha} \leqslant l^{n\alpha}\|f\|_{1,\alpha}, \qquad \|M_{l^{-n}}f\|_{1,\alpha} \leqslant \|f\|_{1,\alpha}, \qquad n = 0, 1, 2,...,$$

in virtue of the relation

$$|l^n x| = l^{-n}|x| \qquad x \in \mathbf{Q}_l^\nu,\ n \in \mathbf{Z}.$$

Let τ be the characteristic function of the compact open subgroup $\mathbf{Z}_l^\nu \subset \mathbf{Q}_l^\nu$. Then τ coincides with its F.t. (cf. (4)). Thus in $\mathscr{F}_\alpha^1(\mathbf{Q}_l^\nu)$—or in any $\mathscr{F}^1_w(\mathbf{Q}^\nu_l)$—there are very simple analogues of the 'trapezium functions' (Chap. 1, § 1.3 (iii)), viz. the functions $t \to \tau(t/l^n)$, $n \in \mathbf{Z}$, which are the Fourier transforms of $M_{l^n}\tau$ (cf. (7)), and their translates.

$\mathscr{F}^1_\alpha(\mathbf{Q}^\nu_l)$ *is a Wiener algebra*: this follows from the general theory, as pointed out further above.†

$\mathscr{F}^1_\alpha(\mathbf{Q}^\nu_l)$ *also has approximate units*: this is true of every Beurling algebra (Chap. 3, § 7.2). We now show that *the condition of Wiener–Ditkin holds for* $\mathscr{F}^1_\alpha(\mathbf{Q}^\nu_l)$, *for all* $\alpha \geqslant 0$ [117]. The proof is based on two lemmas.

LEMMA 1. Let τ be the characteristic function of $\mathbf{Z}^\nu_l$, introduced above. Then we have for $y \in \mathbf{Q}^\nu_l$, $n = 0, 1, 2,...$, and every $\alpha \geqslant 0$

$$\|L_y M_{l^n}\tau - M_{l^n}\tau\|_{1,\alpha} \begin{cases} = 0 & \text{if } |y| \leqslant l^n, \\ \leqslant C|y|^\alpha & \text{if } |y| \geqslant l^n, \end{cases} \tag{8}$$

where C is independent of y and n. Here $\|.\|_{1,\alpha}$ is the norm in $L^1_\alpha(\mathbf{Q}^\nu_l)$.

Since $L_y\tau = \tau$ for $y \in \mathbf{Z}^\nu_l$, we have $L_y M_{l^n}\tau = M_{l^n}\tau$ if $l^n y \in \mathbf{Z}^\nu_l$, and *a fortiori* if $|y| \leqslant l^n$, whence the first part of (8). Now write $I_{y,n}$ for the left-hand side of (8); then for $n = 0, 1, 2,...$,

$$I_{y,n} \leqslant \int \{\tau(x - l^n y) + \tau(x)\}(1 + |l^{-n}x|)^\alpha \, dx$$
$$\leqslant l^{\alpha n}\{\|L_{l^n y}\tau\|_{1,\alpha} + \|\tau\|_{1,\alpha}\}.$$

Moreover,
$$\|L_{l^n y}\tau\|_{1,\alpha} \leqslant (1 + |l^n y|)^\alpha \|\tau\|_{1,\alpha}$$
$$\leqslant (2l^{-n}|y|)^\alpha \|\tau\|_{1,\alpha} \quad \text{if } |y| \geqslant l^n.$$

Hence $I_{y,n} \leqslant (2^\alpha + 1)\|\tau\|_{1,\alpha}|y|^\alpha$ if $|y| \geqslant l^n$.

The analogy between (8) and (2) is worth observing.

LEMMA 2. Let τ be as in Lemma 1. Then for $f \in L^1_\alpha(\mathbf{Q}^\nu_l)$ the relation

$$\left\| f \star M_{l^n}\tau - \left[\int f(x)\, dx\right] . M_{l^n}\tau \right\|_{1,\alpha} \to 0 \qquad n \to \infty$$

holds, for all $\alpha \geqslant 0$.

The expression above is dominated by (cf. Chap. 1, § 2.2)

$$\int |f(x)| . \|L_x M_{l^n}\tau - M_{l^n}\tau\|_{1,\alpha}\, dx \leqslant C \int_{|x| \geqslant l^n} |f(x)| . |x|^\alpha \, dx,$$

by Lemma 1, and the right-hand side $\to 0$ when $n \to \infty$.

Lemma 2 is analogous to Chapter 1, § 2.4. It now follows that $\mathscr{F}^1_\alpha(\mathbf{Q}^\nu_l)$ satisfies the condition of Wiener–Ditkin, *for all* $\alpha \geqslant 0$ (cf. Chap. 1, § 3.2). This contrasts with $\mathscr{F}^1_\alpha(\mathbf{R}^\nu)$ (cf. § 3.3); a closer study of the analogies and differences between these algebras might reveal some interesting facts.

† It is not without interest, though, to observe that the elementary proofs in Chapter 1, §§ 6.4 and 6.5 also apply to $L^1_\alpha(\mathbf{Q}^\nu_l)$ and $\mathscr{F}^1_\alpha(\mathbf{Q}^\nu_l)$, with only minor notational changes.

3.5. As another example we consider the group $\mathbf{Z}$ of integers and put for $\alpha \geqslant 0$ $w_\alpha(n) = (1+|n|)^\alpha$, $n \in \mathbf{Z}$; let $L^1_\alpha(\mathbf{Z})$ be the corresponding Beurling algebra which is analogous to $L^1_\alpha(\mathbf{R})$ (cf. § 3.3); we write $\mathscr{F}^1_\alpha(\mathbf{T})$ for the algebra of Fourier transforms, in analogy with $\mathscr{F}^1_\alpha(\mathbf{R})$. We can define, in the same way, $L^1_\alpha(\mathbf{Z}^\nu)$ and $\mathscr{F}^1_\alpha(\mathbf{T}^\nu)$ for $\nu > 1$.

For each $\alpha \geqslant 0$ *the algebras* $\mathscr{F}^1_\alpha(\mathbf{R})$ *and* $\mathscr{F}^1_\alpha(\mathbf{T})$ *are 'locally isomorphic' in the following sense.* Let K_c, $0 < c < \frac{1}{2}$, be the compact interval $[-c, c]$; let $\hat{f}$ be a continuous function on $\mathbf{R}$ with support in K_c. Then $\hat{f}$ is in $\mathscr{F}^1_\alpha(\mathbf{R})$ if and only if $\mathrm{T}_\mathbf{Z}\hat{f}$ is in $\mathscr{F}^1_\alpha(\mathbf{T})$, where $\mathrm{T}_\mathbf{Z}\hat{f}(\dot{t}) = \sum_{n\in\mathbf{Z}} \hat{f}(t+n)$, $\dot{t} = \pi_\mathbf{Z}(t)$ ($\pi_\mathbf{Z}$: $\mathbf{R} \to \mathbf{R}/\mathbf{Z} \simeq \mathbf{T}$). *Likewise for* $\mathscr{F}^1_\alpha(\mathbf{R}^\nu)$ *and* $\mathscr{F}^1_\alpha(\mathbf{T}^\nu)$. For $\alpha = 0$ and $\nu = 1$ this is a classical result of Wiener's [143, § 11, Lemma 6_7] and the proof for $\alpha > 0$ is almost the same.

Let $\hat{f}$ be the F.t. of $f \in L^1_\alpha(\mathbf{R})$; we may assume f continuous. Let τ be a continuous function in $L^1_\alpha(\mathbf{R})$ such that $\tau(x) = O(|x|^{-(\alpha+2)})$ as $|x| \to \infty$, and

$$\hat{\tau}(t) = 1 \quad \text{for } |t| \leqslant c, \qquad \hat{\tau}(t) = 0 \quad \text{for } |t| \geqslant c' \ (c < c' < \tfrac{1}{2})$$

(cf. Chap. 1, § 1.3 (v)). We have $f = f \star \tau$; hence, putting $w_\alpha(x) = (1+|x|)^\alpha$, $x \in \mathbf{R}$, we get

$$\sum_{n\in\mathbf{Z}} |f(n)|\, w_\alpha(n) \leqslant \int_\mathbf{R} |f(y)|\, w_\alpha(y)\, dy \,.\, \max_{0\leqslant y\leqslant 1} \sum_{n\in\mathbf{Z}} |\tau(n-y)|\, w_\alpha(n-y).$$

Thus the restriction $f_\mathbf{Z}$ of f to $\mathbf{Z}$ is in $L^1_\alpha(\mathbf{Z})$; moreover, $f_\mathbf{Z}$ has the F.t. $\mathrm{T}_\mathbf{Z}\hat{f}$, as we know (Chap. 5, § 5.5 (ii)).

Conversely, let $\mathrm{T}_\mathbf{Z}\hat{f}$ be the F.t. of a function $f_\mathbf{Z}$ in $L^1_\alpha(\mathbf{Z})$. Put $f = \mu \star \tau$, where μ is the measure defined on $\mathbf{R}$ by $f_\mathbf{Z}$ (Chap. 4, § 4.4) and τ is as above. Then

$$\int_\mathbf{R} |f(x)|\, w_\alpha(x)\, dx \leqslant \sum_{n\in\mathbf{Z}} |f_\mathbf{Z}(n)|\, w_\alpha(n) \,.\, \int_\mathbf{R} |\tau(x)|\, w_\alpha(x)\, dx,$$

so that $f \in L^1_\alpha(\mathbf{R})$; also the F.t. of f is $[(\mathrm{T}_\mathbf{Z}\hat{f}) \circ \pi_\mathbf{Z}].\hat{\tau} = \hat{f}$, the given function. We remark that the restriction of f to $\mathbf{Z}$ is, of course, $f_\mathbf{Z}$. For $\mathbf{R}^\nu$, $\mathbf{Z}^\nu$, $\nu > 1$, the proof is the same.

The result above, combined with 3.3, shows that $\mathscr{F}^1_\alpha(\mathbf{T}^\nu)$ satisfies the condition of Wiener–Ditkin, if $0 \leqslant \alpha < 1$.

3.6. We conclude this section with two general results useful for applications.

(i) Let G be any l.c. group (not necessarily abelian) and let $L^1_{w_1}(G)$, $L^1_w(G)$ be two Beurling algebras such that $L^1_{w_1}(G) \subset L^1_w(G)$. Then, for some constant C, $w(x) \leqslant Cw_1(x)$ l.a.e. on G.

(ii) Let G be a l.c.a. group. Let $L^1_{w_1}(G)$, $L^1_w(G)$ be Beurling algebras satisfying the condition of Beurling–Domar (§ 3.1) and such that $L^1_{w_1}(G) \subset L^1_w(G)$. Let I be a [closed] ideal of $L^1_w(G)$ and put $I_1 = I \cap L^1_{w_1}(G)$. Then I_1 is a [closed] ideal of $L^1_{w_1}(G)$ and† $\operatorname{cosp} I_1 = \operatorname{cosp} I$.

† The definition of cospectrum carries over to Beurling algebras.

(i) Put $\|f\| = \|f\|_{1,w_1} + \|f\|_{1,w}$ for $f \in L^1_{w_1}$. With this new norm $L^1_{w_1}$ is still complete: indeed, if (f_n) is a Cauchy sequence for this norm, then (f_n) has a limit, say g_1, in $L^1_{w_1}$ and a limit g in L^1_w, but $g_1(x) = g(x)$ a.e., since $f_n \to g_1$ *and* $f_n \to g$ in the ordinary space L^1. By a general theorem on Banach spaces there is a C such that $\|f\| \leqslant C\|f\|_{1,w_1}$ and hence $\|f\|_{1,w} \leqslant C\|f\|_{1,w_1}$, for all $f \in L^1_{w_1}$. In particular $\int_A w(x)\, dx \leqslant C \int_A w_1(x)\, dx$ for all integrable sets $A \subset G$ with compact closure, whence the assertion. The preceding argument is quite analogous to that in Chapter 2, § 3.5; but it should be observed that Beurling algebras are non-commutative if G is non-abelian (cf. Chap. 3, § 5.1).

(ii) Clearly I_1 is an ideal of $L^1_{w_1}(G)$; by (i) it is closed in $L^1_{w_1}$ if I is closed in L^1_w. Obviously $\operatorname{cosp} I_1 \supset \operatorname{cosp} I$. Now consider any $\hat{a} \in \hat{G} - \operatorname{cosp} I$. There is an $f \in L^1_{w_1}$ such that $\hat{f}(\hat{a}) \neq 0$ and $\operatorname{Supp} \hat{f}$ is compact and disjoint from $\operatorname{cosp} I$. Thus f (being also in L^1_w) belongs to I (cf. Chap. 2, § 1.4 (ii)), hence $f \in I_1$ and $\hat{a} \notin \operatorname{cosp} I_1$.

REMARK. Statement (ii) readily extends to complete normed Wiener algebras (with the same proof). The investigation of examples where the map $I \to I_1$ in (ii) is injective, or perhaps even bijective, would be of interest; compare 2.5 (and also Chap. 7, § 3.10).

7

THE SPECTRUM AND ITS APPLICATIONS

1. Definition and properties of the spectrum

1.1. LET G be a l.c.a. group. If $f \in L^1(G)$ we call the support of the F.t. $\hat{f}$ the *spectrum* of f, $\mathrm{sp} f$, and for an ideal I of $L^1(G)$ we define the spectrum of I, $\mathrm{sp}\, I$, as the smallest closed set in $\hat{G}$ containing $\mathrm{sp} f$ for all $f \in I$.

We have already defined the cospectrum of f, $\mathrm{cosp} f$, the set of zeros of $\hat{f}$, and the cospectrum of I: $\mathrm{cosp}\, I = \bigcap_{f \in I} \mathrm{cosp} f$. We have seen the role of the cospectrum in harmonic analysis (Chap. 6, § 1.3).

Spectrum and cospectrum are closed sets in the dual group $\hat{G}$, attached to each function and each ideal in $L^1(G)$. If G is compact, so that $\hat{G}$ is discrete, these two sets are complementary; but for non-discrete $\hat{G}$ they are, in general, not disjoint.

Now we want to define the spectrum for functions in $L^\infty(G)$. Given any $\phi \in L^\infty(G)$, we introduce a closed ideal I_ϕ of $L^1(G)$ by putting

$$(1) \qquad I_\phi = \{f \mid f \in L^1(G), f^* \star \phi = 0\},$$

where $f^*(x) = \overline{f(x^{-1})}$, and we define

$$(2) \qquad \mathrm{sp}\, \phi = \mathrm{cosp}\, I_\phi.$$

I_ϕ is simply the largest ideal of $L^1(G)$ to which ϕ is orthogonal (cf. Chap. 3, § 6.2 (ii)).

REMARK 1. If $\phi \in L^\infty \cap L^1(G)$, this definition agrees with that given before (i.e. yields the same set in $\hat{G}$).

REMARK 2. Let $\phi(x) = \int_{\hat{G}} \langle x, \hat{x} \rangle \, d\mu(\hat{x})$, where $\mu \in M^1(\hat{G})$. Then $\mathrm{sp}\, \phi = \mathrm{Supp}\, \mu$, as can readily be shown.

REMARK 3. Let Λ be a closed set in $\hat{G}$. For $\phi \in L^\infty(G)$ we have: $\mathrm{sp}\, \phi \subset \Lambda \Leftrightarrow \phi \perp J_\Lambda$, where J_Λ is the smallest closed ideal of $L^1(G)$ with cospectrum Λ.

We can also define the spectrum of a linear subspace of $L^\infty(G)$, in an obvious way.

The notion of spectrum, fundamental in harmonic analysis, is implicit in Carleman's book [20, pp. 91 and 111]; the definition was stated explicitly by Godement [54, p. 130, Definition I].†

REMARK 4. Godement [54, pp. 128–9] also gave another, equivalent definition involving the $\sigma(L^\infty, L^1)$ topology; we shall not need it. It states that $\operatorname{sp}\phi$ is the set of all characters of G contained in the closed linear subspace spanned by $(L_y\phi)_{y\in G}$ in $L^\infty(G)$ (with the topology mentioned). This second definition was influenced by a result proved by Beurling [8] for the line (later extended to l.c.a. groups by Domar, cf. [40]); Beurling coined the term 'spectral synthesis' in this context ([9], [10]). Compare also P. Koosis [85] for some interesting remarks on these matters.

1.2. Using the notion of spectrum, we can now give a very useful criterion for Wiener sets. *A closed set $\Lambda \subset \hat{G}$ is a Wiener set for $\mathscr{F}^1(\hat{G})$ if and only if for every function $f \in L^1(G)$ such that the F.t. $\hat{f}$ vanishes on Λ and for every* CONTINUOUS *function $\phi \in L^\infty(G)$ such that $\operatorname{sp}\phi \subset \Lambda$ the relation $\langle f, \phi\rangle = 0$ holds.*

Proof. Suppose the condition above is fulfilled. Let I_0 be a closed ideal of $L^1(G)$ such that $\operatorname{cosp} I_0 = \Lambda$ and consider any $f \in L^1(G)$ with $\hat{f}$ vanishing on Λ. To show that $f \in I_0$, it is enough to verify that every *continuous* $\phi \perp I_0$ satisfies $\phi \perp f$ (Chap. 3, § 6.3). Now $\phi \perp I_0 \Leftrightarrow f_0^* \star \phi = 0$ for all $f_0 \in I_0$. Thus $I_0 \subset I_\phi$ (defined in (1)) and hence $\operatorname{sp}\phi = \operatorname{cosp} I_\phi$ lies in Λ. This implies, by hypothesis, $\langle f, \phi\rangle = 0$, that is, $\phi \perp f$; thus $f \in I_0$, and Λ is a Wiener set.

Now suppose Λ is a Wiener set and $\operatorname{sp}\phi \subset \Lambda$. Consider the ideal I_ϕ: we have $\Lambda \supset \operatorname{cosp} I_\phi$. If $f \in L^1(G)$ is such that $\hat{f}$ vanishes on Λ, then $f \in I_\phi$ by Chapter 2, § 7.5 (i); thus $f^* \star \phi = 0$ and $\langle f, \phi\rangle = 0$.

REMARK. The criterion above also holds, of course, if the word 'continuous' is omitted.

1.3. We now list some basic properties of the spectrum.

(i) If $\phi \in L^\infty(G)$ has *empty spectrum*, then $\phi(x) = 0$ l.a.e. on G, and hence $\phi(x) = 0$ for all $x \in G$ if ϕ is continuous. This is simply the dual formulation of Wiener's theorem (Chap. 6, § 1.3) (cf. [54, p. 128, Theorem B]; for the line it goes back to Carleman [20, p. 74]).

(ii) $\operatorname{sp} f^* \star \phi \subset \operatorname{sp} f \cap \operatorname{sp}\phi$ for $f \in L^1(G)$, $\phi \in L^\infty(G)$. This can be verified immediately. It follows that, if the F.t. $\hat{f}$ vanishes *near* $\operatorname{sp}\phi$, then $f^* \star \phi = 0$ (note that $f^* \star \phi$ is continuous, cf. Chap. 3, § 6.1). Likewise for $\mu \in M^1(G)$ and *continuous* $\phi \in L^\infty(G)$: if the F.t. $\hat{\mu}$ vanishes near $\operatorname{sp}\phi$, then $\mu^* \star \phi = 0$. (We have $\operatorname{sp}\mu^* \star \phi \subset \operatorname{Supp}\hat{\mu} \cap \operatorname{sp}\phi = \emptyset$; cf. also Chapter 3, § 6.1, Remark 2.)

(iii) If $\phi \in L^\infty(G)$ has *compact spectrum*, then ϕ coincides l.a.e. with a continuous function [54, Theorem E].

† Godement uses $f \star \phi$ rather than $f^* \star \phi$ in (1), but this gives the same spectrum. Herz [65, p. 186] formulates the definition in a slightly different way, easily seen to be equivalent to that of Godement.

Take $\tau \in L^1(G)$ such that $\hat{\tau}$ is 1 near $\operatorname{sp}\phi$. Then $\phi - \tau^* \star \phi$ has empty spectrum, as is readily verified; since $\tau^* \star \phi$ is continuous, (iii) follows (cf. (i)).

REMARK 1. In (iii) we can assert more: ϕ coincides l.a.e. with a *uniformly* continuous function. This is clear from the proof above.

REMARK 2. Suppose, conversely, that $\phi \in L^\infty(G)$ coincides l.a.e. with a uniformly continuous function (cf. Chap. 3, § 6.7). Then $\operatorname{sp}\phi$ is a union of countably many compact sets [65, Theorem 1.9 and p. 192, last example].

For a *uniformly* continuous $\phi \in L^\infty(G)$ there is clearly a sequence $(f_n)_{n \geq 1}$ in $L^1(G)$ such that $\|f_n^* \star \phi - \phi\|_\infty < 1/n$; moreover, we may take f_n such that $\operatorname{sp} f_n$ is compact (cf. Chap. 5, § 4.1 (i)). Now let $\hat{K}$ be a compact nd. of $\hat{e}$ and put $\hat{K}_n = \hat{K}^{-1}.\operatorname{sp} f_n$: then $\operatorname{sp}\phi \subset \bigcup_{n \geq 1} \hat{K}_n$ (observe that, if $\hat{a} \in \hat{G} - \bigcup_{n \geq 1} \hat{K}_n$, then $(\hat{a}\hat{K}) \cap \operatorname{sp} f_n = \varnothing$, $n \geq 1$; choosing $f \in L^1(G)$ such that $\operatorname{sp} f \subset \hat{a}\hat{K}$, $\hat{f}(\hat{a}) \neq 0$, we have $f^* \star f_n^* \star \phi = 0$ for all n, hence $f^* \star \phi = 0$ and $\hat{a} \notin \operatorname{sp}\phi$). Thus $\operatorname{sp}\phi = \bigcup_{n \geq 1} (\hat{K}_n \cap \operatorname{sp}\phi)$.

(iv) (*a*) $\operatorname{sp} L_y \phi = \operatorname{sp}\phi$ $(y \in G)$; (*b*) $\operatorname{sp}(\chi_{\hat{y}} \cdot \phi) = \hat{y}.\operatorname{sp}\phi$ $(\hat{y} \in \hat{G})$;

(*c*) $\operatorname{sp}\bar{\phi} = (\operatorname{sp}\phi)^{-1}$; (*d*) $\operatorname{sp}\tilde{\phi} = \operatorname{sp}\phi$.

(v) If $\phi \in L^\infty(G)$ has *finite spectrum*, say $\operatorname{sp}\phi = (\hat{a}_n)_{1 \leq n \leq N}$, then $\phi = \sum_{1 \leq n \leq N} c_n \chi_{\hat{a}_n}$, with constant coefficients $(\neq 0)$. This is the dual formulation of the proposition that finite sets in $\hat{G}$ are Wiener sets for $\mathscr{F}^1(\hat{G})$ (cf. Chap. 6, § 1.5). For the line it was proved by Carleman [20, p. 78], for l.c.a. groups it was stated as a hypothesis by Godement [54, p. 136] and proved by Kaplansky [82, Theorem 1].

Take $\tau_n \in L^1(G)$, $1 \leq n \leq N$, such that the F.t. $\hat{\tau}_n$ is 1 near $\hat{a}_n$, $\operatorname{Supp}\hat{\tau}_n$ is compact, and the supports are mutually disjoint. Then $\phi = \sum_n \phi_n$, where $\phi_n = \tau_n^* \star \phi$ (cf. the proof of (iii)), and $\operatorname{sp}\phi_n = \{\hat{a}_n\}$. Thus it is enough to prove: if $\phi \in L^\infty(G)$ is continuous and $\operatorname{sp}\phi = \{\hat{e}\}$, then ϕ is constant. Now $\{\hat{e}\}$ is a Wiener set, hence (§ 1.2) $\langle L_y f - f, \phi\rangle = 0$ for all $f \in L^1(G)$, $y \in G$, or

$$\langle f, L_{y^{-1}}\phi - \phi\rangle = 0 \qquad y \in G$$

for all $f \in L^1(G)$: this implies $\phi(yx) = \phi(x)$ for all $x \in G$, since ϕ is continuous.

(vi) $\operatorname{sp}(\phi + \psi) \subset \operatorname{sp}\phi \cup \operatorname{sp}\psi$ for $\phi, \psi \in L^\infty(G)$. If $\operatorname{sp}\phi \cap \operatorname{sp}\psi = \varnothing$, then $\operatorname{sp}(\phi + \psi) = \operatorname{sp}\phi \cup \operatorname{sp}\psi$ [54, Theorem G]; some converse results are given in [111, VI].

(vii) (*a*) $\operatorname{sp}(\phi . f) \subset \overline{(\operatorname{sp}\phi . \operatorname{sp} f)}$ for $\phi \in L^\infty(G)$, $f \in L^1(G)$;

(*b*) $\operatorname{sp}(\phi . \psi) \subset \overline{(\operatorname{sp}\phi . \operatorname{sp}\psi)}$ for $\phi, \psi \in L^\infty(G)$.†

(*a*) We have $(f . \chi_{\hat{x}})^* \star \bar{\phi} = 0$ if $(\hat{x}.\operatorname{sp} f) \cap (\operatorname{sp}\phi)^{-1} = \varnothing$ (cf. (i), (ii), (iv *c*)). This gives $\int \phi(x) f(x) \langle x, \hat{x}\rangle \, dx = 0$ for $\hat{x} \notin (\operatorname{sp}\phi . \operatorname{sp} f)^{-1}$ or $(\phi . f)^\wedge(\hat{x}) = 0$ for $\hat{x} \notin \operatorname{sp}\phi . \operatorname{sp} f$.

† Cf. [65, Lemma 1.5].

(*b*) Let $\hat{U}$ be a compact nd. of $\hat{e}$. Let $f \in L^1(G)$ be such that $\hat{f}(\hat{e}) \neq 0$, $\operatorname{Supp}\hat{f} \subset \hat{U}$. Then (cf. (*a*)) $(\bar{\phi}.L_y f)^\wedge(\hat{x}) = 0$ for all $y \in G$, if $\hat{x} \notin (\operatorname{sp}\phi)^{-1}.\hat{U}$. Thus we have

$$(\bar{\phi}.L_y f.\chi_{\hat{a}})^* \star \psi = 0 \qquad y \in G,$$

if $\{\hat{a}.(\operatorname{sp}\phi)^{-1}.\hat{U}\} \cap \operatorname{sp}\psi = \emptyset$; this yields $(f.\chi_{\hat{a}})^* \star (\phi.\psi) = 0$ for $\hat{a} \notin \hat{U}^{-1}.\operatorname{sp}\phi.\operatorname{sp}\psi$. Hence

$$\operatorname{sp}(\phi.\psi) \subset \hat{U}^{-1}.\operatorname{sp}\phi.\operatorname{sp}\psi;$$

since $\hat{U}$ is arbitrarily small, (*b*) follows.

(viii) Let H be a closed subgroup of G and let $\phi \in L^\infty(G)$ be H-periodic: $\phi = \dot{\phi} \circ \pi_H$, where $\dot{\phi} \in L^\infty(G/H)$ (cf. Chap. 3, § 3.9). Then $\operatorname{sp}\dot{\phi} \subset H^\perp$ coincides with $\operatorname{sp}\phi \subset \hat{G}$.

This follows from the relation

$$f^* \star \phi = (T_H f)^* \star^{G/H} \dot{\phi} \qquad f \in L^1(G),$$

and from Chapter 4, § 4.3.

1.4. Let us now consider a Beurling algebra $L^1_w(G)$; we assume, as always, that w satisfies the condition of Beurling–Domar (Chap. 6, § 3.1). Let $L^\infty_w(G)$ be the dual space of $L^1_w(G)$ (Chap. 3, § 7.3). *The definition of spectrum in* 1.1 *also applies to functions in* $L^\infty_w(G)$, in an entirely analogous way.

Consider now another Beurling algebra $L^1_{w_1}(G)$ contained in $L^1_w(G)$. Then $L^\infty_{w_1}(G)$ contains $L^\infty_w(G)$; moreover, the following *permanence property of the spectrum* holds: the spectrum of $\phi \in L^\infty_w(G)$ is the same whether ϕ is regarded as an element of $L^\infty_w(G)$ or of $L^\infty_{w_1}(G)$.

By Chapter 6, § 3.6 (i) $w(x) \leqslant Cw_1(x)$ l.a.e., hence $\phi \in L^\infty_w \Rightarrow \phi \in L^\infty_{w_1}$; moreover, if $I = \{f \mid f \in L^1_w, f^* \star \phi = 0\}$, $I_1 = I \cap L^1_{w_1}$, then $\operatorname{cosp} I_1 = \operatorname{cosp} I$ by Chapter 6, § 3.6 (ii).

The criterion for Wiener sets in 1.2 *also applies to Beurling algebras.* The proof is the same as in 1.2 (cf. also Chap. 3, § 7.3, Remark 1).

The properties of the spectrum stated in 1.3 *carry over to functions in* $L^\infty_w(G)$ *either as they stand or with some additional conditions.*

(i) goes over unchanged; (ii) remains unchanged, but now only those measures μ on G are considered for which $w.\mu \in M^1(G)$ (cf. also Chap. 3, § 7.3, Remark 2). (iii) carries over unchanged; Remarks 1 and 2 remain valid in $L^\infty_w(G)$ if we replace 'uniformly continuous' by 'w-uniformly continuous': $\phi \in L^\infty_w(G)$ is said to be w-uniformly continuous if, given $\epsilon > 0$, there is a nd. U_ϵ of e such that

$$|\phi(y^{-1}x) - \phi(x)| < \epsilon.w(x) \qquad \text{for all } x \in G,\ y \in U_\epsilon.$$

The properties (iv *a*, *b*, *c*) carry over; also (iv *d*), if w is symmetric. (v) holds in $L^\infty_w(G)$ if w satisfies Shilov's condition (Chap. 6, § 3.2); this will be shown in 2.4. Property (vi) carries over. In (vii *a*) the conditions are now $\phi \in L^\infty_w(G)$, $f \in L^1_w(G)$, whence $\phi.f \in L^1(G)$; (vii *b*) holds if $\phi, \psi \in L^\infty_w(G)$ *and* $\phi.\psi \in L^\infty_w(G)$. Finally, (viii) also carries over (cf. Chap. 3, §§ 7.4 and 7.10).

2. Relativization and the spectrum

2.1. (i) *Let* ϕ *be a continuous function in* $L^\infty(G)$. *Let* H *be a closed subgroup of* G *and put* $\phi_{x,H}(\xi) = \phi(x\xi)$, $\xi \in H$, *for fixed* $x \in G$. The spectrum of $\phi_{x,H} \in L^\infty(H)$ lies in $\hat{H}$, while ϕ itself has its spectrum in $\hat{G}$. The relation between these two spectra is as follows. *The spectrum of* $\phi_{x,H}$ *is contained in the closure of* $\pi_{H^\perp}(\operatorname{sp}\phi)$, *where* $H^\perp$ *is the closed subgroup of* $\hat{G}$ *orthogonal to* H *and* $\pi_{H^\perp}$ *the canonical map* $\hat{G} \to \hat{G}/H^\perp = \hat{H}$.

Let Λ be the closure of $\pi_{H^\perp}(\operatorname{sp}\phi)$ in $\hat{H}$. Let λ' be any element of $\hat{H}$ outside Λ: we want to show $\lambda' \notin \operatorname{sp}\phi_{x,H}$. Take an $f \in L^1(H)$ such that $\hat{f}$, defined on $\hat{H}$, vanishes near Λ and $\hat{f}(\lambda') \neq 0$. The function f defines a measure $\mu_f \in M^1(G)$ and the F.t. $\hat{\mu}_f$ is simply $\hat{f} \circ \pi_{H^\perp}$ (Chap. 4, § 4.4). Thus $\hat{\mu}_f$ vanishes *near* $\operatorname{sp}\phi$ and hence $\mu_f^* \star \phi = 0$ (1.3 (ii); note especially that $\mu_f^* \star \phi$ is continuous, since ϕ is a bounded function, uniformly continuous on compact sets). Explicitly: $\int_H \overline{f(\xi)}\phi(\xi x)\,d\xi = 0$ for *all* $x \in G$. Replacing here x by ηx, $\eta \in H$, we get $f^* \star^H \phi_{x,H} = 0$, hence $\lambda' \notin \operatorname{sp}\phi_{x,H}$, since $\hat{f}(\lambda') \neq 0$.

We shall see some applications of this *'relativization theorem for the spectrum'* in 2.4 and in § 3.

(ii) *Conversely, if* Λ *is a closed set in* $\hat{H}$ *and the continuous function* $\phi \in L^\infty(G)$ *is such that* $\operatorname{sp}\phi_{x,H} \subset \Lambda$ *for all* $x \in G$, *then* $\operatorname{sp}\phi \subset \pi_{H^\perp}^{-1}(\Lambda)$.

We obtain by an obvious modification of the proof of (i): given $\lambda \in \hat{G} - \pi_{H^\perp}^{-1}(\Lambda)$, there is a $\mu \in M^1(G)$ such that $\hat{\mu}(\lambda) \neq 0$ and $\mu^* \star \phi = 0$. Then clearly there is also a $g \in L^1(G)$ such that $\hat{g}(\lambda) \neq 0$ and $g^* \star \phi = 0$, so $\lambda \notin \operatorname{sp}\phi$.

(iii) The subgroup H in (i) and (ii) need not be closed. *Let* H *be* ANY *subgroup of* G *and denote by* H_d *the same abstract group with the discrete topology. Then* (i) *and* (ii) *still hold if* H *and* $\hat{H}$ *are replaced respectively by* H_d *and* $(H_d)^\wedge$, *and* $\pi_{H^\perp}$ *by the canonical morphism of* $\hat{G}$ *into* $(H_d)^\wedge$ (Chap. 4, § 2.8 (ii), Remark 2). The proofs in (i) and (ii) apply unchanged.

References. (i) [111, I, Theorem 1]; (ii) and (iii) are due—in an even more general formulation—to Herz [65, Lemma 5.6].

2.2. It is worth while to state a special case of 2.1 (i) explicitly. *Let* f *be a bounded continuous function in* $L^1(G)$ *such that* f_H, *the restriction of* f *to the closed subgroup* $H \subset G$, *is in* $L^1(H)$. *Then the support of the F.t. of* f_H *is contained in the closure of* $\pi_{H^\perp}(\operatorname{Supp}\hat{f})$.

This is a special case of 2.1 (i), since for functions in $L^\infty \cap L^1$ the spectrum coincides with the support of the F.t. (§ 1.1, Remark 1). The proof in 2.1 (i) applies, of course, also directly. If f satisfies certain additional conditions, then a more precise statement can be made (Chap. 5, § 5.5 (ii)).

2.3. *The results in* 2.1 *remain true for continuous functions in* $L_w^\infty(G)$, *the dual space of a Beurling algebra* $L_w^1(G)$, *if* w *satisfies the condition of*

Beurling–Domar (cf. Chap. 6, § 3.1). We recall that, if $\phi \in L_w^\infty(G)$ is continuous, then $|\phi(x)| \leqslant \|\phi\|_{\infty,w} \cdot w(x)$ for *all* $x \in G$ (Chap. 3, § 7.3). Hence $\phi_{x,H}$, as defined in 2.1, belongs to $L_{w_H}^\infty(H)$ for every $x \in G$, w_H being the restriction of w to H.

The precise statements are as follows.

(i) If ϕ is a continuous function in $L_w^\infty(G)$ and H a closed subgroup of G, then $\phi_{x,H} \in L_{w_H}^\infty(H)$ has a spectrum contained in the closure of $\pi_{H^\perp}(\operatorname{sp}\phi)$ (*relativization theorem for the spectrum*).

(ii) Conversely, if Λ is a closed set in $\hat{H} = \hat{G}/H^\perp$ and ϕ is a continuous function in $L_w^\infty(G)$ such that $\operatorname{sp}\phi_{x,H} \subset \Lambda$ for every $x \in G$, then

$$\operatorname{sp}\phi \subset \pi_{H^\perp}^{-1}(\Lambda).$$

(iii) We may consider an arbitrary subgroup H of G and the corresponding discrete subgroup H_d. Then (i) and (ii) remain true if we replace $H, \hat{H}$ by H_d, $(H_d)^\wedge$ respectively and $\pi_{H^\perp}$ by the canonical morphism of $\hat{G}$ into $(H_d)^\wedge$ (Chap. 4, § 2.8 (ii), Remark 2).

The proofs are outlined below.

(i) The proof in 2.1 (i) still goes through. We remark that, if $\phi \in L_w^\infty(G)$ is continuous and f is in $L_{w_H}^1(H)$, then $\mu_f^* \star \phi: x \to \int_H \overline{f(\xi)}\phi(\xi x)\, d\xi$, $x \in G$, exists and is a *continuous* function in $L_w^\infty(G)$. This is a special case of Chapter 3, § 7.3, Remark 2, and may be proved directly as follows. We have $|\mu_f^* \star \phi(x)| \leqslant \|f\|_{1,w_H} \cdot \|\phi\|_{\infty,w} \cdot w(x)$; moreover, f can be approximated in $L_{w_H}^1(H)$ by functions $k \in \mathscr{K}(H)$, hence $\mu_f^* \star \phi$ can be approximated, uniformly on compact sets, by functions $\mu_k^* \star \phi$, $k \in \mathscr{K}(H)$, and these are continuous (let $K = \operatorname{Supp} k$; for fixed $x \in G$ we have, if $\epsilon > 0$ is given: $|\phi(\xi y x) - \phi(\xi x)| < \epsilon$ for all $\xi \in K$, if $y \in U_\epsilon$, a nd. of e independent of $\xi \in K$, and thus $|\mu_k^* \star \phi(yx) - \mu_k^* \star \phi(x)| \leqslant \epsilon \int_H |k(\xi)|\, d\xi$ for $y \in U_\epsilon$).

(ii) The proof is as indicated in 2.1 (ii). For the very last part we observe: if $f \in L_{w_H}^1(H)$, $h \in L_w^1(G)$ and $g(x) = \int_H f(\xi) h(\xi^{-1}x)\, d\xi$ a.e. on G, then $g \in L_w^1(G)$.

(iii) The proof is as in (i) and (ii).

2.4. As an application we now establish the following result which extends a theorem of Shilov (cf. Chap. 6, § 3.2). *If w is a weight function on G satisfying Shilov's condition, then finite sets in $\hat{G}$ are Wiener sets for $\mathscr{F}_w^1(\hat{G})$ or, dually, every $\phi \in L_w^\infty(G)$ with finite spectrum is of the form $\sum_n c_n \chi_{\hat{a}_n}$ with $\hat{a}_n \in \operatorname{sp}\phi$ and $c_n \in \mathbf{C}$.*

Suppose first the spectrum of $\phi \in L^\infty(G)$ consists of the single element $\hat{e} \in \hat{G}$. For $G = \mathbf{Z}$ the result is true, as proved by Shilov (loc. cit., Chap. 6, § 3.2). For general G observe that we may assume ϕ continuous (§§ 1.3 (iii) and 1.4). Choose any $\xi \in G$ and consider the cyclic subgroup $H_d = (\xi^n)_{n \in \mathbf{Z}}$, with the discrete topology: by 2.3 (iii) the function $\xi^n \to \phi(\xi^n)$ is in $L_{w_{H_d}}^\infty(H_d)$ and has a spectrum containing at most the neutral element of $(H_d)^\wedge$. Now by Shilov's result, combined with 1.3 (i) (cf. also § 1.4), we have: $\phi(\xi^n) = \phi(e)$ for all $n \in \mathbf{Z}$ (the case that $\xi^n = e$ for some $n \neq 0$ is of course obvious). In particular $\phi(\xi) = \phi(e)$ and, since $\xi \in G$ was arbitrary, ϕ is constant.

The general case of a finite spectrum reduces to that of a single point (cf. the proof of § 1.3 (v)).

REMARK. We can also prove 2.4 by using only 2.3 (i) rather than 2.3 (iii). For, the closed subgroup generated in G by a single element ξ is either $\mathbf{Z}$ or else compact (Chap. 4, § 2.9, Remark 1); on a compact subgroup H the continuous function ϕ is bounded and 2.3 (i) then shows directly that ϕ is constant on H, in particular $\phi(\xi) = \phi(e)$.

3. Wiener sets for Beurling algebras

3.1. Let w be a weight function on G satisfying the condition of Beurling–Domar; let H be a closed subgroup of G and w_H the restriction of w to H. We have, then, two Beurling algebras $L^1_w(G)$, $L^1_{w_H}(H)$, and the corresponding function algebras $\mathscr{F}^1_w(\hat{G})$, $\mathscr{F}^1_{w_H}(\hat{H})$. We shall prove that *a closed set* $\Lambda \subset \hat{H}$ *is a Wiener set for* $\mathscr{F}^1_{w_H}(\hat{H})$ *if and only if* $\pi^{-1}_{H^\perp}(\Lambda) \subset \hat{G}$ *is a Wiener set for* $\mathscr{F}^1_w(\hat{G})$, $\pi_{H^\perp}$ being the canonical map of $\hat{G}$ onto $\hat{G}/H^\perp = \hat{H}$. This is the '*inverse projection theorem for Wiener sets*' (cf. [111, II and III], [117]).

3.2. To prove 3.1, we shall use the following *criterion for Wiener sets* given in 1.2 (cf. also § 1.4): a closed set $\Lambda_0 \subset \hat{G}$ is a Wiener set for $\mathscr{F}^1_w(\hat{G})$ if and only if for all functions $f \in L^1_w(G)$ such that $\hat{f}$ vanishes on Λ_0 and for all *continuous* functions $\phi \in L^\infty_w(G)$ such that $\operatorname{sp}\phi \subset \Lambda_0$ the relation $\langle f, \phi\rangle = 0$ or, equivalently, $f^* \star \phi = 0$, holds.

3.3. First we prove the following part of 3.1: *if* $\Lambda \subset \hat{H}$ *is a Wiener set for* $\mathscr{F}^1_{w_H}(\hat{H})$, *then* $\pi^{-1}_{H^\perp}(\Lambda)$ *is a Wiener set for* $\mathscr{F}^1_w(\hat{G})$. The proof is based on Poisson's formula (Chap. 5, § 5) and on the relativization theorem 2.3 (i); it is a slight modification of that in [111, II].

If $f_0 \in L^1_w(G)$, then for any $u \in \mathscr{K}(G)$ the convolution $f = f_0 \star u$ is a continuous function in $L^1_w(G)$; also

$$\int_H |f(x\xi)|w(\xi)\,d\xi \leqslant \|f_0\|_{1,w}\cdot w(x^{-1}).\sup_{z\in G}\int_H |u(z\xi)|w(z\xi)\,d\xi,$$

i.e. $f_{x,H} \in L^1_{w_H}(H)$ for every $x \in G$, where $f_{x,H}(\xi) = f(x\xi)$, $\xi \in H$.

Let here $u = u_1 \star u_2$, with $u_1, u_2 \in \mathscr{K}(G)$; then (cf. Chap. 5, § 5.1, footnote) we may apply Chapter 5, § 5.3, Remark 2, to f: the F.t. of $f_{x,H}$ is $T_{H^\perp}(\hat{f}.\chi_x)$. Since $\hat{f} = \hat{f}_0.\hat{u}$, it follows that, if $\hat{f}_0$ vanishes on $\pi^{-1}_{H^\perp}(\Lambda)$, then $(f_{x,H})^\wedge$ vanishes on Λ, for each $x \in G$.

Now let f_0 be such that $\hat{f}_0$ vanishes on $\pi^{-1}_{H^\perp}(\Lambda)$ and let ϕ be any *continuous* function in $L^\infty_w(G)$ with $\operatorname{sp}\phi \subset \pi^{-1}_{H^\perp}(\Lambda)$. Let f be as above; by the extended Weil formula we can write $\langle f,\phi\rangle = \int_{G/H}\left\{\int_H f(x\xi)\overline{\phi(x\xi)}\,d\xi\right\}d\dot{x}$. But here the inner integral is zero for each $x \in G$: the F.t. of $f_{x,H} \in L^1_{w_H}(H)$ vanishes on Λ (cf. above) and by 2.3 (i) $\phi_{x,H} \in L^\infty_{w_H}(H)$ has spectrum in Λ; by hypothesis Λ is a Wiener set for $\mathscr{F}^1_{w_H}(\hat{H})$, hence $\langle f_{x,H}, \phi_{x,H}\rangle = 0$ by 3.2. Thus $\langle f,\phi\rangle = 0$, or $\langle f_0 \star u_1 \star u_2, \phi\rangle = 0$; this holds for all u_1, u_2 in $\mathscr{K}(G)$, hence $\langle f_0, \phi\rangle = 0$, since we can take u_1, u_2 as approximate units (of bounded L^1_w-norm) for f_0. Thus $\pi^{-1}_{H^\perp}(\Lambda)$ is a Wiener set for $\mathscr{F}^1_w(\hat{G})$ according to 3.2.

REMARK. The proof above also shows the following. Let Λ be an arbitrary closed set in $\hat{H} = \hat{G}/H^\perp$. Let $\phi \in L^\infty_w(G)$ be continuous. If for every $x \in G$ the function $\phi_{x,H}$ is orthogonal to *all* $h \in L^1_{w_H}(H)$ such that $\hat{h}$ vanishes on Λ, then ϕ is orthogonal to *all* $f_0 \in L^1_w(G)$ such that $\hat{f}_0$ vanishes on $\pi^{-1}_{H^\perp}(\Lambda)$. This should be compared with 2.3 (ii) (cf. also § 1.1, Remark 3); it was pointed out by Herz [65, § 5.8].

COROLLARY. If the weight function w satisfies Shilov's condition, then closed subgroups of $\hat{G}$ are Wiener sets for $\mathscr{F}^1_w(\hat{G})$; in particular this holds for $\mathscr{F}^1(\hat{G})$, for $\mathscr{F}^1_\alpha(\mathbf{R}^\nu)$, $0 \leqslant \alpha < 1$, and for $\mathscr{F}^1_\alpha(\mathbf{Q}^\nu_l)$, $\alpha \geqslant 0$ (Chap. 6, §§ 3.3 and 3.4). This corollary follows at once from the result above, combined with Chapter 6, § 3.2 ([111, I, Theorem 3], [117]).

3.4. We shall now prove the other part of 3.1: if $\pi^{-1}_{H^\perp}(\Lambda)$ is a Wiener set for $\mathscr{F}^1_w(\hat{G})$, then Λ is a Wiener set for $\mathscr{F}^1_{w_H}(\hat{H})$. This is equivalent to the following (cf. § 3.2): *if* $f_0^* \star \phi_0 = 0$ *for all* $f_0 \in L^1_w(G)$ *with* $\hat{f}_0$ *vanishing on* $\pi^{-1}_{H^\perp}(\Lambda)$ *and all continuous* $\phi_0 \in L^\infty_w(G)$ *with* $\operatorname{sp}\phi_0 \subset \pi^{-1}_{H^\perp}(\Lambda)$, *then* $f^* \star^H \phi = 0$ *for all* $f \in L^1_{w_H}(H)$ *with* $\hat{f}$ *vanishing on* Λ, *and all continuous* $\phi \in L^\infty_{w_H}(H)$ *with* $\operatorname{sp}\phi \subset \Lambda$.

3.5. To prove 3.4 we introduce some auxiliary functions (cf. [111, III]). Let *any* two functions $\phi \in L^\infty_{w_H}(H)$ and $f \in L^1_{w_H}(H)$ be given, let u be any function in $\mathscr{K}(G)$ and define

(1 a) $\phi_{u^*}(x) = \int_H \phi(\xi)\overline{u(x^{-1}\xi)}\,d\xi,$ (1 b) $f_u(x) = \int_H f(\xi)u(\xi^{-1}x)\,d\xi,$

for $x \in G$. These functions and their properties will be of importance in the sequel.

(A) The function ϕ_{u^*} in (1 a) is continuous and in $L^\infty_w(G)$. This follows at once from the fact that, if $u \in \mathscr{K}(G)$ and H is a closed subgroup of G, then

(i) $\int_H |u(x^{-1}\xi)|\,w(\xi)\,d\xi \leqslant C\,.\,w(x)$ for all $x \in G$, where C is independent of x;

(ii) $\int_H |u(y^{-1}x^{-1}\xi)-u(x^{-1}\xi)|\, w(\xi)\, d\xi < \epsilon . w(x)$ for all $x \in G$, if $y \in U_\epsilon$ ($\epsilon > 0$), a neighbourhood of e independent of x.

(i) Since $w(\xi) \leqslant w(x)w(x^{-1}\xi)$, we may take $C = \sup_{z\in G} \int_H |u(z\xi)|\, w(z\xi)\, d\xi$.

(ii) Let K be a compact nd. of $\operatorname{Supp} u$ and put $A = \sup_{x\in K} w(x)$. Then we have for all $z \in G$

$$|u(y^{-1}z)-u(z)|\, w(z) \leqslant A . |u(y^{-1}z)-u(z)|,$$

if y is in some (small) nd. of e, and (ii) follows from Chapter 3, § 3.2, Lemma.

(B) The function f_u in (1 b) is continuous and in $L^1_w(G)$; moreover, it has properties analogous to (A) (i) and (ii), since

(i) $\int_H |f_u(x\eta)|\, w(\eta)\, d\eta \leqslant \int_H |f(\xi)|\, w(\xi)\, d\xi . \int_H |u(x\eta)|\, w(\eta)\, d\eta,$

(ii) $\int_H |f_u(yx\eta)-f_u(x\eta)|\, w(\eta)\, d\eta$

$$\leqslant \int_H |f(\xi)|\, w(\xi)\, d\xi . \int_H |u(yx\eta)-u(x\eta)|\, w(\eta)\, d\eta.$$

(C) We can take, in particular, a function $u \in \mathscr{K}_+(G)$ such that $\int_H u(\xi)\, d\xi = 1$ and $\operatorname{Supp} u \subset U$, where U is any pre-assigned neighbourhood of e in G. Then, if ϕ_{u^*}, f_u are defined by (1 a, b), we have:

$$\int_H \overline{f_u(\xi^{-1})}\phi_{u^*}(\xi^{-1}\eta)\, d\xi \to f^* \star^H \phi(\eta) \qquad \eta \in H \tag{2}$$

when U 'shrinks to $\{e\}$'.

Let $(f_u)_H$, $(\phi_{u^*})_H$ be the restrictions to H of f_u, ϕ_{u^*} respectively. Then the left-hand side of (2) is the value at $\eta \in H$ of the convolution $(f_u)^*_H \star^H (\phi_{u^*})_H$. But $(f_u)_H = f \star^H u_H$ (u_H = restriction of u to H) and $(\phi_{u^*})_H = \phi \star^H u^*_H$; thus $(f_u)^*_H \star^H (\phi_{u^*})_H = (v \star^H f)^* \star^H \phi$, with $v = u_H \star^H u_H \in \mathscr{K}_+(H)$. Now we observe that $\operatorname{Supp} v \subset U^2 \cap H$ and $\int_H v(\xi)\, d\xi = 1$, so $v \star^H f \to f$ in $L^1_{w_H}(H)$ when $U \to \{e\}$, and (2) follows (cf. Chap. 3, §§ 7.2 and 7.3).

(D) *The spectrum of* ϕ_{u^*} *is contained in* $\pi_{H^\perp}^{-1}(\operatorname{sp}\phi)$. This is an analogue of 2.3 (i), in the opposite direction, and is easy to guess by considering the case when $\phi \in L^\infty(H) \cap L^1(H)$.

(D) follows from 2.3 (ii): for every fixed $x \in G$ the function $\xi \to \phi_{u^*}(x\xi)$, $\xi \in H$, is simply $\phi \star^H (L_x u)^*_H$, where $(\,.\,)_H$ means restriction to H; hence its spectrum is contained in $\operatorname{sp}\phi$ (cf. §§ 1.3 (ii) and 1.4).

3.6. We can now establish 3.4. Let f, ϕ be as stated there and define, for $u \in \mathscr{K}(G)$, the functions ϕ_{u^*}, f_u by (1 a, b). The F.t. of $f_u \in L^1_w(G)$ is $(\hat{f} \circ \pi_{H^\perp}) . \hat{u}$ and thus vanishes on $\pi_{H^\perp}^{-1}(\Lambda)$, since $\hat{f}$ vanishes on Λ. Hence $f^*_u \star \phi_{u^*} = 0$ by 3.5 (D) and the hypothesis of 3.4. Moreover, $\pi_{H^\perp}^{-1}(\Lambda)$ is invariant under translations by elements of $H^\perp \subset \hat{G}$, so we may replace f_u by $f_u \cdot \bar{\chi}_\alpha$, $\alpha \in H^\perp$. Thus for any $\alpha \in H^\perp$

$$\int f_u(y^{-1}x)\,\overline{\chi_\alpha(x)}\,\overline{\phi_{u^*}(x)}\,dx = 0 \qquad y \in G.$$

Since χ_α is H-periodic, we can write this

$$(3)\qquad \int\limits_{G/H}\left\{\int\limits_H f_u(y^{-1}x\xi)\overline{\phi_{u^*}(x\xi)}\,d\xi\right\}\overline{\langle \dot{x}, \alpha\rangle}\,d\dot{x} = 0 \qquad \alpha \in H^\perp.$$

Let here $y = \eta \in H$ and consider (for fixed η) the inner integral as a function of x:

$$(4)\qquad F(x) = \int\limits_H f_u(\eta^{-1}x\xi)\overline{\phi_{u^*}(x\xi)}\,d\xi \qquad x \in G.$$

We will show below that F is continuous. Since F is H-periodic, we have $F = F' \circ \pi_H$, where F' is a continuous function on G/H. Moreover, F' is in $L^1(G/H)$ (cf. Chap. 3, § 4.5; observe that $x \to f_u(\eta^{-1}x)\overline{\phi_{u^*}(x)}$ is in $L^1(G)$). Now (3) says that the F.t. of F' vanishes on $H^\perp = (G/H)^\wedge$, hence, since F' is continuous, $F'(\dot{x}) = 0$ for all $\dot{x} \in G/H$ (not merely a.e.); in particular $\int\limits_H f_u(\eta^{-1}\xi)\overline{\phi_{u^*}(\xi)}\,d\xi = 0$. By 3.5 (C) this implies $f^* \star^H \phi = 0$, since $u \in \mathscr{K}(G)$ was arbitrary.

Proof of the continuity of F (cf. (4)). Write

$$F(yx) - F(x) = A_x(y) + B_x(y) \qquad y \in G,$$

where

$$A_x(y) = \int\limits_H \{f_u(\eta^{-1}yx\xi) - f_u(\eta^{-1}x\xi)\}\overline{\phi_{u^*}(yx\xi)}\,d\xi,$$

$$B_x(y) = \int\limits_H \{\overline{\phi_{u^*}(yx\xi)} - \overline{\phi_{u^*}(x\xi)}\}f_u(\eta^{-1}x\xi)\,d\xi.$$

Then both $|A_x(y)|$ and $|B_x(y)|$ are arbitrarily small, if y is near enough to e. Indeed, ϕ_{u^*} is a continuous function in $L^\infty_w(G)$ (§ 3.5 (A)), thus $|\phi_{u^*}(z)| \leqslant \|\phi_{u^*}\|_{\infty,w} \cdot w(z)$ for all $z \in G$ (Chap. 3, § 7.3). We can now apply 3.5 (B) (ii) to $A_x(y)$ (cf. also § 3.5 (A) (ii)). To $B_x(y)$ we can apply the method used in proving 2.3 (i) (for any $z \in G$ the function $\xi \to f_u(z\xi)$, $\xi \in H$, is in $L^1_{w_H}(H)$, cf. § 3.5 (B) (i)).

This concludes the proof of Theorem 3.1. Other results in this direction have been obtained by Herz and de Leeuw [73].

As Theorem 3.1 shows, there are relations between the closed ideals of $L^1_{w_H}(H)$ with cospectrum $\Lambda \subset \hat{G}/H^\perp$ and the closed ideals of $L^1_w(G)$ with cospectrum $\pi^{-1}_{H^\perp}(\Lambda)$; in particular this applies to $L^1(H)$ and $L^1(G)$. It would be of interest to investigate this further and make these relations entirely clear.

Remark. The method of proof above also leads to the following result. Consider an *arbitrary* closed set Λ in $\hat{H} = \hat{G}/H^\perp$. Let $\phi \in L^\infty_w(G)$ be continuous. If ϕ is orthogonal to *all* $g \in L^1_w(G)$ such that $\hat{g}$ vanishes on $\pi_H^{-1}(\Lambda)$, then for every $x \in G$ the function $\phi_{x,H} \in L^\infty_{w_H}(H)$ is orthogonal to *all* $f \in L^1_{w_H}(H)$ such that $\hat{f}$ vanishes on Λ. This is a converse of 3.3, Remark, and should be compared with 2.3 (i). The proof is quite analogous to that above, ϕ_{u*} being replaced by ϕ in (3) and (4).

3.7. Let us now consider subgroups of the dual group $\hat{G}$. *A closed subgroup* Γ *of* $\hat{G}$ *is a Wiener set for* $\mathscr{F}^1_w(\hat{G})$ *if and only if every* $\phi \in L^\infty_w(G)$ *with* $\operatorname{sp}\phi \subset \Gamma$ *is (l.a.e. equal to) an H-periodic function, H being the subgroup of G orthogonal to* Γ.

Suppose the periodicity condition holds. Let I be any closed ideal of $L^1_w(G)$ with $\operatorname{cosp} I = \Gamma$ and consider any $\phi \in L^\infty_w(G)$ such that $\phi \perp I$. Then $\operatorname{sp}\phi \subset \Gamma$ and hence, by assumption, $\phi = \phi' \circ \pi_H$, where ϕ' is a function on G/H. Thus by the extended Weil formula $\langle f, \phi\rangle = \langle T_H f, \phi'\rangle$ for all $f \in L^1_w(G)$. Now, if f is any function in $L^1_w(G)$ such that the F.t. $\hat{f}$ vanishes on Γ, then $T_H f = 0$ (Chap. 4, § 4.3); hence $\langle f, \phi\rangle = 0$, so $f \in I$. Thus Γ is a Wiener set for $\mathscr{F}^1_w(\hat{G})$.

Conversely, let Γ be a Wiener set for $\mathscr{F}^1_w(\hat{G})$ and let $I_{\Gamma,w}$ be *the* closed ideal of $L^1_w(G)$ with cospectrum Γ. Then *every* $\phi \in L^\infty_w(G)$ with $\operatorname{sp}\phi \subset \Gamma$ satisfies $\phi \perp I_{\Gamma,w}$ (cf. § 1.2, Remark, and § 1.4). Hence ϕ defines a continuous linear functional on $L^1_w(G)/I_{\Gamma,w} \cong L^1_{\dot{w}}(G/H)$ (Chap. 3, § 7.4; note that $I_{\Gamma,w} = J^1_w(G,H)$); this functional is defined by $f' \to \langle f, \phi\rangle$, where $f' \in L^1_{\dot{w}}(G/H)$ and f is any function in $L^1_w(G)$ for which $T_H f = f'$. Thus there is a $\phi' \in L^\infty_{\dot{w}}(G/H)$ such that $\langle f, \phi\rangle = \langle T_H f, \phi'\rangle$, for all $f \in L^1_w(G)$ or, by the extended Weil formula, $\langle f, \phi\rangle = \langle f, \phi' \circ \pi_H\rangle$. Hence $\phi(x) = \phi' \circ \pi_H(x)$ l.a.e. This proof is entirely analogous to that in Chapter 3, § 7.10.

Combining this result with 3.3, Corollary, we obtain:

Corollary. If the weight function w on G satisfies Shilov's condition, then every function in $L^\infty_w(G)$ whose spectrum is contained in a closed subgroup Γ of $\hat{G}$ coincides l.a.e. with an H-periodic function, H being the subgroup of G orthogonal to Γ. In particular this holds for $L^\infty(G)$, $L^\infty_\alpha(\mathbf{R}^\nu)$, $0 \leqslant \alpha < 1$, and $L^\infty_\alpha(\mathbf{Q}^\nu_l)$, $\alpha \geqslant 0$ $(L^\infty_\alpha = (L^1_\alpha)')$.

Remark. This corollary, joined to § 1.1, Remark 2, also contains the result in Chapter 5, § 4.4 (ii).

3.8. Consider again a closed subgroup $\Gamma \subset \hat{G}$ and let H be the subgroup of G orthogonal to Γ, whence $(G/H)^\wedge = \Gamma$. Let w be a weight function on G and $\dot{w}$ the induced weight function on G/H (Chap. 3, § 7.4). If Λ is a closed subset of $\hat{G}$ contained in Γ, then there are two families of closed ideals with cospectrum Λ, one in $L^1_w(G)$ and one in $L^1_{\dot{w}}(G/H)$. We shall prove the following result. *Suppose the closed subgroup* $\Gamma \subset \hat{G}$ *is a Wiener set for* $\mathscr{F}^1_w(\hat{G})$. *Then a closed set* $\Lambda \subset \Gamma$ *is a Wiener set for* $\mathscr{F}^1_{\dot{w}}(\Gamma)$ *if and only if it is a Wiener set for* $\mathscr{F}^1_w(\hat{G})$. *In particular this holds for* $\mathscr{F}^1(\Gamma)$ *and* $\mathscr{F}^1(\hat{G})$, *the ordinary Fourier algebras, or, more generally, whenever w satisfies Shilov's condition.* This may be called the *injection theorem for Wiener sets*: it shows that the 'Wiener property' of a closed set is to a certain extent independent of the group into which the set is injected.

Proof. We show that, given $\Lambda \subset \Gamma$, there is a bijective correspondence between those closed ideals $I \subset L_w^1(G)$ and $I' \subset L_{\dot w}^1(G/H)$ for which $\operatorname{cosp} I = \operatorname{cosp} I' = \Lambda$. Consider the restriction of the mapping T_H to $L_w^1(G)$ (Chap. 3, § 7.4) and call it $T_{H,w}$. Given I' as above, we put $I = T_{H,w}^{-1}(I')$; then clearly $\operatorname{cosp} I = \operatorname{cosp} I'$, i.e. $\operatorname{cosp} I = \Lambda$. Conversely, if I is given, we put $I' = T_{H,w}(I)$; then also $\operatorname{cosp} I' = \Lambda$. Observe that I contains the kernel $J_w^1(G,H)$ of $T_{H,w}$ (cf. Chap. 3, § 7.4): indeed, by hypothesis, Γ is a Wiener set for $\mathscr{F}_w^1(\hat{G})$ and thus we can apply Chapter 2, § 7.5 (i) to I and $J_w^1(G,H)$. It follows that $I' = T_{H,w}(I)$ is a *closed* ideal and $T_{H,w}^{-1}(I') = I$. The special case of $\mathscr{F}^1(\Gamma)$ and $\mathscr{F}^1(\hat{G})$, or of Shilov's condition, then follows by § 3.3, Corollary.

It will be noted that the inverse projection theorem 3.1 holds for all Beurling algebras, but the injection theorem above is of a slightly different nature, requiring an additional hypothesis which holds, for instance, in the case of $\mathscr{F}_\alpha^1(\mathbf{R}^\nu)$, $0 \leqslant \alpha < 1$, or $\mathscr{F}_\alpha^1(\mathbf{Q}_l^\nu)$, $\alpha \geqslant 0$.

3.9. We prove here the remarkable fact, already mentioned in Chapter 2, § 7.3, that *a circle is a Wiener set for the algebra* $\mathscr{F}_\alpha^1(\mathbf{R}^2)$, *if* $0 \leqslant \alpha < \frac{1}{2}$. For $\alpha = 0$ this is a well-known result of Herz's [64] and his method still applies for $\alpha < \frac{1}{2}$.

The precise statement of *Herz's theorem* is as follows. *Every continuous function* $\phi \in L_\alpha^\infty(\mathbf{R}^2)$, $0 \leqslant \alpha < \frac{1}{2}$, *such that* $\operatorname{sp}\phi \subset S_1 = \{t \mid t \in \mathbf{R}^2, |t| = 1\}$ *can be approximated, uniformly on compact sets, by functions* σ *of the form* $\sigma(x) = \int \langle x, t\rangle \, d\mu(t)$, *where* μ *is a bounded measure with support in* S_1, *and* $\|\sigma\|_{\infty,\alpha} \leqslant \|\phi\|_{\infty,\alpha}$. It is readily seen that this implies the result above (cf. § 1.2).

Proof. We shall work as long as possible in $\mathbf{R}^\nu$, $\nu \geqslant 1$, and specialize to $\mathbf{R}^2$ at the critical moment. Define $F_0(x) = \int_{|t| \leqslant \frac{1}{2}} \langle x, t\rangle \, dt$ and put $k(x) = \{F_0(x)/F_0(0)\}^2$; thus $\hat{k}(t) = 0$ for $|t| \geqslant 1$ (cf. Chap. 5, § 1.2). By familiar properties of the Bessel functions, $k(x) = O(|x|^{-(\nu+1)})$ ($|x| \to \infty$), hence $k \in L_\alpha^1(\mathbf{R}^\nu)$ for $\alpha < 1$. Now put for *any* $\phi \in L_\alpha^\infty(\mathbf{R}^\nu)$, $0 \leqslant \alpha < 1$,

$$\Phi_h(t) = \int \phi(x) k(hx) \overline{\langle x, t\rangle} \, dx \qquad h > 0.$$

Then $\Phi_h(t) = 0$ for all t at distance $> h$ from $\operatorname{sp}\phi$ (cf. §§ 1.3 (vii *a*) and 1.4).

Now let $K \subset \mathbf{R}^\nu$ be compact and suppose $\operatorname{sp}\phi \subset K$. Then ϕ is (a.e. equal to) a continuous function (§§ 1.3 (iii) and 1.4) and we can apply the inversion theorem:

$$\phi(x) k(hx) = \int \langle x, t\rangle \Phi_h(t) \, dt.$$

Now for any $f \in L_\alpha^1(\mathbf{R}^\nu)$ we have

$$f^* \star \phi(x) = \lim_{h \downarrow 0} \int \overline{f(y)} \phi(x+y) k(h(x+y)) \, dy = \lim_{h \downarrow 0} \int \overline{\hat{f}(t)} \Phi_h(t) \langle x, t\rangle \, dt,$$

the last integral extending only over the set K_h of points at distance $\leqslant h$ from K. Thus by Schwarz's inequality

$$|f^* \star \phi(x)|^2 \leqslant \liminf_{h \downarrow 0} \left\{ \int_{K_h} |\hat{f}(t)|^2 \, dt \, . \int_{K_h} |\Phi_h(t)|^2 \, dt \right\}.$$

The second factor on the right is

$$\|\Phi_h\|_2{}^2 = \int |\phi(x)k(hx)|^2\,dx = \frac{1}{h^\nu}\int \left|\phi\left(\frac{x}{h}\right)k(x)\right|^2 dx = O\left(\frac{1}{h^{\nu+2\alpha}}\right) \quad (h \to 0).$$

(Here we have used the fact that the function $x \to (1+|x|)^\alpha k(x)$ is in $L^2(\mathbf{R}^\nu)$, since $\alpha < 1$.) Hence

$$(*) \qquad |f^* \star \phi(x)|^2 \leqslant \liminf_{h\downarrow 0}\left\{O\left(\frac{1}{h^{\nu+2\alpha}}\right)\cdot\int_{K_h} |\hat{f}(t)|^2\,dt\right\} \qquad x \in \mathbf{R}^\nu.$$

Next let τ be an indefinitely differentiable function in $L^1_\alpha(\mathbf{R}^\nu)$ such that all partial derivatives of τ are in $L^1_\alpha(\mathbf{R}^\nu)$ and the F.t. $\hat{\tau}$ is 1 near K (cf. e.g. Chap. 1, § 1.3 (v) and (vi)). Now $\phi = \tau^* \star \phi$ (cf. the proof of § 1.3 (iii)), hence ϕ is indefinitely differentiable. Let Δ be the Laplace operator on $\mathbf{R}^\nu$; putting

$$f = \tau + \left(\frac{1}{2\pi}\right)^2 \Delta\tau \in L^1_\alpha(\mathbf{R}^\nu),$$

we have

$$(\sharp) \qquad \phi + \left(\frac{1}{2\pi}\right)^2 \Delta\phi = f^* \star \phi.$$

We note that

$$(**) \qquad \hat{f}(t) = \hat{\tau}(t).(1-|t|^2) \qquad t \in \mathbf{R}^\nu.$$

Now let K be the unit sphere $S_{\nu-1} = \{t \mid |t| = 1\}$. Then (**) yields, if K_h has the former meaning,

$$({}^{\,*}_{**}) \qquad \int_{K_h} |\hat{f}(t)|^2\,dt = O(h^3) \qquad h \downarrow 0.$$

From (*) and $({}^{\,*}_{**})$ we obtain, for the f chosen: $f^* \star \phi = 0$ *if* $\nu = 2$ *and* $0 \leqslant \alpha < \frac{1}{2}$ (or if $\nu = 1$ and $\alpha < 1$, which is not of interest here). We have now obtained the following result (cf. ($\sharp$)): *if* $0 \leqslant \alpha < \frac{1}{2}$, *then every (continuous) function* $\phi \in L^\infty_\alpha(\mathbf{R}^2)$ *with spectrum in* S_1 *satisfies* $\phi + (2\pi)^{-2}\Delta\phi = 0$ (cf. [64, Lemma]; the factor $(2\pi)^{-2}$ is irrelevant and due to a different normalization of the F.t.).

For the remainder of the proof the reader is referred to [64]. We observe that, in the notation used there, we have for each fixed $r > 0$

$$\max_{0\leqslant\theta\leqslant 2\pi} |\sigma_k(\phi; r, \theta)| \leqslant \max_{0\leqslant\theta\leqslant 2\pi} |\phi(r,\theta)| \qquad k = 1, 2,\ldots,$$

hence we obtain for the norms in $L^\infty_\alpha(\mathbf{R}^2)$

$$\|\sigma_k\|_{\infty,\alpha} \leqslant \|\phi\|_{\infty,\alpha} \qquad k = 1, 2,\ldots,$$

and the final result follows from the boundedness of $C_\nu(\phi; r)$, $-k \leqslant \nu \leqslant k$, near $r = 0$.

Thus the proof of the result stated in Chapter 2, § 7.3 is complete.

REMARK. An analysis of the proof shows that the result also holds for algebras $\mathscr{F}^1_w(\mathbf{R}^2) \cong L^1_w(\mathbf{R}^2)$ if the weight function w is a radial function and satisfies $w(x) = O(|x|^\alpha)$ $(|x| \to \infty)$ for some $\alpha < \frac{1}{2}$. Compare Chapter 2, § 7.3, Remark.

3.10. We note, finally, that, if $\alpha \geqslant \frac{1}{2}$, then there are closed ideals in $L^1_\alpha(\mathbf{R}^2)$ which are not the intersection of a closed ideal of $L^1(\mathbf{R}^2)$ with $L^1_\alpha(\mathbf{R}^2)$. It would be of interest to investigate these ideals for $\frac{1}{2} \leqslant \alpha < 1$.

If I is a closed ideal of $L^1(\mathbf{R}^2)$, then $I \cap L^1_\alpha(\mathbf{R}^2)$ is a closed ideal of $L^1_\alpha(\mathbf{R}^2)$ with the *same* cospectrum (Chap. 6, § 3.6 (ii)). By Chapter 2, § 7.3, $L^1_\alpha(\mathbf{R}^2)$ contains distinct closed ideals with the unit circle as cospectrum, if $\alpha \geqslant \frac{1}{2}$; since $L^1(\mathbf{R}^2)$ contains only one such ideal, the assertion follows.

4. Wiener–Ditkin sets on groups

4.1. We shall prove here several results for the ordinary Fourier algebra $\mathscr{F}^1(\hat{G}) \simeq L^1(G)$; references will be given in 4.8 and 4.9. First we show that *closed subgroups of $\hat{G}$ are Wiener–Ditkin sets* (Chap. 2, § 5.2) *for $\mathscr{F}^1(\hat{G})$*. The proof is based on a lemma.

4.2. Lemma. Let Γ be a closed subgroup of the dual group $\hat{G}$. Choose any $A > 1$. Then, given $f \in L^1(G)$ and any $\epsilon > 0$, there is a measure $\mu \in M^1(G)$ such that $\|\mu\| < A$ and

(i) the F.t. $\hat{\mu}$ is 1 near Γ,

(ii) $\|\mu \star f\|_1 < A\,.\,\|T_H f\|_1 + \epsilon$, where H is the subgroup of G orthogonal to Γ and $T_H f \in L^1(G/H)$ is defined as in Chapter 3, § 4.4 (10).

This lemma is a relativization (Chap. 4, § 5) of Chapter 5, § 2.3 (ii), relation (6). It will be proved in 4.4.

Remark. We can replace (ii) by the slightly stronger assertion $\|\mu \star f\|_1 < \|T_H f\|_1 + \epsilon$. (We choose any $A' > 1$ such that $A' \leqslant A$ and $(A'-1)\|T_H f\|_1 < \frac{1}{2}\epsilon$, and apply the lemma with $A', \frac{1}{2}\epsilon$ instead of A, ϵ.)

4.3. It is easy to see that Lemma 4.2 implies Theorem 4.1.

Let the closed subgroup $\Gamma \subset \hat{G}$ be fixed and let $H = \Gamma^\perp \subset G$. Consider any $f \in L^1(G)$ such that $\hat{f}$ vanishes on Γ; then $T_H f = 0$ (Chap. 4, § 4.3). Given $\epsilon > 0$, choose $u \in L^1(G)$ so that $\|f - u \star f\|_1 < \frac{1}{2}\epsilon$. Since $T_H(u \star f) = T_H u \star T_H f = 0$, there is a $\mu \in M^1(G)$ such that the F.t. $\hat{\mu}$ is 1 near Γ and $\|\mu \star u \star f\|_1 < \frac{1}{2}\epsilon$, by Lemma 4.2 (the constant $A > 1$ there is irrelevant here). Then

$$\|f - (u - \mu \star u) \star f\|_1 < \tfrac{1}{2}\epsilon + \tfrac{1}{2}\epsilon = \epsilon$$

and the F.t. of $u - \mu \star u \in L^1(G)$ vanishes near Γ; thus Γ is a Wiener–Ditkin set for $\mathscr{F}^1(\hat{G})$.

4.4. The proof of Lemma 4.2 is based on Chapter 5, § 3.1 and consists of two steps:

(i) Proof for the case of functions $f \in \mathscr{K}(G)$.

(ii) Extension to general $f \in L^1(G)$.

(i) Given $A > 1$, consider the family $\mathfrak{F}_{H,A}$ of all $\tau \in L^1(H)$ such that $\|\tau\|_1 < A$ and $\hat{\tau}$ is 1 near $\hat{e} \in \hat{H}$. We can apply Chapter 5, § 3.1 to $\mathfrak{F}_{H,A}$ (cf. Chap. 5, § 2.2, Remark 1, with $G = H$); thus, given $f \in \mathscr{K}(G)$ and $\epsilon > 0$, there is a $\tau \in \mathfrak{F}_{H,A}$ such that

$$(*) \qquad \int_G \left| \int_H f(x\xi)\tau(\xi^{-1})\,d\xi \right| dx < A\,.\,\|T_H f\|_1 + \epsilon.$$

The function $\tau \in L^1(H)$ defines a measure $\mu = \mu_\tau$ in $M^1(G)$ and $\hat{\mu} = \hat{\tau} \circ \pi_\Gamma$ (π_Γ: $\hat{G} \to \hat{G}/\Gamma = \hat{H}$), by Chapter 4, § 4.4; thus $\hat{\mu}$ is 1 near Γ. Moreover, we have $\|\mu\| = \int_H |\tau(\xi)|\, d\xi < A$ and the left-hand side of (*) is simply $\|\mu \star f\|_1$: for, since G is abelian,

$$\int_H f(x\xi)\tau(\xi^{-1})\, d\xi = \int_H f(\xi^{-1}x)\tau(\xi)\, d\xi = \int_G f(y^{-1}x)\, d\mu(y).$$

Thus Lemma 4.2 is proved for $f \in \mathscr{K}(G)$.

(ii) Let $A > 1$, $f \in L^1(G)$ and $\epsilon > 0$ be given. There is a $k \in \mathscr{K}(G)$ such that $\|f-k\|_1 < \frac{1}{3}(\epsilon/A)$ and (cf. (i)) a $\mu \in M^1(G)$ such that $\|\mu\| < A$, $\hat{\mu}$ is 1 near Γ and $\|\mu \star k\|_1 < A\,.\|T_H k\|_1 + \frac{1}{3}\epsilon$. Then we have

$$\|\mu \star f\|_1 \leqslant \|\mu \star k\|_1 + \|\mu \star (f-k)\|_1 < A\,.\|T_H k\|_1 + \tfrac{1}{3}\epsilon + \tfrac{1}{3}\epsilon < A\,.\|T_H f\|_1 + \epsilon.$$

Thus Lemma 4.2 is proved and the proof of Theorem 4.1 is complete (cf. § 4.3).

Theorem 4.1 implies, in particular, the proposition that *closed subgroups of $\hat{G}$ are Wiener sets for $\mathscr{F}^1(\hat{G})$*, which is also contained in 3.3, Corollary. The proof given here is more direct than the one in 3.3 in that it works only with $L^1(G)$, making no use of $L^\infty(G)$; but the corollary in 3.3 is more general, since it also applies to a certain class of Beurling algebras. It would be of interest to investigate whether Theorem 4.1 holds for some Beurling algebras, say $\mathscr{F}^1_\alpha(\mathbf{R}^\nu)$, $0 < \alpha < 1$, or $\mathscr{F}^1_\alpha(\mathbf{Q}^\nu_l)$, $\alpha > 0$ (cf. Chap. 6, §§ 3.3 and 3.4).

A *compact open* subgroup of $\hat{G}$ is a Wiener–Ditkin set for $\mathscr{F}^1_w(\hat{G})$, for *any* weight function w; this is nearly trivial, as is readily seen.

4.5. Next we prove the following *injection theorem for Wiener–Ditkin sets*. Let Γ be a closed subgroup of $\hat{G}$, let Δ be a closed subset of Γ. Then Δ is a Wiener–Ditkin set for $\mathscr{F}^1(\Gamma)$ if and only if Δ is a Wiener–Ditkin set for $\mathscr{F}^1(\hat{G})$. This is entirely analogous to the case of Wiener sets (§ 3.8).

4.6. One half of Theorem 4.5 is easy to prove: if Δ is a Wiener–Ditkin set for $\mathscr{F}^1(\hat{G})$, then Δ is a Wiener–Ditkin set for $\mathscr{F}^1(\Gamma)$.

Put $H = \Gamma^\perp \subset G$; thus $(G/H)^\wedge = \Gamma$. Suppose $\dot{f} \in L^1(G/H)$ is such that the F.t. $\dot{f}^\wedge$—defined on Γ—vanishes on Δ. Take any $f \in L^1(G)$ for which $T_H f = \dot{f}$ (cf. Chap. 3, § 4.4): then the F.t. $\hat{f}$ vanishes on Δ, since the restriction of $\hat{f}$ to Γ is $\dot{f}^\wedge$ (Chap. 4, § 4.3). Given $\epsilon > 0$, there is, by hypothesis, a $v \in L^1(G)$ such that the F.t. $\hat{v}$ vanishes on some nd. Ω of Δ in $\hat{G}$ and $\|f - v \star f\|_1 < \epsilon$. Put $\dot{v} = T_H v$; then $\|\dot{f} - \dot{v} \star \dot{f}\|_1 < \epsilon$ and $\dot{v}^\wedge$ vanishes on a nd. of Δ in Γ, viz. $\Omega \cap \Gamma$.

4.7. The proof of the other half of Theorem 4.5 is based on 4.2, Lemma and Remark, by means of which we show first:

(i) Let $H \subset G$ and $\Gamma \subset \hat{G}$ be closed orthogonal subgroups. Given any $g \in L^1(G)$, there is to every $\epsilon > 0$ an $h_0 \in L^1(G)$ such that the F.t. $\hat{h}_0$ vanishes near Γ and

$$\|g - h_0 \star g\|_1 < \|T_H g\|_1 + \epsilon.$$

Using (i) we can then prove:

(ii) If $\Delta \subset \Gamma$ is a Wiener–Ditkin set for $\mathscr{F}^1(\Gamma)$, then Δ is a Wiener–Ditkin set for $\mathscr{F}^1(\hat{G})$.

(i) Take $u \in L^1(G)$ such that $\|u\|_1 = 1$ and $\|g - u \star g\|_1 < \frac{1}{2}\epsilon$. By 4.2, Lemma and Remark, there is a $\mu \in M^1(G)$ such that $\hat{\mu}$ is 1 near Γ and

$$\|\mu \star u \star g\|_1 < \|T_H(u \star g)\|_1 + \tfrac{1}{2}\epsilon.$$

Then
$$\|g - (u - \mu \star u) \star g\|_1 < \|T_H(u \star g)\|_1 + \epsilon \leqslant \|T_H g\|_1 + \epsilon.$$

Thus we may take $h_0 = u - \mu \star u$. This proof, of course, is simply a generalization of that in 4.3.

(ii) Let $f \in L^1(G)$ be such that $\hat{f}$ vanishes on Δ. Put $T_H f = \dot{f}$ $[H = \Gamma^\perp \subset G]$. Now $\dot{f}^\wedge$ vanishes on Δ; hence, given $\epsilon > 0$, there is by hypothesis a $\dot{u} \in L^1(G/H)$ such that $\dot{u}^\wedge$ vanishes on a nd. of Δ in Γ and $\|\dot{f} - \dot{u} \star \dot{f}\|_1 < \epsilon$. Take $u \in L^1(G)$ such that $T_H u = \dot{u}$; thus the F.t. $\hat{u}$ vanishes on Δ (even on a relative nd. of Δ in Γ).

Δ is, in particular, a Wiener set for $\mathscr{F}^1(\Gamma)$ (Chap. 2, § 5.2) and hence a Wiener set for $\mathscr{F}^1(\hat{G})$ (§ 3.8). Thus there is a $v \in L^1(G)$ such that $\hat{v}$ vanishes on a nd. of Δ in $\hat{G}$ and $\|u - v\|_1 < \epsilon / \|f\|_1$ (we may suppose $\|f\|_1 > 0$); put $\dot{v} = T_H v \in L^1(G/H)$. Then $\|\dot{f} - \dot{v} \star \dot{f}\|_1 < 2\epsilon$.

Now we apply (i) to $g = f - v \star f$: there is an $h_0 \in L^1(G)$ such that $\hat{h}_0$ vanishes near Γ and

$$\|(f - v \star f) - h_0 \star (f - v \star f)\|_1 < \|\dot{f} - \dot{v} \star \dot{f}\|_1 + \epsilon.$$

Thus
$$\|f - (v + h_0 - h_0 \star v) \star f\|_1 < 3\epsilon.$$

The F.t. of $v + h_0 - h_0 \star v \in L^1(G)$ vanishes on a nd. of Δ in $\hat{G}$, since $\hat{v}$ vanishes near Δ and $\hat{h}_0$ vanishes near Γ. Hence Δ is a Wiener–Ditkin set for $\mathscr{F}^1(\hat{G})$, since $\epsilon > 0$ was arbitrary.

Thus the proof of Theorem 4.5 is complete.

4.8. Theorem 4.5 can be applied to polyhedral sets in $\mathbf{R}^\nu$. These are defined inductively: every closed set in $\mathbf{R}$ with a boundary consisting of countably many points is a polyhedral set; in $\mathbf{R}^\nu$, $\nu \geqslant 2$, a polyhedral set is any closed set whose boundary is a union of countably many (translates of) polyhedral sets in $\mathbf{R}^{\nu'}$, $\nu' < \nu$. It follows from 4.5, combined with Chapter 2, § 5.3 and the obvious fact that a translate of a Wiener–Ditkin set is again a Wiener–Ditkin set: *polyhedral sets in* $\mathbf{R}^\nu$ *are Wiener–Ditkin sets for* $\mathscr{F}^1(\mathbf{R}^\nu)$. For $\nu = 1$ this is, of course, already contained in Chapter 6, § 1.6.

Whether this result also holds for $\mathscr{F}^1_\alpha(\mathbf{R}^\nu)$, $0 < \alpha < 1$, does not seem to have been investigated.

Examples of polyhedral sets are the intersections of finitely many closed half-spaces, in particular the compact convex sets generated by finitely many points (cf. [2], [140]).

References. The results in 4.1–4.8 are a combination of theorems due to Calderón [19], Herz [65, Theorem 6.2],† and Rudin [120, Theorem 7.5.2].

4.9. There is an interesting *criterion for Wiener–Ditkin sets* which applies not only to $\mathscr{F}^1(\hat{G})$ but also to Beurling algebras $\mathscr{F}^1_w(\hat{G}) \simeq L^1_w(G)$. A necessary and sufficient condition for a closed set $\Lambda \subset \hat{G}$ to be a Wiener–Ditkin set for $\mathscr{F}^1_w(\hat{G})$ is the following: whenever $f \in L^1_w(G)$ is such that $\hat{f}$ vanishes on Λ and $\phi \in L^\infty_w(G)$ is such that $\operatorname{sp} f^* \star \phi \subset \Lambda$, then $f^* \star \phi = 0$; in other words, if $\hat{f}$ vanishes on Λ and $\psi \in L^\infty_w(G)$ has spectrum in Λ, then the integral equation $f^* \star \phi = \psi$ has no solution $\phi \in L^\infty_w(G)$ unless $\psi = 0$.

Necessity. Consider any $f \in L^1_w(G)$ with $\hat{f}$ vanishing on Λ. By hypothesis there is a sequence $(j_n)_{n \geqslant 1}$ in $L^1_w(G)$ such that each F.t. $\hat{j}_n$ vanishes near Λ and $j_n \star f \to f$ in $L^1_w(G)$ $(n \to \infty)$. If $\operatorname{sp} f^* \star \phi \subset \Lambda$, then $j_n^* \star (f^* \star \phi) = 0$, $n \geqslant 1$ (cf. §§ 1.3 (ii) and 1.4), and for $n \to \infty$ we obtain $f^* \star \phi = 0$.

Sufficiency. Let J_0 be the ideal of all $j_0 \in L^1_w(G)$ with $\hat{j}_0$ vanishing near Λ. Let $f \in L^1_w(G)$ be such that $\hat{f}$ vanishes on Λ; let I_f be the ideal $\{j_0 \star f \mid j_0 \in J_0\}$. We want to show: f is in the closure of I_f. Let $\phi \in L^\infty_w(G)$ be orthogonal to I_f; thus

$$(j_0 \star f)^* \star \phi = 0 \quad \text{for all } j_0 \in J_0$$

(cf. Chap. 3, §§ 6.2 (ii) and 7.3, Remark 1), whence $\operatorname{sp} f^* \star \phi \subset \Lambda$. By hypothesis, this implies $f^* \star \phi = 0$, thus $\phi \perp f$ and the desired result follows.

It is instructive to consider the case when Λ reduces to a single point (cf. in this connexion [109, Lemma 1.1.2]); compare also the criterion for Wiener sets in 3.2.

The criterion above is due to Herz [65, p. 227] and Glicksberg.‡

4.10. Let us finally call attention to a problem which, in special cases, we have already met before. Stated in rather general terms, the problem is this: to investigate the existence of approximate units in closed ideals of $L^1(G)$, considered as Banach algebras in their own right; likewise for Beurling algebras $L^1_w(G)$. The problem also applies to general locally compact groups if we consider approximate left, or right, units.

If G is abelian, we can pass to $\mathscr{F}^1(\hat{G})$ or $\mathscr{F}^1_w(\hat{G})$ and then the whole question can be considered for, say, Wiener algebras with approximate units, as was already done in Chapter 2, § 5.2 (ii), Remark. The problem

† It is of interest to observe that Herz's proof is based on the existence of invariant means on l.c.a. groups (cf. Chap. 8, § 5).

‡ Herz [loc. cit.] *defined* a certain class of sets by a property which is essentially that stated in the criterion. In the spring of 1962 Glicksberg discovered that these sets are precisely the Wiener–Ditkin sets and communicated the proof to the author who is indebted to him for permission to publish it here.

leads, in particular, to the condition of Wiener–Ditkin (cf. Chap. 2, § 5.1) and to the consideration of Wiener–Ditkin sets (cf. Chap. 2, § 5.2 (iii) and Chap. 6, § 1.8); compare also Chapter 2, § 6.2, Example 3.

Other special cases of the problem were mentioned in Chapter 2, § 7.4 (ii), Remark, and, in the present chapter, in 4.4 (cf. also 4.8).

It is a remarkable aspect of harmonic analysis that it offers problems which can be stated in a simple way and are of considerable interest not only in classical analysis, but in a far wider setting.

8

FUNCTIONS ON GENERAL LOCALLY COMPACT GROUPS

1. Quasi-invariant measures on quotient spaces

1.1. LET H be a closed subgroup of a locally compact group G and consider the quotient space G/H (Chap. 3, § 1.6). If H is normal, then G/H carries a Haar measure and Weil's formula holds (Chap. 3, § 3.3 (3)). We shall now discuss a generalization, due to Mackey [94, I, § 1] and Bruhat [17, I, § 1, 1], for arbitrary closed subgroups.

In Chapter 3, § 3.2 we defined a linear map T_H of $\mathscr{K}(G)$ into $\mathscr{K}(G/H)$ and we showed that T_H is actually surjective (cf. Chap. 3, § 4.2, footnote). Now suppose that μ and $\dot{\mu}$ are (complex) measures on G and G/H, respectively, satisfying the relation

$$\int_{G/H}\left\{\int_H f(x\xi)\,d\xi\right\}d\dot{\mu}(\dot{x}) = \int_G f(x)\,d\mu(x) \qquad f\in\mathscr{K}(G), \tag{1}$$

which is analogous to Weil's formula. Let us apply (1) to the function f_η: $f_\eta(x) = f(x\eta^{-1})$, with $\eta\in H$. Since $\int_H f(x\xi\eta^{-1})\,d\xi = \Delta_H(\eta)\int_H f(x\xi)\,d\xi$, we obtain from (1)

$$\int_G f(x\eta^{-1})\,d\mu(x) = \Delta_H(\eta)\int_G f(x)\,d\mu(x) \quad \eta\in H, f\in\mathscr{K}(G). \tag{2}$$

Conversely, let μ be a measure on G satisfying (2). Then we can show that there is a (necessarily unique) measure $\dot{\mu}$ on G/H satisfying (1).

Fix $f\in\mathscr{K}(G)$. Then for any $g\in\mathscr{K}(G)$ we have, if (2) holds:

$$\int_G f(x)\left\{\int_H g(x\xi)\,d\xi\right\}d\mu(x) = \int_H\left\{\int_G f(x)g(x\xi)\,d\mu(x)\right\}d\xi$$

$$= \int_H \Delta_H(\xi^{-1})\left\{\int_G f(x\xi^{-1})g(x)\,d\mu(x)\right\}d\xi = \int_G g(x)\left\{\int_H f(x\xi)\,d\xi\right\}d\mu(x).$$

We can choose $g\in\mathscr{K}(G)$ so that $\int_H g(x\xi)\,d\xi = 1$ for all $x\in \operatorname{Supp} f$ (cf. Chap. 3, § 4.2). Hence, if $\int_H f(x\xi)\,d\xi = 0$ for all $x\in G$, then also $\int_G f(x)\,d\mu(x) = 0$ (if (2) holds). Thus we can put $\dot{\mu}(f') = \int_G f(x)\,d\mu(x)$ for $f'\in\mathscr{K}(G/H)$, where f is *any* function in $\mathscr{K}(G)$ such that $f' = T_H f$, and the number $\dot{\mu}(f')$ is *uniquely* defined. Also $f'\to\dot{\mu}(f')$ is a *measure* on G/H; for, given $K'\subset G/H$ compact, let $f_1\in\mathscr{K}(G)$ be such that $T_H f_1$ is 1 on K'; then, whenever $f' \in \mathscr{K}(G/H)$ has support in K', we can put $f = (f' \circ \pi_H).f_1$, whence $|\dot{\mu}(f')| = \left|\int\limits_G f\, d\mu\right| \leqslant C_{K'}.\|f'\|_\infty$, with $C_{K'}$ independent of f'. This measure $\dot{\mu}$ satisfies (1).

Thus we have: *let μ be a measure on G; then* (2) *is a necessary and sufficient condition for the existence of a measure $\dot{\mu}$ on G/H satisfying* (1).

1.2. Consider, in particular, a measure μ of the form $d\mu(x) = q(x)\, dx$, where dx is a left Haar measure on G and q a continuous function. The relation (2) then becomes

$$\int f(x\eta^{-1}) q(x)\, dx = \Delta_H(\eta) \int f(x) q(x)\, dx \qquad \eta \in H.$$

But we also have: $\int f(x\eta^{-1}) q(x)\, dx = \Delta_G(\eta) \int f(x) q(x\eta)\, dx$. Since q is continuous and $f \in \mathscr{K}(G)$ arbitrary, it follows that *for measures of the form $q(x)dx$ the relation* (2) *is equivalent to the functional equation*

$$q(x\eta) = q(x)\frac{\Delta_H(\eta)}{\Delta_G(\eta)} \qquad \textit{for all } x \in G,\ \eta \in H. \tag{3}$$

We shall see later that, given H, there is always a continuous, strictly positive solution q of (3) (cf. § 1.6); let us assume it for the moment. Thus, *if H is any closed subgroup of G, there is a continuous, strictly positive function q on G and a positive measure $d_q\dot{x}$ on G/H such that*

$$\int\limits_{G/H} \left\{ \int\limits_H f(x\xi)\, d\xi \right\} d_q\dot{x} = \int\limits_G f(x) q(x)\, dx \qquad f \in \mathscr{K}(G). \tag{4}$$

If H is normal, (4) reduces to Weil's formula for $q = 1$ and $d_q\dot{x} = d\dot{x}$.

1.3. Suppose again that q is a continuous, strictly positive function on G satisfying (3). We can then put

$$\lambda_y(\dot{x}) = \frac{q(yx)}{q(x)} \qquad \dot{x} = \pi_H(x), \tag{5}$$

since the right-hand side is invariant under $x \to x\xi$ ($\xi \in H$).

The group G operates on G/H by means of the transformations

$$\Lambda_y\colon\ \Lambda_y(xH) = y^{-1}xH \qquad y \in G$$

which are homeomorphisms of G/H with itself. The measure $d_q\dot{x}$ on G/H has the property: *for all $f' \in \mathscr{K}(G/H)$*

$$\int\limits_{G/H} f'(\Lambda_y \dot{x})\, d_q\dot{x} = \int\limits_{G/H} f'(\dot{x})\lambda_y(\dot{x})\, d_q\dot{x} \qquad y \in G, \tag{6}$$

λ_y being given by (5).

Choose $f \in \mathscr{K}(G)$ such that $T_H f = f'$. Then by (4)

$$\int\limits_{G/H} f'(\Lambda_y\dot{x})\, d_q\dot{x} = \int\limits_G f(y^{-1}x) q(x)\, dx = \int\limits_G f(x)\frac{q(yx)}{q(x)} q(x)\, dx = \int\limits_{G/H} f'(\dot{x})\lambda_y(\dot{x})\, d_q\dot{x}.$$

We express (6) by saying that the measure $d_q \dot{x}$ on G/H is *quasi-invariant* under G. If the function λ_y on G/H reduces, for each $y \in G$, to a constant (depending on y), we call $d_q \dot{x}$ *relatively invariant* under G, and *invariant* if $\lambda_y = 1$ for all $y \in G$. This extends the definitions in Chapter 3, §§ 3.1 (i) and 3.4; examples will be given later.

The concepts of quasi-invariance, relative invariance, and invariance of measures can be defined in a more general way (cf. [13 *c*, Chap. VII, § 1]) but we shall not need this.

One can also show that the measure $d_q \dot{x}$ corresponding to a continuous, strictly positive solution q of (3) is, in a certain sense, the most general quasi-invariant measure on G/H (relative to G); cf. [94, Theorem 1.1], [13 *c*, Chap. VII, § 2, n° 5, Theorem 1].

1.4. Suppose next that there is a continuous, strictly positive function r on G satisfying

$$(7\,\text{a}) \qquad r(xy) = r(x)r(y) \quad x, y \in G; \qquad (7\,\text{b}) \quad r(\xi) = \frac{\Delta_H(\xi)}{\Delta_G(\xi)} \qquad \xi \in H.$$

If we put $q(x) = r(x)$, then (3) holds; moreover, $\lambda_y(\dot{x})$ in (5) becomes $r(y)$ and thus the corresponding measure $d_r \dot{x}$ on G/H is relatively invariant. Hence, *if there is a continuous, strictly positive function r on G satisfying* (7 a, b), *then there is a relatively invariant positive measure $d_r \dot{x}$ on G/H such that for all $f' \in \mathscr{K}(G/H)$*

$$(8) \qquad \int_{G/H} f'(\Lambda_y \dot{x})\, d_r \dot{x} = r(y) \int_{G/H} f'(\dot{x})\, d_r \dot{x} \qquad y \in G$$

and (4) *holds with $q = r$. In particular, if*

$$(9) \qquad \Delta_H(\xi) = \Delta_G(\xi) \qquad \xi \in H,$$

then there exists an invariant positive measure on G/H.

It is easy to show that, *conversely, the existence of a strictly positive function r on G satisfying* (7 a, b) *is also necessary in order that G/H should carry a relatively invariant positive measure* (not identically zero). In particular, (9) *is also necessary for the existence of an invariant positive measure on G/H.*

Suppose that G/H carries a positive measure $d_r \dot{x}$ not identically zero and satisfying (8). Then we can define a positive measure μ on G by

$$\mu(f) = \int_{G/H} \left\{ \int_H f(x\xi)\, d\xi \right\} d_r \dot{x}, \qquad f \in \mathscr{K}(G),$$

and

$$\int f(y^{-1}x)\, d\mu(x) = r(y) \int f(x)\, d\mu(x) \qquad y \in G,$$

that is, μ is relatively (left) invariant. Hence (Chap. 3, § 3.4) r is continuous, satisfies (7 a) and is strictly positive; moreover, $\{1/r(x)\}\, d\mu(x) = dx$, a (left) Haar measure on G. Thus

$$\int_{G/H} \left\{ \int_H f(x\xi)\, d\xi \right\} d_r \dot{x} = \int_G f(x) r(x)\, dx.$$

The relation (7 b) then follows, as a special case of (3). In particular, if $r = 1$ (i.e. if $d_r \dot{x}$ is invariant), then (9) holds; for normal subgroups H this was already shown in Chapter 3, § 3.6.

REMARK. We may, of course, also consider complex-valued functions r in (7) and corresponding complex measures $d_r \dot{x}$.

These results were given by A. Weil [138, § 9].

Example. Let H be any closed unimodular subgroup of G, that is, $\Delta_H = 1$. Then we can put $r = 1/\Delta_G$, thus G/H carries a relatively invariant measure. If H is compact, then there is an invariant measure on G/H, by (9) and Chapter 3, § 3.6 (ii).

1.5. We shall now consider a case where a function q satisfying (3) can be obtained in a rather simple way. Suppose that G is of the form $G = G_1 H$, where G_1 and H are closed subgroups and $G_1 \cap H = \{e\}$; this means that every $x \in G$ can be written uniquely in the form $x = gh$ with $g \in G_1$, $h \in H$. Suppose further that g and h are continuous functions of x: thus the map $x \to (g, h)$ of G onto the product $G_1 \times H$ is continuous and hence a homeomorphism, since the inverse is also continuous.

Example. If $G = SL(n, \mathbf{R})$, then $G = KH$, where $K = SO(n, \mathbf{R})$, the compact group of all $n \times n$ real orthogonal matrices of determinant 1, and $H = ST_+(n, \mathbf{R})$ (Chap. 3, § 3.8, Example (iii)); moreover, the conditions above are satisfied. An analogous result holds for $SL(n, \mathbf{C})$. This is called the *Iwasawa decomposition* of $SL(n, \mathbf{R})$ and $SL(n, \mathbf{C})$; see [49, § 6, 1], [13 c, Chap. VII, § 3, n° 3, Example 7], and the first footnote on page 187.

Now define the function q as follows: for $h \in H$ let $q(h) = \Delta_H(h)/\Delta_G(h)$ and for general $x \in G$ put $q(x) = q(h)$ if $x = gh$ ($g \in G_1$, $h \in H$). Then q is (single-valued and) continuous, strictly positive, and satisfies (3). Moreover, since $q(gx) = q(x)$ for all $g \in G_1$, $x \in G$, the corresponding quasi-invariant measure $d_q \dot{x}$ on G/H is actually invariant under the *subgroup* G_1 (cf. § 1.3). Now, by our topological assumption, G/H is isomorphic, as a topological space, to G_1. It follows that $d_q \dot{x}$ is a (left) Haar measure on G_1. The general formula (4) becomes here

$$\int_{G_1} \left\{ \int_H f(gh)\, dh \right\} dg = \int_G f(x) q(x)\, dx \qquad f \in \mathscr{K}(G),$$

or, if we replace f by f/q,

$$\int_G f(x)\, dx = \int_{G_1} \left\{ \int_H f(gh) \frac{\Delta_G(h)}{\Delta_H(h)}\, dh \right\} dg \qquad f \in \mathscr{K}(G), \tag{10}$$

where dx, dh, dg are (left) Haar measures on G, H, G_1 respectively. Thus we have expressed dx in terms of dh and dg. Special cases of (10) are:

(i) If G is unimodular, then

$$\int_G f(x)\,dx = \int_{G_1} \left\{ \int_H f(gh)\,\Delta_H(h^{-1})\,dh \right\} dg \qquad f \in \mathscr{K}(G). \tag{11}$$

We note that $\Delta_H(h^{-1})\,dh$ is simply a *right* Haar measure on H. It is instructive to consider the example $SL(n, \mathbf{R})$ mentioned above, especially for $n = 2$.

(ii) If H is normal, then $\int_G f(x)\,dx = \int_{G_1} \left\{ \int_H f(gh)\,dh \right\} dg$ (cf. (9)); this can of course also be obtained directly from Weil's formula.

Reference. [25, Theorem 3.]

1.6. We shall now prove that, *given any closed subgroup $H \subset G$, the functional equation* (3) *has a continuous, strictly positive solution q.* Suppose that we can obtain a function F on G such that $\int_H F(x\xi)\,d_R\xi$, where $d_R\xi$ is a *right* Haar measure on H, exists for all $x \in G$ and is a strictly positive and continuous function of x. Then clearly

$$q(x) = \left\{ \int_H F(x\xi)\,d_R\xi \right\} \Big/ \Delta_G(x) \tag{12}$$

satisfies (3).

The construction of a function F with the properties above is based on the following purely topological proposition.

1.7. *Let H be a closed subgroup of G. Let Ω be a* (*non-empty*) *open set in G with compact closure. Then there is a subset $Y \subset G$ such that the family $(\Omega y H)_{y \in Y}$ covers G and is locally finite.*† In other words, the family $(\pi_H(\Omega y))_{y \in Y}$ forms a locally finite covering of G/H. When H reduces to the neutral element, this proposition becomes rather intuitive.

The condition that Ω have compact closure cannot be omitted.

For the proof we assume, without loss of generality, that $e \in \Omega$. Put

$$K = \bar{\Omega} \cup \bar{\Omega}^{-1},$$

so that K is compact. Let $G_1 = \bigcup_{r \geqslant 1} K^r$ be the subgroup generated by K; then G_1 is open and hence closed in G. We show first:

(i) *There is a sequence $(z_m)_{m \geqslant 1}$, possibly finite, of distinct elements of G_1 such that $(\Omega z_m H)_{m \geqslant 1}$ is a locally finite covering of $G_1 H$.*

Proof. If $\pi_H(G_1)$ is compact, then $\pi_H(G_1)$ can be covered by a finite sequence

† A family $\mathfrak{F}$ of subsets of a topological space X is said to be *locally finite* if every point of X has a nd. meeting at most finitely many sets of $\mathfrak{F}$. If X is locally compact, this means that any compact set in X meets at most finitely many sets of $\mathfrak{F}$.

$(\pi_H(\Omega z_n))_{1 \leqslant n \leqslant N}$, $z_n \in G_1$ (observe that $(\pi_H(\Omega z))_{z \in G_1}$ is an *open* covering of $\pi_H(G_1)$). Now suppose that $\pi_H(G_1)$ is not compact. Put $K'_1 = \pi_H(K)$, with K as above, and† let K'_r be the closure of the 'annulus' $\pi_H(K^r) - \pi_H(K^{r-1})$ for $r = 2, 3, \ldots$. Then $K'_r \neq \emptyset$ for all $r \geqslant 1$ (if $K^r H = K^{r-1} H$ for some $r \geqslant 2$, it follows that $K^{r+1} H = K^r H$ for all r sufficiently large, hence $\pi_H(G_1)$ would be compact); moreover, K'_r is compact and $\bigcup_{1 \leqslant s \leqslant r} K'_s = \pi_H(K^r)$, $r \geqslant 1$. We further observe that

$$(*) \qquad K'_s \cap \pi_H(K^r) = \emptyset \quad \text{if } s > r+1,$$

for $\pi_H(K^r)$ is contained in the interior of $\pi_H(K^{r+1})$, since π_H is open.

Each K'_r, $r \geqslant 1$, is compact and is covered by the family of open sets

$$(\pi_H(\Omega y))_{y \in G_1, \pi_H(y) \in K'_r}.$$

Hence K'_r may be covered by finitely many of these sets. Thus there is a strictly increasing sequence of integers $(N_r)_{r \geqslant 0}$, starting with $N_0 = 0$, and a sequence of (not necessarily distinct) points $(y_n)_{n \geqslant 1}$ in G_1 such that $\pi_H(y_n)$ lies in K'_r for $N_{r-1} < n \leqslant N_r$ and $(\pi_H(\Omega y_n))_{N_{r-1} < n \leqslant N_r}$ covers K'_r, for each $r \geqslant 1$.

Now $\bigcup_{r \geqslant 1} K'_r = \pi_H(G_1)$, thus each point of $\pi_H(G_1)$ lies in some $\pi_H(\Omega y_n)$. Moreover,

$$(**) \qquad \pi_H(\Omega y_n) \cap \pi_H(K^r) = \emptyset \quad \text{if } n > N_{r+2},$$

which is shown as follows: if $n > N_{r+2}$, then $\pi_H(y_n) \in K'_s$ for some $s > r+2$ and (*) now implies $K'_s \cap \pi_H(K^{r+1}) = \emptyset$; thus $(y_n H) \cap (K^{r+1} H) = \emptyset$ or (since $K = K^{-1}$) $(K y_n H) \cap (K^r H) = \emptyset$, whence (**).

There is, finally, a subsequence $(z_m)_{m \geqslant 1} = (y_{n_m})_{m \geqslant 1}$ of *distinct* elements of G_1 such that $(\pi_H(\Omega z_m))_{m \geqslant 1}$ covers $\pi_H(G_1)$, and this covering is locally finite by (**). Thus (i) is proved.

(ii) If $G_1 H = G$, then the proof is finished. Otherwise let A be a complete system of representatives of G mod(G_1, H), that is, a subset of G such that the double cosets $(G_1 a H)_{a \in A}$ are mutually disjoint and their union is G. For each $a \in A$ we apply (i) to G_1 and the subgroup aHa^{-1} of G: thus there is a locally finite covering $(\Omega z_{a,m} aHa^{-1})_{m \geqslant 1}$ of $G_1 aHa^{-1}$, with distinct elements $z_{a,m} \in G_1$. Then $(\Omega z_{a,m} aH)_{m \geqslant 1}$ is a locally finite covering of $G_1 aH$. Let Y be the set of all $y = z_{a,m} a$, $a \in A$, $m \geqslant 1$; then $(\Omega y H)_{y \in Y}$ is a locally finite covering of G.

(iii) To see that the proposition may fail if Ω does *not* have compact closure, let $G = \mathbf{R}$, $H = \{0\}$; let $\Omega \subset \mathbf{R}$ be any open set containing all rational numbers, but of finite Lebesgue measure. Then only an infinite number of translates of Ω can form a covering of $\mathbf{R}$, and any interval meets all of them. Likewise for any l.c. group which is neither compact nor discrete and contains a countable dense subset.

1.8. Using the result in 1.7, we can now obtain, in a very simple way, a function F of the kind required in 1.6. Take any $k \neq 0$ in $\mathscr{K}_+(G)$; put $\Omega = \{x \mid x \in G, k(x) > 0\}$. There is a subset Y of G with the following properties:

(*a*) For each compact set $K \subset G$ there are only finitely many $y \in Y$ such that $(KH) \cap (\Omega y) \neq \emptyset$.

† In reading the argument that follows, it may be helpful to plot some figures.

(*b*) For every $x \in G$ there is a $y \in Y$ such that $(xH) \cap (\Omega y) \neq \emptyset$.

This is the result in 1.7, stated in a slightly different way. Now put $k_y(x) = k(xy^{-1})$, $y \in Y$. Then the function F given by

$$F(x) = \sum_{y \in Y} k_y(x) \qquad x \in G$$

is not only well-defined and continuous, but also has the following properties arising respectively from (*a*) and (*b*):

(*a'*) If K is any compact set in G, then F coincides *on the 'strip'* KH with a function in $\mathscr{K}_+(G)$ (viz., with the sum of finitely many functions k_y).

(*b'*) For every $x \in G$ there is a $\xi \in H$ such that $F(x\xi) > 0$.

Hence, as is readily seen, F satisfies the requirements laid down in 1.6. Thus we have a solution (12) of the functional equation (3); we shall, however, modify this a little, also for the sake of later applications. Put

$$F_1(x) = \int_H F(x\xi)\, d\xi \qquad x \in G.$$

Then F_1 is continuous, on account of (*a'*) (cf. Chap. 3, § 3.2); also $F_1(x) > 0$ for all $x \in G$ by (*b'*). Now put

$$\beta(x) = \frac{F(x)}{F_1(x)}.$$

Then $\int_H \beta(x\xi)\, d\xi = 1$ for all $x \in G$.

1.9. We have now obtained the result that *for every closed subgroup* H *there exists a function* β on G with the following properties:

(i) If K is any compact set in G, then β coincides on the 'strip' KH with a function in $\mathscr{K}_+(G)$.

(ii) $\int_H \beta(x\xi)\, d\xi = 1$ for all $x \in G$.

We call β a *Bruhat function* for H. From (i) it follows, in particular, that for every $x \in G$ the function $\xi \to \beta(x\xi)$, $\xi \in H$, is in $\mathscr{K}_+(H)$ and (ii) implies that $\beta(x\xi) > 0$ for some $\xi \in H$ (depending on x).

Remark. In all arguments involving only strips KH, with $K \subset G$ compact, or subsets of such strips, we can treat β as though it were a function in $\mathscr{K}_+(G)$; this is the significance of (i).

We can now define, instead of (12),

$$q(x) = \int_H \beta(x\xi)\Delta_G(\xi)\Delta_H(\xi^{-1})\, d\xi. \tag{13}$$

Then q is continuous (cf. (i) and Chap. 3, § 3.2, Remark); also $q(x) > 0$

for all $x \in G$ and q satisfies the functional equation (3). The definition (13) has, compared to (12), the advantage that (cf. (ii)) $q = 1$ whenever G/H carries an invariant measure (in particular, when H is normal).

References. The existence of the function β was first shown by Bruhat [17, pp. 102–3] who proved the existence of a *continuous* solution of (3) by means of (13); previously Mackey [94, Theorem 1.1] had shown the existence of a *measurable* solution of (3) for l.c. groups with a countable basis for the open sets (both Mackey and Bruhat considered the analogue of (3) for *right* cosets). Later, but independently, the function β was also introduced by Macbeath and Świerczkowski [92, Theorem C] who likewise discussed quasi-invariant measures. See also [13 *c*, Chap. VII, § 2, nº 4, Prop. 7 and nº 5, Theorem 2].

2. The space $L^1(G/H)$

In § 1 we showed that, if H is *any* closed subgroup of a locally compact group G, then there exists a 'quasi-invariant' measure on the quotient space G/H. We can now extend several results proved in Chapter 3 for closed normal subgroups to closed subgroups in general.

2.1. Let q be any continuous, strictly positive solution of the functional equation 1.2 (3), for a given closed subgroup $H \subset G$. Then we define for every $f \in \mathscr{K}(G)$ a function $T_{H,q}f$ on G/H by

(1) $$T_{H,q}f(\dot{x}) = \int_H \frac{f(x\xi)}{q(x\xi)}\,d\xi \qquad \dot{x} = \pi_H(x),$$

$d\xi$ being a left Haar measure on H; observe that the integral is indeed a function of $\dot{x} \in G/H$. When $q = 1$, this is simply formula (1) in Chapter 3, § 3.2. We note that $T_{H,q}f$ is in $\mathscr{K}(G/H)$, since $f/q \in \mathscr{K}(G)$ (cf. Chap. 3, § 3.2).

We can now rewrite the relation 1.2 (4) in the form

(2) $$\int_{G/H} T_{H,q}f(\dot{x})\,d_q\dot{x} = \int_G f(x)\,dx \qquad f \in \mathscr{K}(G).$$

This will be called the *formula of Mackey–Bruhat* for quasi-invariant measures; it generalizes Weil's formula.†

We can extend (2) to $\mathscr{I}_+(G)$ by the same method as in Chapter 3, § 3.3 (iii): if $F \in \mathscr{I}_+(G)$, then the function $\dot{x} \to \int_H^\times \frac{F(x\xi)}{q(x\xi)}\,d\xi$ is in $\mathscr{I}_+(G/H)$ and

(3) $$\int_{G/H}^\times \left\{\int_H^\times \frac{F(x\xi)}{q(x\xi)}\,d\xi\right\} d_q\dot{x} = \int_G^\times F(x)\,dx.$$

† Cf. Chapter 3, § 3.3 (i); for relatively invariant measures (2) is already contained in [138, p. 45].

2.2. By means of (3) we can show: the result in Chapter 3, § 3.9 holds if H is *any* closed subgroup of G. Statement and proof simply carry over, with obvious modifications, the measure on G/H being now $d_q\dot{x}$.

Reference. [13 *c*, Chap. VII, App. 2, p. 98.]

2.3. We put
$$\|T_{H,q}f\|_1 = \int_{G/H} |T_{H,q}f(\dot{x})|\, d_q\dot{x}$$
(cf. (1)). Then
$$\|T_{H,q}f\|_1 \leqslant \|f\|_1, \tag{4}$$
as in Chapter 3, § 4.2. Moreover, the whole reasoning in Chapter 3, § 4.2 now extends to the general case of an arbitrary closed subgroup $H \subset G$ and we obtain: *the mapping $T_{H,q}$ of $\mathscr{K}(G)$ into $\mathscr{K}(G/H)$ is surjective and, if $\mathscr{J}(G,H)$ is the closed linear subspace of $\mathscr{K}(G)$ defined by*
$$\mathscr{J}(G,H) = \{k_0 \mid k_0 \in \mathscr{K}(G),\ T_{H,q}k_0 = 0\}, \tag{5}$$
then
$$\mathscr{K}(G/H) \cong \mathscr{K}(G)/\mathscr{J}(G,H),$$
the isomorphism being algebraic and isometric. Here $\mathscr{K}(G)/\mathscr{J}(G,H)$ is provided with the usual quotient norm.

REMARK. *$\mathscr{J}(G,H)$ is independent of the particular choice of q* (if q is one solution of equation 1.2 (3) such that $q(x) \neq 0$ for all $x \in G$, then any other solution is of the form $q.f'$, where f' satisfies $f'(x\xi) = f'(x)$ for all $x \in G$, $\xi \in H$). This also shows that *$\|T_{H,q}f\|_1$ is independent of q.*

Let $L^1(G/H)$ be the L^1-space attached to the measure $d_q\dot{x}$, with the usual norm; then we have (cf. Chap. 3, §§ 4.3 and 4.4):
$$L^1(G/H) \cong L^1(G)/J^1(G,H), \tag{6}$$
where $J^1(G,H)$ is the closure of $\mathscr{J}(G,H)$ in $L^1(G)$. The isomorphism (6) *is algebraic and isometric.*† It is defined by extending the map $T_{H,q}$ from $\mathscr{K}(G)$ to $L^1(G)$ by continuity. We write again, for the extended mapping,
$$T_{H,q}f(\dot{x}) = \int_H \frac{f(x\xi)}{q(x\xi)}\, d\xi \qquad f \in L^1(G) \tag{7}$$
and this notation is not merely formal: the result in Chapter 3, § 4.5 and its proof carry over, with obvious changes of notation. Thus (2) extends to all $f \in L^1(G)$ (*'extended formula of Mackey–Bruhat'*); so does (4). Also $J^1(G,H)$ is simply the kernel of (7) (cf. Chap. 3, § 4.6).

2.4. In Chapter 3, § 5.5 we defined the operator R_a in $L^1(G)$. It is sometimes more convenient to use the operator $A_y = R_{y^{-1}}$:
$$A_y f(x) = f(xy)\Delta_G(y) \qquad y \in G. \tag{8}$$

† In the 'extreme' cases $H = G$ or $H = \{e\}$ this is obvious.

This operator satisfies $$A_{y_1y_2} = A_{y_1}A_{y_2}.$$

$T_{H,q}$ has the property that for $f \in L^1(G)$

$$T_{H,q}A_\eta f = T_{H,q}f \qquad \eta \in H, \tag{9}$$

which can be verified by means of 1.2 (3).

2.5. The subspace $J^1(G, H)$ in (6), defined as the closure of $\mathscr{J}(G, H)$ (cf. (5))—or, in view of (7), simply as the kernel of $T_{H,q}$ in $L^1(G)$—may be characterized in a different way, independently of $T_{H,q}$: *the closed linear subspace of* $L^1(G)$ *spanned by the family of all functions* $A_\eta f - f$ *with* $f \in \mathscr{K}(G)$ *and* $\eta \in H$ (cf. (8)) *coincides with* $J^1(G, H)$.

We denote this subspace, for the moment, by J. In view of (9) we have $J \subset J^1(G, H)$. To show the opposite inclusion, we observe that J is left invariant; hence we can apply Chapter 3, § 6.3. The proof in Chapter 3, § 6.4 then carries over to the present case, if we replace the operator L_η there by A_η $(\eta \in H)$ and use the formula (2) of Mackey–Bruhat instead of Weil's formula.

Thus we have an analogue of Chapter 3, § 6.4 valid for general closed subgroups.† It yields an intrinsic definition of $J^1(G, H)$, entirely independent of q (cf. also 2.3, Remark). Thus *the isomorphism* (6) *provides a direct characterization of* $L^1(G/H)$ *in terms of* $L^1(G)$ without reference to the quasi-invariant measure $d_q\dot{x}$ or the function q.

References. For H such that G/H carries an invariant measure the results in 2.3 and 2.5 were given in [110, I], but the proof contained a gap (cf. [111, I, p. 260, footnote], but also 2.6 below). The generalization obtained here is a modification of that given by Świerczkowski in [131, Theorems 3 and 4], where the notation $L^1(G)$ has another meaning, varying with the subgroup H, and the mapping is defined differently.

2.6. Let Φ be a complex-valued measurable function on G. We call an element $\eta \in G$ an *essential right period* of Φ if $\Phi(x\eta) = \Phi(x)$ l.a.e. on G; the set of all $x \in G$ for which equality does not hold may of course depend on η. We shall prove: *the essential right periods of* Φ *form a closed subgroup* H *of* G, *and* Φ *coincides locally almost everywhere with a function* $\Phi' \circ \pi_H$, *where* Φ' *is a function on* G/H. The proof is based on the existence of a quasi-invariant measure on G/H (and Φ' is measurable with respect to that measure). This is an extension of Chapter 3, § 6.5.

Clearly the essential right periods of Φ form a subgroup. Now consider first an essentially bounded measurable function ϕ on G; let H be its group of essential

† If H is normal, the above characterization of $J^1(G, H)$ is not exactly the same as that in Chapter 3, § 6.4, but is related to it by involution: we have

$$(A_\eta f - f)^* = L_\eta(f^*) - f^* \quad \text{(Chap. 3, § 5.5 } (\lambda_7))$$

and the involution maps $J^1(G, H)$ onto itself for normal H (Chap. 3, § 5.3 (8)).

right periods: then for each $\eta \in H$

$$(*) \qquad \langle R_\eta f - f, \phi \rangle = 0 \quad \text{for all } f \in L^1(G).$$

Here the left-hand side is a continuous function of η (cf. Chap. 3, § 6.1), so (*) holds for all $\bar{\eta}$ in the closure of H; conversely, if (*) holds for $\bar{\eta}$ in the place of η, then $\bar{\eta}$ is an essential right period of ϕ. Thus H is closed. Now we can show that there is an essentially bounded measurable function ϕ' on G/H such that

$$\phi(x) = \phi' \circ \pi_H(x) \text{ l.a.e. on } G.$$

The proof is as in Chapter 3, § 6.5, using 2.2 and the extended formula of Mackey–Bruhat (§ 2.3). The general case of a measurable function Φ reduces to that of a bounded function (cf. Chap. 3, § 6.5).

References. For H such that G/H carries a relatively invariant measure the result above dates back to 1955: it was obtained to fill a gap in the proof referred to at the end of 2.5, but was not published. It was given in a still more general form than here by Bourbaki [13 *c*, Chap. VII, § 2, nº 3, Cor. 2], with a proof differing only in appearance. For $\Phi \in L^p(G)$, $1 \leqslant p < \infty$, see also [138, pp. 41–42], and Chapter III, § 6.1, first footnote.

2.7. Consider now the Banach algebra $M^1(G)$ (Chap. 3, § 5.4, Remark). Given a closed subgroup $H \subset G$, we can consider the Banach space $M^1(G/H)$. Let μ be in $M^1(G)$; then we define $\dot{\mu} \in M^1(G/H)$ by

$$(10) \qquad \int_{G/H} f'(\dot{x})\, d\dot{\mu}(\dot{x}) = \int_G f' \circ \pi_H(x)\, d\mu(x) \qquad f' \in \mathscr{K}(G/H).$$

Since here $\|\dot{\mu}\| \leqslant \|\mu\|$, *the linear map* $\mu \to \dot{\mu}$ *is continuous*; moreover, we can show that *it maps* $M^1(G)$ *onto* $M^1(G/H)$.

Let β be a Bruhat function for H (§ 1.9). Consider the map

$$k \to T_{H,\beta}k\colon\ T_{H,\beta}k(\dot{x}) = \int_H k(x\xi)\beta(x\xi)\, d\xi \qquad k \in \mathscr{K}(G).$$

This maps $\mathscr{K}(G)$ into $\mathscr{K}(G/H)$ (in fact, the mapping is surjective, as is not difficult to see); moreover, $\|T_{H,\beta}k\|_\infty \leqslant \|k\|_\infty$. Thus, given $\mu' \in M^1(G/H)$, we can define a measure $\mu \in M^1(G)$ by

$$(*) \qquad \int_G k(x)\, d\mu(x) = \int_{G/H} T_{H,\beta}k(\dot{x})\, d\mu'(\dot{x}) \qquad k \in \mathscr{K}(G).$$

Now, if we apply the map $\mu \to \dot{\mu}$ defined by (10) to this μ, then we obtain precisely the original measure μ': indeed, (*) may be extended to $\mathscr{C}^b(G)$ (to show this, it suffices, by a familiar decomposition, to consider positive measures and hence also positive functions, and then we can apply the same method as in Chapter 3, § 3.3 (iii)). In particular we have

$$\int_G f' \circ \pi_H(x)\, d\mu(x) = \int_{G/H} f'(\dot{x})\, d\mu'(\dot{x}) \qquad f' \in \mathscr{K}(G/H),$$

that is, $\mu' = \dot{\mu}$.

REMARK. The map $\mu' \to \mu$ defined by (*) is the dual of $T_{H,\beta}$ (extended to $\mathscr{C}^0(G)$) and yields an isometric injection of $M^1(G/H)$ into $M^1(G)$. We also have

$$(**) \qquad \mathscr{C}^0(G/H) \cong \mathscr{C}^0(G)/\mathscr{C}^0_\beta(G, H),$$

where $\mathscr{C}^0_\beta(G,H) = \{f \mid f \in \mathscr{C}^0(G),\ T_{H,\beta} f = 0\}$, and the isomorphism (**) is algebraic and isometric as is readily seen.

We observe that, if $d\mu(x) = f(x)\,dx$, with $f \in L^1(G)$, then the measure $\dot{\mu}$ defined by (10) is given by $d\dot{\mu}(\dot{x}) = T_{H,q} f(\dot{x})\, d_q \dot{x}$; thus *the map* $\mu \to \dot{\mu}$ *defined by* (10) *is an extension of the map* $T_{H,q}$ in 2.1.

Now let H be a normal subgroup. Then $(\mu_1 \star^G \mu_2)^{\cdot} = \dot{\mu}_1 \star^{G/H} \dot{\mu}_2$, which is analogous to Chapter 3, § 5.3 (7) and follows quite directly from the definition of convolution for measures (Chap. 3, § 5.4, Remark); relation (8) in Chapter 3, § 5.3 likewise carries over, together with its proof. *If G is abelian, then the* F.t. $\dot{\mu}^\wedge$ *is the restriction of* $\hat{\mu}$ *to* $H^\perp$; this is analogous to Chapter 4, § 4.3 and follows directly from the definition of $\dot{\mu}$.

References. [120, § 2.7.2] (cf. also the review of [120] in *Zentralblatt f. Math.* **107,** 96–97 (1964)); [56, Lemma 5.1.8]; [118, Satz 1]; for far more general results see [65, § 5] and [6].

3. The property P_p

3.1. A locally compact group G is said to have the *property* P_p $(1 \leqslant p < \infty)$ if it satisfies the following condition: given any compact set $K \subset G$, there is, for every $\epsilon > 0$, a positive function† s $(= s_{K,\epsilon})$ in $L^p(G)$ such that $\|s\|_p = 1$ and $\|L_y s - s\|_p < \epsilon$ for all $y \in K$.

In this definition we may clearly replace $L^p(G)$ by $\mathscr{K}(G)$. It was introduced by Dieudonné [33, p. 284]. We can now express the content of Chapter 5, § 2.1 (i) by saying that *every locally compact abelian group has the property* P_1.

3.2. It is sufficient to consider the property P_1, for P_p *is equivalent to* P_1 *for every* $p > 1$. This was shown by Stegeman [127] after some partial results had been obtained by Dieudonné [33, p. 285] and in [115].

Let a compact set $K \subset G$ and $\epsilon > 0$ be given.

To prove $P_1 \Rightarrow P_p$, let $s_1 \in L^1(G)$ be such that $s_1 \geqslant 0$, $\|s_1\|_1 = 1$, $\|L_y s_1 - s_1\|_1 < \epsilon$ $(y \in K)$, and put $s_p = s_1^{1/p}$. Then $s_p \in L^p(G)$, $s_p \geqslant 0$, $\|s_p\|_p = 1$, and the inequality

$$|a-b|^p \leqslant |a^p - b^p| \qquad (a \geqslant 0,\ b \geqslant 0,\ p \geqslant 1)$$

implies $\|L_y s_p - s_p\|_p < \epsilon^{1/p}$ $(y \in K)$.

To prove $P_p \Rightarrow P_1$, let $s_p \in L^p(G)$ be such that $s_p \geqslant 0$, $\|s_p\|_p = 1$, $\|L_y s_p - s_p\|_p < \epsilon$ $(y \in K)$ and put $s_1 = s_p^p$. Thus $s_1 \in L^1(G)$, $s_1 \geqslant 0$, $\|s_1\|_1 = 1$; moreover, the inequality

$$|a^p - b^p| \leqslant p|a-b|(a^{p-1} + b^{p-1}) \quad (a \geqslant 0,\ b \geqslant 0,\ p \geqslant 1),$$

together with Hölder's inequality, yields

$$\|L_y s_1 - s_1\|_1 \leqslant p\|L_y s_p - s_p\|_p (\|L_y s_p^{p-1}\|_{p'} + \|s_p^{p-1}\|_{p'}),$$

where $1/p + 1/p' = 1$ and $s_p^{p-1} \in L^{p'}(G)$. Hence

$$\|L_y s_1 - s_1\|_1 \leqslant 2p\|L_y s_p - s_p\|_p < 2p\epsilon \quad \text{for } y \in K.$$

† Cf. Chapter 3, § 1.9.

A compact group obviously has the property P_1: we may take $s = 1$. On a l.c. non-compact group the Haar measure is unbounded (Chap. 3, § 3.1 (iv)); P_1 signifies, intuitively, that the group carries *bounded* positive measures ($\neq 0$) which are 'approximately left invariant'; but we shall see that not all l.c. groups satisfy P_1.

3.3. The property P_1 holds for abelian l.c. groups (cf. § 3.1). To extend this result to a larger class of groups, we now prove: *if H is a closed normal subgroup of a locally compact group G and if P_1 holds for H and for G/H, then P_1 holds for G.*

Let a compact set $K \subset G$ be given; choose $\epsilon > 0$.

Since G/H satisfies P_1, there is a positive $s' \in \mathscr{K}(G/H)$ such that $\int_{G/H} s'(\dot{x})\, d\dot{x} = 1$ and

$$(*) \qquad \int_{G/H} |s'(\dot{y}^{-1}\dot{x}) - s'(\dot{x})|\, d\dot{x} < \tfrac{1}{2}\epsilon \quad \text{for all } \dot{y} \in \pi_H(K).$$

Let β be a Bruhat function on G for H (§ 1.9) and put

$$f_1 = (s' \circ \pi_H).\beta.$$

Then $f_1 \in \mathscr{K}_+(G)$ and $\int_G f_1(x)\, dx = 1$; put $C_1 = \operatorname{Supp} f_1$. Consider

$$k(x) = f_1(y^{-1}x) - f_1(x) \qquad \text{for arbitrary } y \in K.$$

The functions k all vanish outside some *compact* set K_0 and satisfy $\max_{x \in G} |k(x)| \leqslant M_0$, where K_0 *and* M_0 *are independent of* $y \in K$: we can take $K_0 = (KC_1) \cup C_1$ and $M_0 = 2 \max_{x \in G} f_1(x)$.

H has the property P_1, hence we can apply Chapter 5, § 3.1 with

$$\mathfrak{F}_H = \left\{ s_1 \mid s_1 \in \mathscr{K}(H),\ s_1 \geqslant 0,\ \int_H s_1(\xi)\, d\xi = 1 \right\}$$

and obtain: there is an $s_1 \in \mathfrak{F}_H$ such that for *all* functions k above the inequality

$$(**) \qquad \int_G \left| \int_H k(x\xi) s_1(\xi^{-1})\, d\xi \right| dx < \int_{G/H} \left| \int_H k(x\xi)\, d\xi \right| d\dot{x} + \tfrac{1}{2}\epsilon$$

holds. Now, putting $\dot{y} = \pi_H(y)$, $\dot{x} = \pi_H(x)$, we have

$$\int_H k(x\xi)\, d\xi = s'(\dot{y}^{-1}\dot{x}) \int_H \beta(y^{-1}x\xi)\, d\xi - s'(\dot{x}) \int_H \beta(x\xi)\, d\xi = s'(\dot{y}^{-1}\dot{x}) - s'(\dot{x}).$$

Hence (*) and (**) yield

$$\int_G \left| \int_H k(x\xi) s_1(\xi^{-1})\, d\xi \right| dx < \tfrac{1}{2}\epsilon + \tfrac{1}{2}\epsilon = \epsilon.$$

Thus, if we put

$$s(x) = \int_H f_1(x\xi) s_1(\xi^{-1})\, d\xi,$$

then $s \in \mathscr{K}(G)$ and $\|L_y s - s\|_1 < \epsilon$ for all $y \in K$. Moreover, s is positive and

$$\int_G s(x)\, dx = \int_G f_1(x)\, dx . \int_H \Delta_G(\xi^{-1}) s_1(\xi^{-1})\, d\xi = 1,$$

since $\Delta_G(\xi) = \Delta_H(\xi)$ for $\xi \in H$.

The compact set $K \subset G$ and $\epsilon > 0$ were arbitrary; thus G has the property P_1.

Hence *all soluble locally compact groups have the property* P_1; likewise all locally compact groups G containing a closed normal abelian (or soluble) subgroup H such that G/H is compact, for instance the group of motions of euclidean space (cf. [33, Theorem 2 and p. 289], [113]).

REMARK. A *topological* group G is said to be *soluble* if it contains a finite chain of *closed* subgroups

$$G = G_0 \supset G_1 \supset \ldots \supset G_n \supset G_{n+1} \supset \ldots \supset G_{N+1} = \{e\}$$

such that G_{n+1} is normal in G_n and G_n/G_{n+1} is abelian $(0 \leqslant n \leqslant N)$. We have: *if G is soluble as an abstract (i.e. discrete) group, then it is also soluble as a topological group.*

If in the chain above not all subgroups G_n are closed, consider their closures $\overline{G}_n$. For $x \in G_n$ we have $xG_{n+1}x^{-1} \subset G_{n+1}$, hence $x\overline{G}_{n+1}x^{-1} \subset \overline{G}_{n+1}$ and, since $x \to xax^{-1}$ is continuous, this holds also for $x \in \overline{G}_n$, that is, $\overline{G}_{n+1}$ is normal in $\overline{G}_n$. Moreover, for $x, y \in G_n$ we have $xyG_{n+1} = yxG_{n+1}$ and it readily follows that $xy\overline{G}_{n+1} = yx\overline{G}_{n+1}$ for all $x, y \in \overline{G}_n$, that is, $\overline{G}_n/\overline{G}_{n+1}$ is abelian. See also [83, § 32 and Theorem 8.3].

Example. The groups $ST_+(n, \mathbf{R})$ (Chap. 3, § 3.8, Example (iii)) are soluble.

3.4. We now discuss an interesting application of the property P_p due, essentially, to Dieudonné [33, pp. 283–5]. *Let G have the property P_1 and let μ be a positive measure on G. Suppose that for some p, $1 < p < \infty$, the following holds: for every $f \in \mathscr{K}(G)$ the function $\mu \star f$ is in*† $L^p(G)$ *and $A_\mu: f \to \mu \star f$, $f \in \mathscr{K}(G)$, is a bounded operator in $L^p(G)$, i.e.*

$$\|\mu \star f\|_p \leqslant C\|f\|_p \qquad f \in \mathscr{K}(G).$$

Then μ is bounded and $|A_\mu| = \|\mu\|$. For $p = 1$ this is true for all l.c. groups, independently of the property P_1; the case $p = \infty$ is of course trivial.

The proof is based on the property P_p which, as we know, is equivalent to P_1 for every p, $1 < p < \infty$ (§ 3.2); nevertheless, P_p plays an independent role.

Take any $f_1 \in \mathscr{K}_+(G)$ such that $f_1 \leqslant 1$; then the positive measure $\mu_1 = f_1.\mu$ is bounded and even has compact support contained in $K_1 = \operatorname{Supp} f_1$. Consider the *bounded* operator A_{μ_1}: we will show that

$$(*) \qquad |A_{\mu_1}| = \|\mu_1\|.$$

Then we observe that $|A_{\mu_1}| \leqslant |A_\mu|$, for

$$\|A_{\mu_1} f\|_p \leqslant \|A_{\mu_1}|f|\|_p \leqslant \|A_\mu |f|\|_p \leqslant |A_\mu| . \|f\|_p.$$

Since $\sup\limits_{\mu_1}\|\mu_1\| = \sup\limits_{f_1} \int f_1\, d\mu$ $(f_1 \in \mathscr{K}_+(G), f_1 \leqslant 1)$, it will follow that μ is bounded

† We observe that $\mu \star f$ is certainly a continuous function, even if μ is not bounded (cf. Chap. 3, § 3.2, Remark).

and $\|\mu\| \leqslant |A_\mu|$. The opposite inequality is clear (cf. Chap. 3, § 5.4, 2nd footnote), so we shall have $\|\mu\| = |A_\mu|$.

Proof of (*). If $p = 1$, this is trivial. For $1 < p < \infty$ we use the property P_p as follows. For $f, g \in \mathscr{K}(G)$ we have

$$(**) \qquad \int \Big\{ \int f(y^{-1}x)\, d\mu_1(y) \Big\} g(x)\, dx = \int \Big\{ \int f(y^{-1}x) g(x)\, dx \Big\}\, d\mu_1(y).$$

Now, given any $\epsilon > 0$, there is an $s \in \mathscr{K}_+(G)$ such that $\int s^p = 1$, $\|L_y s - s\|_p < \epsilon$ for all $y \in K_1$ (defined above). Put in (**) $f = s$, $g = s^{p-1}$, so that $g \in \mathscr{K}_+(G)$, $\int g^{p'} = 1$ $(1/p + 1/p' = 1)$. Then $\int fg = 1$ and by Hölder's inequality

$$\int f(y^{-1}x) g(x)\, dx > 1 - \epsilon \qquad \text{for all } y \in K_1.$$

Hence, with this choice of f, g, (**) yields

$$\int A_{\mu_1} f(x) . g(x)\, dx \geqslant (1-\epsilon) \int d\mu_1.$$

Thus $\|A_{\mu_1} f\|_p \geqslant (1-\epsilon)\|\mu_1\|$, since $\|g\|_{p'} = 1$. Since $\|f\|_p = 1$, we have

$$|A_{\mu_1}| \geqslant (1-\epsilon)\|\mu_1\|.$$

Hence (*) holds (since obviously $|A_{\mu_1}| \leqslant \|\mu_1\|$) and the proof is complete.

We shall prove a converse to the proposition above in 3.7.

REMARK.† Let G be an arbitrary l.c. group and μ a positive measure on G. Suppose that for some p, $1 \leqslant p \leqslant \infty$, the functions $\mu \star f$ are in $L^p(G)$ for all $f \in \mathscr{K}(G)$. Then the operator $A_\mu : f \to \mu \star f$ is bounded in $L^p(G)$ if and only if the following condition holds: (♯) for every sequence $(f_n)_{n \geqslant 1}$ in $\mathscr{K}(G)$ such that $|f_n| \leqslant g$, $n \geqslant 1$, where $g \in L^p(G)$, we have $\sup_{n \geqslant 1} \|A_\mu f_n\|_p < \infty$.

If A_μ is L^p-bounded on $\mathscr{K}(G)$, it has a continuous extension to $L^p(G)$ and (♯) holds. Now let (♯) hold. Suppose that there is a sequence $(f_n)_{n \geqslant 1}$ in $\mathscr{K}(G)$ with $\|f_n\|_p \leqslant 1$ for all $n \geqslant 1$ and $\|A_\mu f_n\|_p \to \infty$ $(n \to \infty)$. Then clearly there is also a sequence $(g_n)_{n \geqslant 1}$ in $\mathscr{K}(G)$ such that $\sum_{n \geqslant 1} \|g_n\|_p < \infty$ and $\|A_\mu g_n\|_p \to \infty$ $(n \to \infty)$. Now $g = \sum_{n \geqslant 1} |g_n|$ is in $L^p(G)$ and $|g_n| \leqslant g$, $n \geqslant 1$; thus $\|A_\mu g_n\|_p \leqslant C < \infty$, $n \geqslant 1$, a contradiction. Hence A_μ is bounded.

We note that the particular form of the operator A_μ did not enter into the argument—only the fact that $f \in \mathscr{K}(G) \Rightarrow A_\mu f \in L^p(G)$ and $f \geqslant 0 \Rightarrow A_\mu f \geqslant 0$. Thus the remark above holds for such operators in general.

Dieudonné [34] has shown that on F_2, the free non-abelian group with two generators, there is a positive measure μ such that $\|\mu\| = 3$, but the corresponding operator A_μ in $L^2(F_2)$ has norm $\sqrt{8}$. Thus *the discrete group F_2 does not have the property P_1*. We shall later study the question for which l.c. groups the property P_1 holds (§ 7).

3.5. Let us now consider another property of groups, of importance in the theory of group representations. It was introduced by Godement

† Compare [87].

[55, Problems 4 and 5], but its origin goes back to A. Weil's book [138, p. 60].

We say that a l.c. group G has the *property* P' if the constant 1 can be approximated uniformly on compact sets by functions of the form $k \star \tilde{k}$, where $k \in \mathscr{K}(G)$ and $\tilde{k}(x) = \overline{k(x^{-1})}$. We shall discuss the significance of this property presently, but first we mention a *second equivalent formulation of* P' which is useful in the applications: G has the property P' if the constant 1 can be approximated uniformly on compact sets by positive-definite functions in $\mathscr{K}(G)$.

The equivalence with the first definition follows from an important theorem of Godement [55, Theorem 17]: this theorem implies that a positive-definite function in $\mathscr{K}(G)$ can be approximated uniformly by functions $k \star \tilde{k}$, with $k \in \mathscr{K}(G)$.

From the second definition of P' it is clear that, *if G has the property P', then so has every closed subgroup of G* [78, Proposition 7.4].

If G has the property P', then *every* continuous positive-definite function on G can be approximated uniformly on compact sets by functions of the form $k \star \tilde{k}$, $k \in \mathscr{K}(G)$, which are positive-definite functions of a particularly simple type [55, p. 77].

Let ϕ be a continuous positive-definite function on G. If $1-k \star \tilde{k}$, $k \in \mathscr{K}(G)$, is small on a given compact set, then so is $\phi-(k \star \tilde{k})\phi$; moreover, $(k \star \tilde{k})\phi$ is positive-definite [55, Theorem 13] and in $\mathscr{K}(G)$, hence can be approximated uniformly by $k_1 \star \tilde{k}_1$, $k_1 \in \mathscr{K}(G)$.

Thus P' means, in effect, that the structure of the continuous positive-definite functions on G is rather simple. For the connexion of P' with the unitary representations of G see [55, p. 76, footnote], [132, pp. 153–4], [44, Chap. I], [38, § 18].

3.6. It is a surprising fact that *the properties P' and P_1 are equivalent*, but the proof is very simple [115].

We shall prove $P' \Leftrightarrow P_2$; as $P_2 \Leftrightarrow P_1$ (§ 3.2), this suffices. Let any compact set $K \subset G$ and $\epsilon > 0$ be given.

To prove $P' \Rightarrow P_2$, take $k \in \mathscr{K}(G)$ such that $|1-k \star \tilde{k}(y)| < \epsilon$, $y \in K$; we can clearly choose k so that also $k \star \tilde{k}(e) = 1$. Then

$$\|L_y k - k\|_2^2 = 2-2\,\mathrm{Re}\{k \star \tilde{k}(y)\} \leqslant 2\,.\,|1-k \star \tilde{k}(y)| < 2\epsilon \quad (y \in K)$$

and we obtain P_2 by putting $s = |k|$.

To prove $P_2 \Rightarrow P'$, take $s \in \mathscr{K}(G)$ such that $s \geqslant 0$, $\|s\|_2 = 1$ and $\|L_y s - s\|_2 < \epsilon$ $(y \in K)$; then

$$|1-s \star \tilde{s}(y)| = |s \star \tilde{s}(e)-s \star \tilde{s}(y)| \leqslant \|s\|_2 \|s-L_y s\|_2 < \epsilon \quad (y \in K).$$

3.7. As a first application of the equivalence of P' and P_1 (§ 3.6) we now prove the following converse of 3.4. Let G be a l.c. group. For

$f \in L^1(G)$ and any p, $1 \leqslant p \leqslant \infty$, let $A_{f,p}$ be the operator $g \to f \star g$, $g \in L^p(G)$, and $|A_{f,p}|$ its norm. *If*† *for some* p, $1 < p < \infty$, *we have*

$$(1) \qquad |A_{f,p}| = \int f(x)\,dx \text{ for all POSITIVE } f \in L^1(G),$$

then G *has the property* P_1. We note that for $p = 1$ and $p = \infty$ the relation (1) is obviously true for *all* l.c. groups. We also remark that $|A_{f,p}|$ may be considered as a new norm on $L^1(G)$.

The proof is in three parts:

(i) If (1) holds for *some* $p = p_0$, $1 < p_0 < \infty$, then (1) holds for *all* p, $1 \leqslant p \leqslant \infty$.

(ii) For every fixed p, $1 \leqslant p \leqslant \infty$, the relation (1) implies

$$(2) \qquad |A_{f,p}| \geqslant \left|\int f(x)\,dx\right| \quad \text{for ALL } f \in L^1(G).$$

(iii) If the relation (2) holds for $p = 2$, then G has the property P'.

From (i), (ii), (iii) it will follow that, for *any* p, $1 < p < \infty$, the relation (1) implies P' and hence P_1; the proof of (i)–(iii) is given below.

(i) results from the fact that $\log|A_{f,1/\lambda}|$ is a convex function of λ, $0 \leqslant \lambda \leqslant 1$; cf. [33, pp. 283–4] and [41].

(ii) First let $f \in L^1(G)$ be *real*; put $f = g-h$ ($g \geqslant 0$, $h \geqslant 0$). Then, if (1) holds,

$$\left|\int f\right| = \left|\int g - \int h\right| = \big||A_{g,p}|-|A_{h,p}|\big| \leqslant |A_{g,p}-A_{h,p}| = |A_{f,p}|.$$

Now consider the normed algebra $\mathscr{A}_p$ of all operators $A_{f,p}$, $f \in L^1(G)$, with norm $|A_{f,p}|$. We define a real-linear functional ρ on $\mathscr{A}_p$ by putting $\rho(A_{f,p}) = \int \mathrm{Re}(f)$: this functional has norm 1, for by the inequality above we have

$$|\rho(A_{f,p})| \leqslant |A_{\mathrm{Re}(f),p}| = |A_{\frac{1}{2}(f+\bar{f}),p}| \leqslant \tfrac{1}{2}(|A_{f,p}|+|A_{\bar{f},p}|) = |A_{f,p}|.$$

But then the complex linear functional κ defined by $\kappa(A_{f,p}) = \rho(A_{f,p}) - i\rho(iA_{f,p})$ also has norm 1, as is well known, and $\kappa(A_{f,p}) = \int f$ for all $f \in L^1(G)$.

(iii) results from a theorem of Takenouchi [132, §§ 1.3, 1.4, and 2.2, especially Theorem 1′ (applied to the constant function 1)]; [38, §§ 18.1.4 and 18.3.5].

3.8. *If a l.c. group* G *has the property* P_1, *then so has every closed subgroup and every quotient group of* G. For quotient groups this results almost immediately from the very definition of P_1 and from Weil's formula (cf. also Chap. 3, § 1.8 (ii)). For subgroups it follows via the equivalence of P_1 with P' (§ 3.6), since it is true for P' (§ 3.5).‡

4. The invariant convex hull of a function in $L^1(G)$

4.1. Let H be a closed subgroup of G (which may coincide with G). Given $f \in L^1(G)$, consider the finite linear combinations $\sum_n c_n A_{\xi_n} f$, where

† Cf. H. Leptin [88]. The author is indebted to Professor Leptin for an advance copy of the manuscript prior to publication and for permission to include the above result here.

‡ A direct proof for P_1 is more subtle and has been published only for the case of discrete subgroups [114]; see also 5.5 (i) and § 6.

$A_{\xi_n}f(x) = f(x\xi_n)\Delta_G(\xi_n)$, with $\xi_n \in H$ and coefficients $c_n > 0$ satisfying $\sum\limits_n c_n = 1$. We denote the set of all these linear combinations by $C_H(f)$: this is the smallest convex set in $L^1(G)$ containing f and invariant under all operators A_ξ, $\xi \in H$.

For a certain class of subgroups H the distance of $C_H(f)$ from the 'origin' in $L^1(G)$ can be given explicitly: *if the closed subgroup H of G has the property P_1 (§ 3) then for every $f \in L^1(G)$ we have*

$$\inf \Big\| \sum_n c_n A_{\xi_n} f \Big\|_1 = \|T_{H,q} f\|_1$$

or, explicitly,

$$\inf \int_G \Big| \sum_n c_n f(x\xi_n)\Delta_G(\xi_n) \Big| \, dx = \int_{G/H} \Big| \int_H \frac{f(x\xi)}{q(x\xi)} \, d\xi \Big| \, d_q \dot{x}, \tag{1}$$

the infimum being taken over all finite sums with elements $\xi_n \in H$ and coefficients $c_n > 0$ satisfying

$$\sum_n c_n = 1. \tag{2}$$

The function q appearing on the right in (1) is any strictly positive continuous solution of the functional equation 1.2 (3) and $d_q \dot{x}$ is the corresponding quasi-invariant measure on the quotient space G/H (cf. § 1; also 2.3, Remark). The proof of (1) will be given in 4.2.

Remark. The relation (1) becomes trivial if the equality sign is replaced by '$\geqslant$'. Moreover, the coefficients c_n in (1) may be allowed to take arbitrary complex values, subject to (2).

We have $\|T_{H,q} g\|_1 \leqslant \|g\|_1$ for $g \in L^1(G)$ (cf. 2.3); put $g = \sum\limits_n c_n A_{\xi_n} f$ and observe that $T_{H,q} g = T_{H,q} f$, by 2.4 (9) and the condition (2) above.

Example. It is useful to consider (1) in more detail in the case of $G = SL(n, \mathbf{R})$, $H = ST_+(n, \mathbf{R})$ (cf. §§ 1.5 and 3.3, Example).

4.2. The proof of (1) proceeds in three stages.

(i) First we prove the following extension of Chapter 5, § 3.1. Let H be any closed subgroup of G with the property P_1; let $d_q \dot{x}$ be a quasi-invariant measure on G/H, corresponding to a strictly positive, continuous solution q of 1.2 (3). Then for any given $k \in \mathscr{K}(G)$ we have: for every $\epsilon > 0$ there is an $s \in \mathscr{K}(H)$ such that $s \geqslant 0$, $\int\limits_H s(\xi)\, d\xi = 1$ and

$$\int_G \Big| \int_H k(x\xi) s(\xi^{-1}) \, d\xi \Big| \, q(x) \, dx < \int_{G/H} \Big| \int_H k(x\xi) \, d\xi \Big| \, d_q \dot{x} + \epsilon. \tag{3}$$

(ii) From (3) we can deduce the relation (1) for functions in $\mathscr{K}(G)$.

(iii) By an approximation argument we obtain (1) for all $f \in L^1(G)$.

(i) Since H has the property P_1, we may apply Chapter 5, § 3.2 to the family

$$\mathfrak{F}_H = \left\{ s \mid s \in \mathscr{K}(H),\ s \geqslant 0,\ \int_H s(\xi)\,d\xi = 1 \right\}$$

and obtain: given $k \in \mathscr{K}(G)$ vanishing outside a compact set K, say, there is for every $\epsilon > 0$ an $s \in \mathscr{K}(H)$ such that $s \geqslant 0$, $\int_H s(\xi)\,d\xi = 1$ and (cf. Chap. 5, § 3.2 (2))

$$(*) \qquad \int_H \left| \int_H k(x\eta\xi)s(\xi^{-1})\,d\xi \right| d\eta \leqslant \left| \int_H k(x\xi)\,d\xi \right| + \epsilon . \phi_{KH}(x),$$

for all $x \in G$. Then we proceed exactly as in Chapter 5, § 3.3, but use the quasi-invariant measure $d_q \dot{x}$ on G/H instead of Haar measure, and 1.2 (4) instead of Weil's formula: thus, integrating (*) over G/H, we obtain (3), after a change of ϵ in (*).

(ii) Replacing k in (3) by k/q and applying the functional equation 1.2 (3), we obtain: there is an $s \in \mathscr{K}(H)$ such that $s \geqslant 0$, $\int_H s(\xi)\,d\xi = 1$ and

$$\int_G \left| \int_H k(x\xi)\Delta_G(\xi)s(\xi^{-1})\Delta_H(\xi^{-1})\,d\xi \right| dx < \int_{G/H} |T_{H,q}k(\dot{x})|\,d_q\dot{x} + \epsilon.$$

Simplifying the left-hand side, we get

$$\int_G \left| \int_H k(x\xi^{-1})\Delta_G(\xi^{-1})s(\xi)\,d\xi \right| dx < \int_{G/H} |T_{H,q}k(\dot{x})|\,d_q\dot{x} + \epsilon.$$

The relation (1) now follows, for $f = k \in \mathscr{K}(G)$, if we apply Chapter 3, § 5.8 (19) (cf. also the remark in 4.1 above).

(iii) Given $f \in L^1(G)$ and $\epsilon > 0$, take $k \in \mathscr{K}(G)$ such that $\|f-k\|_1 < \epsilon$. There is a linear combination $\sum_n c_n A_{\xi_n} k$ ($c_n > 0$, $\sum_n c_n = 1$, $\xi_n \in H$) such that

$$\left\| \sum_n c_n A_{\xi_n} k \right\|_1 < \|T_{H,q}k\|_1 + \epsilon.$$

Then

$$\left\| \sum_n c_n A_{\xi_n} f \right\|_1 \leqslant \left\| \sum_n c_n A_{\xi_n} k \right\|_1 + \|f-k\|_1 < \|T_{H,q}k\|_1 + 2\epsilon < \|T_{H,q}f\|_1 + 3\epsilon.$$

The proof of (1), as given here, is entirely elementary, in the sense that it uses only the basic properties of Haar measure (and quasi-invariant measures).

The formula (1) is an extension of various results given in [109, Theorem 1.1], [110, II], [112], and [53, relation (4.7)]. Some applications will be considered in 4.6 and a more abstract version will be discussed in § 6.

4.3. If H is a closed *normal* subgroup with the property P_1, then not only does the formula (1) hold (with $q = 1$), but also the following one:

$$(4) \qquad \inf \int_G \left| \sum_n c_n f(\xi_n^{-1}x) \right| dx = \int_{G/H} \left| \int_H f(x\xi)\,d\xi \right| d\dot{x},$$

where the infimum is taken in the same way as in (1) and $d\dot{x}$ is the Haar measure on G/H, normalized according to Weil's formula.

Apply (1) to f^*: $$\inf\left\|\sum_n c_n A_{\xi_n}(f^*)\right\|_1 = \|T_H(f^*)\|_1.$$

Now $A_{\xi_n}(f^*) = [L_{\xi_n}f]^*$, by Chapter 3, § 5.5 ($\lambda_7$), thus

$$\left\|\sum_n c_n A_{\xi_n}(f^*)\right\|_1 = \left\|\left(\sum_n c_n L_{\xi_n}f\right)^*\right\|_1 = \left\|\sum_n c_n L_{\xi_n}f\right\|_1,$$

and by Chapter 3, § 5.3 (8) we obtain (4).

4.4. As a particular case of (1) we have, on taking $H = G$: if G is any l.c. group with the property P_1, then for each $f \in L^1(G)$

(5) $$\inf \int_G \left|\sum_n c_n f(xy_n)\Delta_G(y_n)\right| dx = \left|\int_G f(x)\,dx\right|,$$

the infimum being taken over all finite sums with $c_n > 0$, $\sum_n c_n = 1$, $y_n \in G$.

The proof of (5) is of course much simpler than that of the general formula (1) which is a relativization (cf. Chap. 4, § 5) of (5).†

4.5. We now show that, conversely, *if a l.c. group G is such that* (5) *holds for all $f \in L^1(G)$, then G has the property P_1.*

Let a compact set $K \subset G$ and $\epsilon > 0$ be given. Choose any $h \in L^1(G)$ such that $h \geqslant 0$ and $\int h = 1$. There is an open nd. U of e such that

(*) $$\|L_y h - h\|_1 < \epsilon \qquad y \in U.$$

Next there are finitely many points of K, say $(a_m)_{1\leqslant m\leqslant M}$, such that the translates $(a_m U)_{1\leqslant m\leqslant M}$ cover K. Now consider the M functions $L_{a_m}h - h$; we shall show: there are finitely many numbers $c_n > 0$ satisfying $\sum_n c_n = 1$ and elements $y_n \in G$ such that for $A = \sum_n c_n A_{y_n}$ (cf. § 2.4 (8)) we have

(**) $$\|A(L_{a_m}h - h)\|_1 < \epsilon \qquad 1 \leqslant m \leqslant M.$$

If $M = 1$, this is true by hypothesis, since $\int (L_{a_1}h - h) = 0$. Now we use induction: we put for simplicity of notation $L_{a_m}h - h = f_m$, $1 \leqslant m \leqslant M$, and suppose that for $1 \leqslant m \leqslant M-1$ we have $\|A'f_m\|_1 < \epsilon$ for some $A' = \sum_j c'_j A_{y'_j}$ with $c'_j > 0$, $\sum_j c'_j = 1$, $y'_j \in G$. Put $g = A'f_M$. We have $\int g = 0$, thus there is an $A'' = \sum_k d_k A_{z_k}$ with $d_k > 0$, $\sum_k d_k = 1$, $z_k \in G$, such that $\|A''g\|_1 < \epsilon$.

Put $A = A''A'$. Then $\|Af_m\|_1 < \epsilon$ for $1 \leqslant m \leqslant M$ (for $1 \leqslant m \leqslant M-1$ we observe that $\|Af_m\|_1 \leqslant \|A'f_m\|_1$); also

$$A = \sum_n c_n A_{y_n}, \quad \text{where} \quad c_n > 0,\ \sum_n c_n = 1,\ y_n \in G.$$

† It is instructive to consider the details of the simplification arising in 4.2 when $H = G$.

Now we put $s = Ah$. Then $s \geqslant 0$, $\int s = 1$ and (as we shall verify)

$$(\ast_{\ast\ast}) \qquad \|L_y s - s\|_1 < 2\epsilon \qquad y \in K.$$

Indeed, if $y \in K$, we can put $y = a_m y'$ for some m, $1 \leqslant m \leqslant M$, and some $y' \in U$, with U as above. Then we can write

$$L_y s - s = L_{a_m} L_{y'} Ah - Ah = L_{a_m} A(L_{y'} h - h) + A(L_{a_m} h - h),$$

since L and A commute. Also L is isometric and A is a contraction, thus (*) and (**) give

$$\|L_y s - s\|_1 \leqslant \|L_{y'} h - h\|_1 + \|A(L_{a_m} h - h)\|_1 < 2\epsilon,$$

which proves $(\ast_{\ast\ast})$ and hence shows that P_1 holds for G.

Thus P_1 *is equivalent to the validity of the relation* (5) *for all* $f \in L^1(G)$. This fact will be of importance later (§§ 6 and 7).

4.6. We now discuss some applications of the relation (1). Consider any l.c. group G.

(i) Let H be a closed subgroup having the property P_1. Let I be a *closed* linear subspace of $L^1(G)$, right invariant under H (i.e. if $f \in I$, then $A_\xi f \in I$ for all $\xi \in H$). Then the image of I under the map $T_{H,q}$ (§ 2) is a *closed* linear subspace of $L^1(G/H)$.

Let $I' = T_{H,q}(I)$ and suppose that $f' \in L^1(G/H)$ is in the closure of I'. Then there is a sequence $(f'_n)_{n \geqslant 1}$ in I' such that $\|f'_n\|_1 > 0$, $\sum_n \|f'_n\|_1 < \infty$, and $\sum_n f'_n = f'$. Let $f_n \in I$ be such that $T_{H,q} f_n = f'_n$. Since H has the property P_1, there are, by (1), linear combinations $g_n = \sum_j c_{j,n} A_{\xi_{j,n}} f_n$ such that $\sum_j c_{j,n} = 1$ and

$$\|g_n\|_1 < 2\|f'_n\|_1 \qquad n = 1, 2, \ldots.$$

Now g_n is in I and (cf. § 2.4 (9)) $T_{H,q} g_n = f'_n$. Put $f = \sum_n g_n$ (a convergent series!); then $f \in I$ and $T_{H,q} f = f'$, thus $f' \in I'$.

(ii) If H is a closed *normal* subgroup with the property P_1, then the image of a *closed* right [left] *ideal* of $L^1(G)$ under the morphism T_H (Chap. 3, § 5.3) is a *closed* right [left] *ideal* of $L^1(G/H)$.

Let $I \subset L^1(G)$ be any right [left] ideal and put $I' = T_H(I)$: then I' is a right [left] ideal of $L^1(G/H)$, since T_H is surjective. That I' is closed if I is closed follows from (i) for right ideals. For left ideals we then get it by involution which interchanges closed left and closed right ideals and also satisfies $T_H(I^*) = (T_H I)^*$ (Chap. 3, §§ 5.2 and 5.3 (8)).

(iii) Let H_1, H_2 be closed subgroups of G. Suppose that one of them is normal and one has the property P_1. Then

$$(6) \qquad J^1(G, \overline{H_1 H_2}) = J^1(G, H_1) + J^1(G, H_2),$$

where $J^1(G, H)$ denotes the kernel of the map $T_{H,q}$ (§ 2.3).

First we show that $J^1(G, H_1) + J^1(G, H_2)$ is closed in $L^1(G)$. If, say, H_1 is normal and has the property P_1 as well, then by (ii) the image of the closed left ideal

$J^1(G, H_2)$ under the map T_{H_1} is closed in $L^1(G/H_1)$ and hence so is its pre-image in $L^1(G)$ which is $J^1(G, H_1)+J^1(G, H_2)$. But if, say, H_1 is normal and it is H_2 which has the property P_1, then we observe that $J^1(G, H_1)$ is also right invariant (Chap. 3, §§ 5.3 and 5.7) and apply (i) to the mapping $T_{H_2,q}$: this shows again that the right-hand side of (6) is closed.

Now we can prove (6) as follows. The kernel $J^1(G, H)$ has been characterized in another way in 2.5 which yields at once $J^1(G, H_j) \subset J^1(G, \overline{H_1 H_2})$, $j = 1, 2$. Hence also

$$J^1(G, H_1)+J^1(G, H_2) \subset J^1(G, \overline{H_1 H_2}).$$

To prove the opposite inclusion, we note that for $f \in L^1(G)$ and $\alpha \in H_1$, $\beta \in H_2$,

$$A_{\alpha\beta}f-f = \{A_\alpha A_\beta f-A_\beta f\}+\{A_\beta f-f\} \in J^1(G, H_1)+J^1(G, H_2).$$

It follows that $A_\gamma f-f$ is in $J^1(G, H_1)+J^1(G, H_2)$ for *every* $\gamma \in \overline{H_1 H_2}$, since $\gamma \to A_\gamma f$ is continuous and $J^1(G, H_1)+J^1(G, H_2)$ is closed. Hence we get

$$J^1(G, \overline{H_1 H_2}) \subset J^1(G, H_1)+J^1(G, H_2)$$

which completes the proof.

It is an unsolved problem whether the results above are true for subgroups that do not possess the property P_1; this problem is of particular significance in the case of (ii).

References. (i) and (ii) go back to the final remarks in [110, II, p. 180]; cf. also [111, VI, Lemma 1.1].

5. The property ($\mathscr{M}$)

5.1. Let G be a topological group. We denote, as usual, by $\mathscr{C}^b_{\mathbf{R}}(G)$ [$\mathscr{C}^b_{\mathbf{C}}(G)$] the space of real [complex] continuous bounded functions on G and use $\mathscr{C}^b(G)$ for either $\mathscr{C}^b_{\mathbf{R}}(G)$ or $\mathscr{C}^b_{\mathbf{C}}(G)$.

A *mean* $\mathscr{M}$ on $\mathscr{C}^b(G)$ is a linear functional $\phi \to \mathscr{M}\{\phi\}$ (also denoted by $\mathscr{M}_x\{\phi(x)\}$), $\phi \in \mathscr{C}^b(G)$, such that $\mathscr{M}\{\phi\} \geqslant 0$ if ϕ is real and $\phi \geqslant 0$, and $\mathscr{M}\{1\} = 1$. We call $\mathscr{M}$ *left invariant* if $\mathscr{M}_x\{\phi(ax)\} = \mathscr{M}_x\{\phi(x)\}$, *right invariant* if $\mathscr{M}_x\{\phi(xa)\} = \mathscr{M}_x\{\phi(x)\}$, for all $a \in G$, and *bi-invariant* if $\mathscr{M}$ is both left and right invariant.

The definition of a mean is entirely analogous to that of a positive measure (Chap. 3, § 2.1 (i)–(v)) and we can show in the same way as before: $|\mathscr{M}\{\phi\}| \leqslant \mathscr{M}\{|\phi|\}$ and hence $|\mathscr{M}\{\phi\}| \leqslant \|\phi\|_\infty$ for all $\phi \in \mathscr{C}^b(G)$. Also a mean defined on $\mathscr{C}^b_{\mathbf{R}}(G)$ may be uniquely extended to a mean on $\mathscr{C}^b_{\mathbf{C}}(G)$ in the obvious way.

A left invariant mean is analogous to left Haar measure and coincides with it for compact groups. The essential difference, if G is a locally compact, but not compact, group, lies in the class of functions for which the mean is defined: in fact, for such a G any left invariant mean vanishes for all functions in $\mathscr{K}(G)$.

Cf. the proof in Chapter 3, § 3.1 (iv). Thus, while the restriction of *any* mean $\mathscr{M}$

to $\mathscr{K}(G)$ is, of course, a bounded positive measure μ, it is not necessarily true that $\int \phi \, d\mu = \mathscr{M}\{\phi\}$ for all $\phi \in \mathscr{C}^b(G)$.

5.2. For any topological group the existence of a left invariant mean implies the existence of a right invariant mean and vice versa: if $\mathscr{M}$ is left [right] invariant, then $\phi \to \mathscr{M}\{\check{\phi}\}$ is right [left] invariant. Thus we may say simply that a topological group is an ($\mathscr{M}$)-*group*, or has the *property* ($\mathscr{M}$), if there exists either a left or a right invariant mean on $\mathscr{C}^b(G)$. If a mean $\mathscr{M}$ is bi-invariant, then $\phi \to \mathscr{M}\{\frac{1}{2}(\phi+\check{\phi})\}$, $\phi \in \mathscr{C}^b(G)$, is a bi-invariant mean which is also invariant under $x \to x^{-1}$.

5.3. If G is a l.c. group, it is convenient, in practice, to consider means on $L^\infty(G)$ rather than on $\mathscr{C}^b(G)$: a mean on $L^\infty(G)$ is defined as for $\mathscr{C}^b(G)$; we note that $\mathscr{M}\{\phi\} = 0$ if $\phi(x) = 0$ l.a.e. The definitions and properties in 5.1 remain unchanged; in particular we note that $|\mathscr{M}\{\phi\}| \leqslant \|\phi\|_\infty$ for $\phi \in L^\infty(G)$.

Let G be a topological group with the property ($\mathscr{M}$) (§ 5.2): *if G is locally compact, then there exists also a left* [*or right*] *invariant mean on* $L^\infty(G)$.

Choose $u \in \mathscr{K}_+(G)$ such that $\int u = 1$. Let $\mathscr{M}$ be a left invariant mean on $\mathscr{C}^b(G)$ and put

$$\mathscr{M}'\{\phi\} = \mathscr{M}_x\Big\{\int_G \phi(xy)u(y)\,dy\Big\} \qquad \phi \in L^\infty(G).$$

Then $\mathscr{M}'$ is a left invariant mean on $L^\infty(G)$.

5.4. We shall now prove that, *if a l.c. group G has the property P_1* (§ 3.1), *then G is an* ($\mathscr{M}$)-*group. Moreover, there exists a bi-invariant mean $\mathscr{M}$ on $L^\infty(G)$ with the additional property that $\mathscr{M}\{\check{\phi}\} = \mathscr{M}\{\phi\}$ and*

(1) $\qquad \mathscr{M}\{f \star \phi\} = \int f(x)\,dx \,.\, \mathscr{M}\{\phi\}$ *for all $f \in L^1(G)$, $\phi \in L^\infty(G)$.*

Let $\mathfrak{F} = \{s \mid s \in L^1(G),\ s \geqslant 0,\ \int s = 1\}$ and put for each *symmetric* compact set $K \subset G$ and each $\epsilon > 0$

$$\mathfrak{F}_{K,\epsilon} = \{s \mid s \in \mathfrak{F},\ \|L_a s - s\|_1 < \epsilon \text{ for all } a \in K\}.$$

Let $\mathfrak{A}_{K,\epsilon}$ be the family of functionals

$$\phi \to \iint \phi(xy^{-1})s(x)s(y)\,dxdy \qquad s \in \mathfrak{F}_{K,\epsilon}$$

on $L^\infty(G)$ and let $\mathfrak{B}_{K,\epsilon}$ be the *closure* of $\mathfrak{A}_{K,\epsilon}$ in $L^\infty(G)'$, the dual of $L^\infty(G)$, *in the* $\sigma((L^\infty)', L^\infty)$-*topology*. Each functional $\mathscr{M}' \in \mathfrak{A}_{K,\epsilon}$ is a mean on $L^\infty(G)$ and satisfies for all $a \in K = K^{-1}$

(*) $\qquad |\mathscr{M}'\{{}_a\phi\} - \mathscr{M}'\{\phi\}| \leqslant \epsilon\|\phi\|_\infty, \qquad |\mathscr{M}'\{\phi_a\} - \mathscr{M}'\{\phi\}| \leqslant \epsilon\|\phi\|_\infty,$

where ${}_a\phi(t) = \phi(at)$, $\phi_a(t) = \phi(ta)$, $t \in G$. By continuity (*) holds also for all $\mathscr{M}' \in \mathfrak{B}_{K,\epsilon}$. The closed sets $\mathfrak{B}_{K,\epsilon}$ are all contained in the closed unit ball of $L^\infty(G)'$ which is compact in the topology considered. The intersection of finitely many sets $\mathfrak{B}_{K,\epsilon}$ is nonempty, since this is true for the sets $\mathfrak{A}_{K,\epsilon}$. Hence

$$\bigcap_{K,\epsilon} \mathfrak{B}_{K,\epsilon} \quad (K = K^{-1} \subset G,\ K \text{ compact},\ \epsilon > 0)$$

is nonempty. Any functional $\mathscr{M}$ in this intersection is clearly a bi-invariant mean on $L^\infty(G)$; it also satisfies $\mathscr{M}\{\check{\phi}\} = \mathscr{M}\{\phi\}$, since this holds for each $\mathscr{M}'$.

To show that $\mathscr{M}$ also satisfies (1), suppose first that $f \in L^1(G)$ vanishes outside some *symmetric* compact set C. Given $\phi \in L^\infty(G)$, put $\psi = f \star \phi$ and choose any $\epsilon > 0$. There is an $s \in L^1(G)$ such that $s \geqslant 0$, $\int s = 1$ and

$$(\alpha) \qquad \|L_a s - s\|_1 < \epsilon \quad \text{for all } a \in C,$$

$$(\beta_1) \qquad \left|\iint \phi(xy^{-1})s(x)s(y)\,dxdy - \mathscr{M}\{\phi\}\right| < \epsilon,$$

$$(\beta_2) \qquad \left|\iint \psi(xy^{-1})s(x)s(y)\,dxdy - \mathscr{M}\{\psi\}\right| < \epsilon.$$

(Observe that $\mathscr{M} \in \mathfrak{B}_{C,\epsilon}$, thus each nd. of $\mathscr{M}$ contains a member of $\mathfrak{A}_{C,\epsilon}$; we take the nd. $U_{\phi,\psi;\epsilon}$ of $\mathscr{M}$.) By (α) and (β_1),

$$\left|\iint \phi(axy^{-1})s(x)s(y)\,dxdy - \mathscr{M}\{\phi\}\right| < \epsilon + \epsilon.\|\phi\|_\infty \qquad \text{for all } a \in C.$$

Hence, since f vanishes outside $C = C^{-1}$,

$$\left|\int f(t)\left\{\iint \phi(t^{-1}xy^{-1})s(x)s(y)\,dxdy\right\}dt - \int f(t)\,dt.\mathscr{M}\{\phi\}\right| < \epsilon.(1+\|\phi\|_\infty).\|f\|_1$$

or

$$\left|\iint \psi(xy^{-1})s(x)s(y)\,dxdy - \int f(t)\,dt.\mathscr{M}\{\phi\}\right| < \epsilon.(1+\|\phi\|_\infty).\|f\|_1$$

and this, combined with (β_2), yields (1) for all $f \in L^1(G)$ with compact support, since $\epsilon > 0$ was arbitrary. Then (1) follows for all $f \in L^1(G)$ by approximation.

We will show later that the converse also holds (§ 6).

5.5. (i) If a l.c. group G has the property $(\mathscr{M})$, then so has every quotient group and every closed subgroup of G.

(ii) If a l.c. group G has a closed normal subgroup H such that H and G/H have the property $(\mathscr{M})$, then so has G.

(i) The case of a quotient group is obvious (and applies even to topological groups in general): if $\mathscr{M}$ is a left invariant mean on $\mathscr{C}^b(G)$, then $\phi \to \mathscr{M}\{\phi \circ \pi_H\}$, $\phi \in \mathscr{C}^b(G/H)$, is a left invariant mean on $\mathscr{C}^b(G/H)$. Now let H be any closed subgroup of G. Let β be a Bruhat function on G for H (§ 1.9). For $\phi \in \mathscr{C}^b(H)$ define $\tau_H\phi$ by $\tau_H\phi(x) = \int_H \beta(x^{-1}\xi)\phi(\xi)\,d\xi$, $x \in G$. Then $\tau_H\phi$ is in $\mathscr{C}^b(G)$: this follows from the properties of β (cf. § 1.9, Remark) and the lemma in Chapter 3, § 3.2. Also $\tau_H L_\eta\phi = L_\eta\tau_H\phi$, $\eta \in H$. It is then readily verified that $\phi \to \mathscr{M}\{\tau_H\phi\}$ is a left invariant mean on $\mathscr{C}^b(H)$.

(ii) Let $\phi \in L^\infty(G)$ be given. For any $f \in L^1(G)$ the function $\xi \to \langle L_\xi f, \phi\rangle$, $\xi \in H$, is in $\mathscr{C}^b(H)$. Let $\mathscr{M}^H$ be a *right* invariant mean on $\mathscr{C}^b(H)$. Then $f \to \mathscr{M}^H_\xi\{\langle L_\xi f, \phi\rangle\}$ is a continuous linear functional on $L^1(G)$. Hence there is a (unique) $\phi' \in L^\infty(G)$ such that

$$(*) \qquad \mathscr{M}^H_\xi\{\langle L_\xi f, \phi\rangle\} = \langle f, \phi'\rangle \qquad \text{for all } f \in L^1(G).$$

Moreover, the left-hand side of (*) is invariant under $f \to L_\eta f$, $\eta \in H$, since $\mathscr{M}^H$ is right invariant. Hence $\langle L_\eta f, \phi'\rangle = \langle f, \phi'\rangle$ for all $f \in L^1(G)$, or $\phi'(\eta x) = \phi'(x)$ l.a.e., for each $\eta \in H$. Thus ϕ' is (l.a.e. equal to) an H-periodic function: $\phi' = \dot{\phi} \circ \pi_H$, where $\dot{\phi} \in L^\infty(G/H)$ (Chap. 3, §§ 6.5 and 3.9, Cor.). Thus we obtain a map $\phi \to \dot{\phi}$ of $L^\infty(G)$ onto $L^\infty(G/H)$ with the following properties: (i) it is linear; (ii) if $\phi(x) \geqslant 0$ l.a.e. on G, then $\dot{\phi}(\dot{x}) \geqslant 0$ l.a.e. on G/H; (iii) if $\phi = 1$, then $\dot{\phi} = 1$; (iv) $[\phi_a]^{\cdot} = \dot{\phi}_{\pi_H(a)}$

for $a \in G$, where $\phi_a(x) = \phi(xa)$ (this is seen by replacing f in (*) by $R_a f$). Now let $\mathscr{M}^{G/H}$ be a *right* invariant mean on $L^\infty(G/H)$. Then (i)–(iv) show that

$$\mathscr{M}\{\phi\} = \mathscr{M}^{G/H}\{\dot{\phi}\} \qquad \phi \in L^\infty(G)$$

is a right invariant mean on $L^\infty(G)$.

REMARK. The map $\phi \to \dot{\phi}$ is the analogue for $L^\infty(G)$ of T_H and there is likewise an analogy between the construction of $\mathscr{M}$ and Weil's formula (cf. Chap. 3, §§ 3.2 (1) and 3.3 (3)).

5.6. For the theory of invariant means on topological groups see [36]. For discrete abelian groups the existence of an invariant mean value can be established in a very simple way and it easily follows that discrete soluble groups are ($\mathscr{M}$)-groups (cf. [103] and the references given there). It is also easy to show that on a discrete ($\mathscr{M}$)-group there exists even a bi-invariant mean (cf. e.g. [45, p. 243]). See also the references at the end of § 7.

The method used in 5.4 is essentially that of [109, Lemma 2.1.1]. On discrete ($\mathscr{M}$)-groups the relation (1) obviously holds for *every* left invariant mean. Invariant means are implicit in [110, II]; see also Chapter 7, § 4.8, footnote.

6. The equivalence of the properties ($\mathscr{M}$) and P_1

6.1. As we have seen in § 5, P_1 implies ($\mathscr{M}$); we shall now prove the converse. For this purpose we first establish a result, due to Glicksberg, on groups of contraction operators in Banach spaces.

(i) Let G be a l.c. group† with the property ($\mathscr{M}$).

Let $y \to A_y$, $y \in G$, be a weakly continuous representation of G by linear contraction operators on a (real or complex) Banach space B; explicitly:

(ii) $A_{y_1 y_2} = A_{y_1} A_{y_2}$ $(y_1, y_2 \in G)$;

(iii) $\|A_y f\| \leqslant \|f\|$ $(f \in B)$;

(iv) the function $y \to \langle A_y f, \phi\rangle$, $y \in G$, is continuous for each $f \in B$ and each $\phi \in B'$, the dual of B.

Define $C_G(f)$, for $f \in B$, as the convex set of all finite linear combinations $\sum_n c_n A_{y_n} f$ with $c_n > 0$, $\sum_n c_n = 1$, $y_n \in G$.

Let J_G be the closed linear subspace of B spanned by the family of *all* vectors $A_y g - g$ $(y \in G, g \in B)$.

Glicksberg [53, § 2.5] has shown that *under the conditions* (i)–(iv) *above the distance of* $C_G(f)$ *from the zero vector coincides with that of the linear*

† Actually it is enough to assume that G is a topological group, or semi-group (possibly without neutral element).

variety $f+J_G$, or

(1) $$\operatorname{dist}\{C_G(f), 0\} = \|Tf\|_{B/J_G} \qquad f \in B,$$

where T is the canonical map $B \to B/J_G$.

REMARK. (1) becomes trivial if '$=$' is replaced by '$\geqslant$', since $C_G(f) \subset f+J_G$.

The proof of (1) is based on the methods of functional analysis; it is a modification of that given by Glicksberg (loc. cit.).

6.2. LEMMA† 1. Let C be a convex set in a real or complex Banach space B and put $d = \operatorname{dist}\{C, 0\} = \inf_{v \in C} \|v\|$. If $d > 0$, then there is a functional $\phi \in B'$, the dual of B, such that

(2) (a) $\operatorname{Re}\langle v, \phi\rangle \geqslant 1$ for all $v \in C$; (b) $\|\phi\|_{B'} = 1/d$.

Here $\langle v, \phi\rangle$ is the value of ϕ for $v \in B$, 'Re' denotes the real part, and $\|.\|_{B'}$ is the norm in B'.

Assume first that B is real. Put $B_d = \{b \mid b \in B,\ \|b\| < d\}$. Then the sum $A = C+B_d$ is open, convex, and does not contain the vector zero. Hence there is a $\psi \in B'$ such that $\langle a, \psi\rangle > 0$ for all $a \in A$ [13 *b*, Chap. II, § 5, n° 1, Remarque 1]. Then $\langle c, \psi\rangle > |\langle b, \psi\rangle|$ for all $c \in C$, $b \in B_d$. Thus for each $c \in C$ we have

$$\langle c, \psi\rangle \geqslant \sup_{b \in B_d} |\langle b, \psi\rangle| = d\,.\,\|\psi\|_{B'}.$$

Put $\phi = \psi/(d\,.\,\|\psi\|_{B'})$: then $\langle c, \phi\rangle \geqslant 1$ for all $c \in C$, and $\|\phi\|_{B'} = 1/d$. Now, if B is complex, we can consider B also as a Banach space over **R**. Then, by the preceding, there is an **R**-linear continuous functional ϕ_r on B such that $\langle c, \phi_r\rangle \geqslant 1$ for all $c \in C$ and the norm of ϕ_r is $1/d$. As is well known, there is a **C**-linear continuous functional ϕ on B such that $\operatorname{Re}\phi = \phi_r$ and $\|\phi\|_{B'} = 1/d$.

Now suppose that the conditions 6.1 (i)–(iv) hold. Let us call a functional $\dot{\phi} \in B'$ *invariant* (with respect to the given family of operators A_y, $y \in G$) if, for every $v \in B$, we have $\langle A_y v, \dot{\phi}\rangle = \langle v, \dot{\phi}\rangle$ for all $y \in G$. Using Lemma 1, we can now prove:

LEMMA‡ 2. Under the conditions 6.1 (i)–(iv) the following holds. If $f \in B$ is such that $\inf \left\| \sum_n c_n A_{y_n} f \right\| = d > 0$, where the infimum is taken over all finite sums with $c_n > 0$, $\sum_n c_n = 1$, $y_n \in G$, then there exists an INVARIANT functional $\dot{\phi} \in B'$ such that

$$\langle f, \dot{\phi}\rangle = 1 \quad \text{and} \quad \|\dot{\phi}\|_{B'} = 1/d.$$

Let $\mathscr{M}$ be a *right* invariant mean on $\mathscr{C}^b(G)$. For fixed $v \in B$, $\phi \in B'$, the function $x \to \langle A_x v, \phi\rangle$, $x \in G$, is continuous; also $|\langle A_x v, \phi\rangle| \leqslant \|v\|\,.\,\|\phi\|_{B'}$ for all $x \in G$. Thus

† Cf. [53, p. 100, relation (2.6) and footnote 6].

‡ Cf. [53, Lemma 2.1]; the proof there is based on a theorem of M. M. Day.

$v \to \mathscr{M}_x\{\langle A_x v, \phi\rangle\}$ is a linear functional on B, and $|\mathscr{M}_x\{\langle A_x v, \phi\rangle\}| \leqslant \|\phi\|_{B'}.\|v\|$. Hence there is a $\dot\phi \in B'$ such that

$$(\alpha) \qquad \langle v, \dot\phi\rangle = \mathscr{M}_x\{\langle A_x v, \phi\rangle\} \quad \text{for all } v \in B;$$

$$(\beta) \qquad \|\dot\phi\|_{B'} \leqslant \|\phi\|_{B'}.$$

Moreover, if we replace v by $A_y v$ in (α), we obtain, since $\mathscr{M}$ is right invariant, $\langle A_y v, \dot\phi\rangle = \langle v, \dot\phi\rangle$ for all $y \in G$, that is, $\dot\phi$ is an *invariant* functional. We may express (α) by saying that the value of $\dot\phi$ at v is the mean value of ϕ along the orbit $A_x v$, $x \in G$. Thus to every $\phi \in B'$ there corresponds an *invariant* $\dot\phi \in B'$, defined by (α) (the same method was used in the proof of 5.5 (ii)).

By Lemma 1 there is a $\phi \in B'$ satisfying (2 a, b) for $C = C_G(f)$, as defined in 6.1. Consider the corresponding invariant functional $\dot\phi$: then

$$(*) \qquad \operatorname{Re}\langle c, \dot\phi\rangle \geqslant 1 \quad \text{for all } c \in C_G(f);$$

$$(**) \qquad \|\dot\phi\|_{B'} \leqslant 1/d.$$

((*) follows from (2 a), with $C = C_G(f)$, and (α); (**) results from (2 b) and (β).) Thus for $c \in C_G(f)$ we have

$$1 \leqslant \operatorname{Re}\langle c, \dot\phi\rangle \leqslant |\langle c, \dot\phi\rangle| \leqslant \|c\|.1/d.$$

But here $\|c\|$ can be arbitrarily close to d, while $\langle c, \dot\phi\rangle$ is constant on $C_G(f)$. It follows that

$$1 = \operatorname{Re}\langle c, \dot\phi\rangle = |\langle c, \dot\phi\rangle|,$$

hence even $\langle c, \dot\phi\rangle = 1$ for $c \in C_G(f)$; moreover, there must actually be equality in (**). Thus Lemma 2 is proved.

6.3. By means of Lemma 2 (§ 6.2) we can now prove the result in 6.1. Put (with the notation used in 6.1)

$$d = \operatorname{dist}\{C_G(f), 0\}, \qquad d' = \operatorname{dist}\{f + J_G, 0\}.$$

We want to show $d = d'$.

Clearly $d \geqslant d'$ (cf. § 6.1, Remark). Thus we need only show $d \leqslant d'$ and here we may assume $d > 0$. But if $d > 0$, then by Lemma 2 there is an *invariant* $\dot\phi \in B'$ such that $\langle f, \dot\phi\rangle = 1$ and $\|\dot\phi\|_{B'} = 1/d$. Since $\dot\phi$ is invariant, it vanishes on J_G, whence $\langle f+v, \dot\phi\rangle = 1$ for all $v \in J_G$. But $\|f+v\|$, $v \in J_G$, can be arbitrarily close to d' by proper choice of v. Hence $d'.(1/d) \geqslant 1$ or $d \leqslant d'$. Thus $d = d'$ and 6.1 is proved.

Remark. We can also show that, conversely, the relation (1) implies the assertion of Lemma 2. Indeed, suppose $(A_y)_{y \in G}$ is *any* family of operators on a Banach space B (G need not be a group) and (1) holds. Then we have for fixed $f \in B$: if $\operatorname{dist}\{C_G(f), 0\} = \|Tf\|_{B/J_G} = d$, say, and if $d > 0$, then there is a continuous linear functional ϕ' on B/J_G such that $\langle Tf, \phi'\rangle = 1$ and the norm of ϕ' is $1/d$. Now ϕ' defines a continuous linear functional $\dot\phi$ on B if we put

$$\langle v, \dot\phi\rangle = \langle Tv, \phi'\rangle \qquad v \in B,$$

i.e. $\dot\phi = \phi' \circ T$, and the norm of $\dot\phi$ is the same as that of ϕ'. Moreover, $\dot\phi$ vanishes on J_G, thus $\dot\phi$ is invariant; also $\langle f, \dot\phi\rangle = 1$, $\|\dot\phi\|_{B'} = 1/d$.

6.4. Using (1) we can prove: *for locally compact groups the property* ($\mathscr{M}$) *implies the property* P_1.

Suppose that G is a l.c. group with the property ($\mathscr{M}$). Let $B = L^1(G)$ and let A_y, $y \in G$, be the usual operators (§ 2.4): then the conditions 6.1 (i)–(iv) are satisfied, hence (1) holds and yields, in the present case,

$$\inf\left\|\sum_n c_n A_{y_n} f\right\|_1 = \left|\int f(x)\,dx\right| \qquad f \in L^1(G), \tag{3}$$

the infimum being taken over all finite sums with $c_n > 0$, $\sum_n c_n = 1$, $y_n \in G$.

The subspace J_G in (1) consists here of *all* $f_0 \in L^1(G)$ such that $\int_G f_0(x)\,dx = 0$: this follows from 2.5 if we put there $H = G$. Thus (1) yields (3) (cf. also the footnote in § 2.3, p. 165).

But (3) implies that G has the property P_1 (§ 4.5).

6.5. Thus *for locally compact groups the properties P_1 and ($\mathscr{M}$) are equivalent*; this results by combining 6.4 with 5.4.

Glicksberg's theorem (§ 6.1) now appears as an abstract extension of Theorem 4.1 (the subgroup H in 4.1 taking the place of the group G in 6.1). Clearly there is also an analogue of 4.6 (i) in the abstract setting of 6.1.

As the equivalence of P_1 and ($\mathscr{M}$) shows, there is some connexion between Haar measure and invariant means even for non-compact l.c. groups.

References. For countable discrete groups the equivalence of P_1 and ($\mathscr{M}$) is contained in a result of Følner's [45, § 2]; this was pointed out by Day who also proved this equivalence (using a somewhat different terminology) for arbitrary discrete groups and semi-groups [29, pp. 524–5]. For l.c. groups it was proved in [116]; cf. also [78, Proposition 4.1] for a similar result (which actually amounts to the same by § 5.4 (1)†).

6.6. We add a remark concerning the right invariant convex hull of a function in $L^p(G)$, $1 < p < \infty$. Define the operators A_y, $y \in G$, in $L^p(G)$ by $A_y = R^{(p)}_{y^{-1}}$ (Chap. 3, § 5.5, Remark). Let H be a closed subgroup of G. Then for every $f \in L^p(G)$ we have‡

$$\inf\left\|\sum_n c_n A_{\xi_n} f\right\|_p = 0 \qquad p > 1, \tag{4}$$

if H is not compact, the infimum being taken over all finite sums such that $c_n > 0$, $\sum_n c_n = 1$, $\xi_n \in H$; but

$$\inf\left\|\sum_n c_n A_{\xi_n} f\right\|_p = \left\{\int_{G/H}\left|\int_H f(x\xi)\,d\xi\right|^p d\dot{x}\right\}^{1/p} \qquad p \geqslant 1, \tag{5}$$

† Another method of showing this was given in [102]; the reader should observe that some historical remarks made there are in error.

‡ Cf. [53, § 5]; the proof above avoids some questions of measure theory.

if H is compact (and its Haar measure is normalized so that $m_H(H) = 1$). We note that for compact H there is always an invariant measure $d\dot{x}$ on G/H (§ 1.4, Example), and that the right-hand side of (5) is defined also if $p > 1$.

Proof of (4). If the left-hand side of (4) were > 0, then by 6.2, Lemma 1, there would be a $g \in L^{p'}(G)$ $(1/p + 1/p' = 1)$ such that

$$(*) \qquad \operatorname{Re}\langle A_\xi f, g\rangle \geqslant 1 \quad \text{for all } \xi \in H.$$

Put $\psi(y) = \langle A_y f, g\rangle$, $y \in G$; then $\psi \in \mathscr{C}^0(G)$ (for f, $g \in \mathscr{K}(G)$ we have $\psi \in \mathscr{K}(G)$ and for $f \in L^p(G)$, $g \in L^{p'}(G)$ the assertion follows by approximation). Thus (*) is impossible for non-compact H; hence (4) holds.

Proof of (5). For compact H the relation (5) certainly holds with '$\geqslant$' instead of '$=$', as can readily be verified (cf. § 4.1, Remark; if $p > 1$, then Hölder's inequality is also applied, on H). On the other hand, let $g(x) = \int_H f(x\xi)\, d\xi$, $x \in G$; then g is in $L^p(G)$ and can be approximated in $L^p(G)$ by $\sum_n c_n A_{\xi_n} f$, with $c_n > 0$, $\sum_n c_n = 1$, $\xi_n \in H$ (Chap. 3, § 5.9, Example; cf. also Chap. 3, § 3.6 (ii)). Thus $\inf \left\| \sum_n c_n A_{\xi_n} f \right\|_p \leqslant \|g\|_p$ and (5) follows (note that the right-hand side of (5) is equal to $\|g\|_p$).

7. The structure of groups with the property P_1

7.1. Suppose a locally compact group has the property P_1 (cf. § 3); what does this imply about its structure? We can answer this question to the following extent. *Let G be a locally compact group, G_0 the connected component of the neutral element in G;*† *suppose that the quotient group G/G_0 is compact or abelian or, more generally, soluble. Then G has the property P_1 if and only if G_0/R_0 is compact, R_0 being the radical of G_0.*

The *radical* of a connected l.c. group is the largest connected soluble normal subgroup; it is necessarily closed (cf. [80, Theorem 15]; also [83, p. 54, Cor. 2]). For the structure of connected l.c. groups G_0 such that G_0/R_0 is compact, see [80, Theorem 19].

7.2. That the condition stated in Theorem 7.1 is sufficient follows from the result in 3.3. To prove the necessity, we observe that, if G has the property P_1, then so has G_0 (§ 3.8). The proof is thus reduced to connected groups G and proceeds in three stages:

(i) First we prove a lemma (§ 7.3) from which we can deduce (§ 7.4) that certain l.c. groups do not have the property P_1. Among these groups are, in particular, all connected, semi-simple, non-compact Lie groups with finite centre (§ 7.5).

† G_0 is a closed normal subgroup ([138, p. 13], [100, Theorem 1.25]).

(ii) Next we show (§ 7.6) that, if G is any connected Lie group with the property P_1, then G/R must be compact, R being the radical of G.

(iii) Lastly we extend (ii) to connected l.c. groups in general (§ 7.7).

7.3. Lemma. Let G be a l.c. group satisfying the following conditions.†

(*a*) G contains a compact subgroup K and a closed subgroup H such that $K \cap H = \{e\}$ and $G = KH$.

(*b*) In the (unique) decomposition $x = kh$ ($k \in K$, $h \in H$) for each $x \in G$, the elements k and h are continuous functions of x.

(*c*) The group G is unimodular, but the subgroup H is not.

Then there is a constant $c_0 > 0$ and an element $h_0 \in G$ with the following property: for all functions $f \in \mathscr{K}(G)$ such that $\int\limits_G f(x)\,dx = 1$ and‡ $f(kx) = f(x)$ for all $k \in K$ and all $x \in G$, the inequality

$$\|L_{h_0}f - f\|_1 \geqslant c_0$$

holds.

Proof. The Haar measures on G, K, H are related by

$$\int\limits_G f(x)\,dx = \int\limits_K \left\{ \int\limits_H f(kh)\Delta_H(h^{-1})\,dh \right\} dk,$$

where dh is a left Haar measure (cf. § 1.5 (11)). Choose any $h_0 \in H$ such that $\Delta_H(h_0) \neq 1$. Given $k \in K$, define $h_1 \in H$ by the relation $k_1 h_1 = h_0^{-1}k$ ($k_1 \in K$): then $h_1 = h_1(k)$ is a (single-valued and) continuous function of $k \in K$. Consider any $f \in \mathscr{K}(G)$ such that $f(k_1 x) = f(x)$ for all $k_1 \in K$, $x \in G$. Then

$$\|L_{h_0}f - f\|_1 \geqslant \int\limits_K \left| \int\limits_H \{f(h_0^{-1}kh) - f(kh)\}\Delta_H(h^{-1})\,dh \right| dk$$

and the right-hand side is equal to

$$\int\limits_K |\Delta_H(h_1(k)) - 1| . \left| \int\limits_H f(h)\Delta_H(h^{-1})\,dh \right| dk.$$

Here $|\Delta_H(h_1(k)) - 1|$ is a positive continuous function of $k \in K$ which for $k = e$ has the value $|\Delta_H(h_0^{-1}) - 1| > 0$; moreover, $\int\limits_H f(h)\Delta_H(h^{-1})\,dh = \int\limits_G f(x)\,dx$ (we may assume $\int\limits_K dk = 1$). Thus the lemma holds with h_0 as above and with

$$c_0 = \int\limits_K |\Delta_H(h_1(k)) - 1|\,dk > 0.$$

7.4. From Lemma 7.3 we can deduce that, *if a l.c. group G satisfies the conditions* 7.3 (a, b, c), *then G does not have the property P_1.*

It will be enough to show: if G is a l.c. group with the property P_1 and K a compact subgroup of G, then for every compact set $C \subset G$ and

† Cf. 1.5 for an example.

‡ Clearly there are such functions in $\mathscr{K}(G)$, since K is compact.

every $\epsilon > 0$ there is a function $s \in \mathscr{K}(G)$ such that $s \geqslant 0$, $\int\limits_G s(x)\,dx = 1$ and $\|L_y s - s\|_1 < \epsilon$ for all $y \in C$, and with the property that $s(kx) = s(x)$ for all $k \in K$ and all $x \in G$. Combining this with 7.3, we shall obtain the result above.

Put $C' = (CK) \cup K$. There is an $s' \in \mathscr{K}(G)$ such that $s' \geqslant 0$, $\int s' = 1$ and $\|L_{y'}s' - s'\|_1 < \frac{1}{2}\epsilon$ for all $y' \in C'$. Then for $k \in K$, $y \in C$ we have

$$\int |s'(ky^{-1}x) - s'(x)|\,dx < \tfrac{1}{2}\epsilon, \qquad \int |s'(kx) - s'(x)|\,dx < \tfrac{1}{2}\epsilon$$

and thus
$$\int |s'(ky^{-1}x) - s'(kx)|\,dx < \epsilon \qquad k \in K,\ y \in C.$$

Hence, putting
$$s(x) = \int\limits_K s'(kx)\,dk \qquad x \in G$$

$\big($where $\int\limits_K dk = 1\big)$, we obtain a function s satisfying all conditions above.

Remark. We can also make another application of 7.3 as follows. *Let G be any l.c. group satisfying the conditions* 7.3 (a, b, c). *Then there are functions $f_0 \in L^1(G)$ such that*

(1) $$\int f_0(x)\,dx = 0, \quad \textit{but} \quad \inf\Big\|\sum_n c_n A_{y_n} f_0\Big\|_1 > 0,$$

the infimum being taken over all finite sums with c_n complex, $\sum\limits_n c_n = 1$, $y_n \in G$. *In particular, the relation* 4.4 (5)—which, as we know, is equivalent to P_1 (§§ 4.4 and 4.5)—*does not hold for G*. By using Lemma 7.3 we can give a simple proof of these assertions.

Let $G = KH$, as in 7.3, and take any $g \in \mathscr{K}(G)$ such that $\int\limits_G g(x)\,dx = 1$ and $g(kx) = g(x)$ for all $k \in K$, $x \in G$. Put $f_0 = L_{h_0}g - g$, with $h_0 \in H$ as in 7.3. We can write

$$\sum_n c_n A_{y_n} f_0 = L_{h_0}f - f, \quad \text{where } f = \sum_n c_n A_{y_n} g.$$

Here f has the properties required in Lemma 7.3 $\big($note that $\int f = \sum\limits_n c_n . \int g = 1\big)$. Hence f_0 satisfies (1).

7.5. Every connected, semi-simple, non-compact Lie group with finite centre has the properties 7 (a, b, c): this is the content of the *Iwasawa decomposition* of such a group.† Thus 7.4 shows: *a connected, semi-simple Lie group with finite centre does not have the property P_1 unless it is compact.*

Example 1. The groups $SL(n, \mathbf{R})$, $n \geqslant 2$, do not have the property P_1 (cf. § 1.5, Example). It follows (§ 3.8) that the same is true for $GL(n, \mathbf{R})$ and for $SL(n, \mathbf{C})$, $GL(n, \mathbf{C})$.

Example‡ 2. Let $\mathbf{Q}_p$ be the field of p-adic numbers; let $GL(n, \mathbf{Q}_p)$ and

† Cf. [80, Lemma 3.12], [59, Chap. VI, § 5, and Chap. X, Proposition 1.11].

‡ For this example, and the construction below, the author is indebted to T. A. Springer.

$SL(n, \mathbf{Q}_p)$, $n \geqslant 2$, be the multiplicative groups of $n \times n$ matrices with entries from $\mathbf{Q}_p$ and determinant $\neq 0$ and $= 1$ resp. These are l.c. groups in the topology induced by that of $\mathbf{Q}_p^{n^2}$; they are totally disconnected, but not discrete. The groups $GL(n, \mathbf{Q}_p)$, $SL(n, \mathbf{Q}_p)$, $n \geqslant 2$, do not have the property P_1. This is the p-adic analogue of Example 1.

Let first $n = 2$: put $G = GL(2, \mathbf{Q}_p)$. Choose any $\omega \in \mathbf{Q}_p$ that is not a square (e.g. $\omega = p$); put

$$G_1 = \left\{\begin{bmatrix} a & b \\ b\omega & a \end{bmatrix} \middle| a, b \in \mathbf{Q}_p, \text{ not both } 0\right\}, \qquad H = \left\{\begin{bmatrix} u & v \\ 0 & 1 \end{bmatrix} \middle| u, v \in \mathbf{Q}_p, u \neq 0\right\}.$$

G_1 and H are closed subgroups of G and $G_1 \cap H$ contains only the unit matrix. It is readily seen that every $x \in G$ can be represented in the form $x = gh$ with $g \in G_1$, $h \in H$ and that g and h are continuous functions of x.

G_1 contains the closed subgroup $Z = \left\{\begin{bmatrix} \lambda & 0 \\ 0 & \lambda \end{bmatrix} \middle| \lambda \in \mathbf{Q}_p, \lambda \neq 0\right\}$ (which is the centre of G) and G_1/Z is clearly compact. We note that G_1 is isomorphic to the multiplicative group of the extension field $\mathbf{Q}_p(\sqrt{\omega})$.

Now consider $G' = G/Z$: we have $G' = KH$ with $K = G_1/Z$ and H as above, and it is readily seen that the conditions 7.3 (a, b, c) are fulfilled (for (c) we observe that G is unimodular and Z the centre). Thus G' does not have the property P_1. By 3.8 it follows that $G = GL(2, \mathbf{Q}_p)$ does not have the property P_1, and since $G \subset GL(n, \mathbf{Q}_p)$ for $n > 2$, we obtain the same result for all $GL(n, \mathbf{Q}_p)$. Now $SL(n, \mathbf{Q}_p)$ is a normal subgroup of $GL(n, \mathbf{Q}_p)$ with abelian quotient group; hence (§ 3.3) $SL(n, \mathbf{Q}_p)$ does not have the property P_1 either.

The construction above also works for $GL(2, \mathbf{R})$, say with $\omega = -1$, and then yields Example 1.

7.6. We now show first: if G is any connected *semi-simple* Lie group with the property P_1, then G is compact. Let Z be the centre of G. The quotient group G/Z also has the property P_1 (§ 3.8) and is a connected, semi-simple Lie group with centre reduced to $\{e\}$ (it is the adjoint group of a real, semi-simple Lie algebra); hence (§ 7.5) G/Z must be compact. By a theorem of H. Weyl ([59, Chap. II, Theorem 6.9] or [25, Cor. 1 of Theorem 2]) G itself is also compact.

Now let G and R be as stated in 7.2 (ii): then G/R is a connected, semi-simple Lie group and also has the property P_1, hence by the above G/R is compact. Thus 7.2 (ii) is proved.

REMARK.† Using Example 2 in 7.5 and applying results from the theory of algebraic groups, one can establish a proposition entirely analogous to 7.2 (ii) when G is the group of rational points of an algebraic group A defined over $\mathbf{Q}_p$ (thus G is a l.c. group and is totally disconnected, but not discrete). For example, if A is a linear algebraic group and is semi-simple, then G does not have the property P_1 unless G is compact; this covers the case of all so-called classical groups over $\mathbf{Q}_p$.

† This remark is due to T. A. Springer.

7.7. We can now prove 7.2 (iii) by using the structure theory of connected l.c. groups: a group G of this type contains a family of (compact) normal subgroups H_α such that $\bigcap_\alpha H_\alpha = \{e\}$ and each G/H_α is a connected Lie group [100, Theorem 4.6]. If G has the property P_1, then so has each G/H_α (§ 3.8). It follows from 7.2 (ii) and a theorem of Iwasawa [80, Theorem 17] that G/R is compact.

Thus Theorem 7.1 is proved.

The proof above is a modification of that of Takenouchi [132, § 3] who proved 7.1 for G such that G/G_0 is compact, using the equivalent property P' (§ 3.5).

Theorem 7.1 reduces the question of the structure of l.c. groups with the property P_1 to the case of totally disconnected groups. For such groups we have indicated some results in 7.5, Example 2, and 7.6, Remark. For results concerning discrete groups see [36, Theorem 4] and [29, § 4], where the equivalent property ($\mathscr{M}$) is considered, and also [34] and [30]. See also a recent paper by N. W. Rickert in *Trans. Amer. Math. Soc.* **127**, 221–32 (1967), especially §§ 4 and 5, where another property equivalent to P_1 is disussed. Further results and references will be found in [88].

BIBLIOGRAPHY

Abbreviations of names of Journals are as in *Mathematical Reviews*

1. ACHIESER, N. I. *Lectures on approximation theory*. In Russian, Moscow (1947);† German translation, Akademie-Verlag, Berlin (1953); English translation, Ungar, New York (1956).
2. ALEXANDROFF, P. and HOPF, H. *Topologie*, pp. 609–14. Springer, Berlin (1935).
3. ARTIN, E. *Collected papers*, p. 523. Addison–Wesley, Reading, Mass., U.S.A. (1965).
4. BARGMANN, V. *Ann. of Math.* **48,** 568–640 (1947).
5. BAUER, H. *Arch. Math.* **9,** 389–93 (1958).
6. —— *Rev. Roumaine Math. Pures Appl.* **7,** 747–52 (1966).
7. BEURLING, A. Sur les intégrales de Fourier absolument convergentes. *IXe Congrès Math. Scand.*, Helsinki, pp. 345–66 (1938).
8. —— Un théorème sur les fonctions bornées et uniformément continues sur l'axe réel. *Acta Math.* **77,** 127–36 (1945).
9. —— Sur les spectres des fonctions. *Colloque sur l'analyse harmonique*, Nancy, pp. 9–29 (1947).
10. —— On the spectral synthesis of bounded functions. *Acta Math.* **81,** 225–38 (1949).
11. BOCHNER, S. *Vorlesungen über Fouriersche Integrale*. Akad. Verlagsges., Leipzig (1932); English translation, Princeton University Press (1959).
12. —— and CHANDRASEKHARAN, K. *Fourier transforms*. Princeton University Press (1949).
13. BOURBAKI, N. *Éléments de mathématique*. Hermann, Paris. (*a*) *Algèbre*, Chap. III (1948). (*b*) *Espaces vectoriels topologiques*, Chaps. I and II, 2nd edn (1966). (*c*) *Intégration*, Chaps. I–IV, 2nd edn (1965); Chap. V, 2nd edn (1967); Chaps. VII and VIII (1963). (*d*) *Théories spectrales*, Chaps. I and II (1967). (*e*) *Topologie générale*, Chaps. III and IV, 3rd edn (1960); Chaps. V–VIII, 3rd edn (1963).
14. BRACONNIER, J. Sur les groupes topologiques localement compacts. *J. Math. Pures Appl.* **27,** 1–85 (1948).
15. —— L'analyse harmonique dans les groupes abéliens, I, II. *Enseignement Math.* **2,** 12–41, 257–73 (1956).
16. BREDON, G. E. *Michigan Math. J.* **10,** 365–73 (1963).
17. BRUHAT, F. Sur les représentations induites des groupes de Lie. *Bull. Soc. Math. France* **84,** 97–205 (1956).
18. BUCY, R. S. and MALTESE, G. M. *J. Math. Anal. Appl.* **12,** 371–7 (1966).
19. CALDERÓN, A. P. Ideals in abelian group algebras. *Symposium on harmonic analysis*, Cornell University (1956) [mimeographed lecture notes].
20. CARLEMAN, T. *L'intégrale de Fourier et questions qui s'y rattachent*. Inst. Mittag–Leffler Publ. Scient., Uppsala (1944).‡

† A second Russian edition has now appeared: cf. Review No. 6108, *Math. Rev.* **32,** 1039 (1966).

‡ Lectures given at the Mittag–Leffler Institute in 1935.

21. CARTAN, H. *C. R. Acad. Sci. Paris* **211,** 759–62 (1940).
22. —— *Théorie élémentaire des fonctions.* Hermann, Paris (1961).
23. —— and GODEMENT, R. Théorie de la dualité et analyse harmonique dans les groupes abéliens localement compacts. *Ann. Sci. École. Norm. Sup.* **64,** 79–99 (1947).
24. CARTIER, P. Über einige Integralformeln in der Theorie der quadratischen Formen. *Math. Z.* **84,** 93–100 (1964).
25. —— Structure topologique des groupes de Lie généraux. *Sém. Sophus Lie,* Exposé No. 22. Secrétariat Math., Paris (1965).
26. CHEVALLEY, C. *Theory of Lie groups,* vol. i. Princeton University Press (1946).
27. VAN DANTZIG, D. *Fund. Math.* **15,** 102–25 (1930).
28. DAVIS, H. F. *Proc. Amer. Math. Soc.* **6,** 318–21 (1955).
29. DAY, M. M. Amenable semigroups. *Illinois J. Math.* **1,** 509–44 (1957).
30. —— Convolutions, means and spectra. *Illinois J. Math.* **8,** 100–11 (1964).
31. DEDEKIND, R. *Math. Werke,* Vol. II, pp. 420–1. Vieweg, Braunschweig (1931).
32. DIEUDONNÉ, J. Sur les espaces de Köthe. *J. Analyse Math.* **1,** 81–115 (1951).
33. —— Sur le produit de composition, II. *J. Math. Pures Appl.* **39,** 275–92 (1960).
34. —— Sur une propriété des groupes libres. *J. Reine Angew. Math.* **204,** 30–4 (1960).
35. DITKIN, V. A. On the structure of ideals in certain normed rings (in Russian, English summary). *Uchen. Zap. Mosk. Gos. Univ. Matem.* **30,** 83–130 (1939).
36. DIXMIER, J. Les moyennes invariantes dans les semi-groupes et leurs applications. *Acta Sci. Math. Szeged* **12**A, 213–27 (1950).
37. —— Quelques exemples concernant la synthèse spectrale. *C.R. Acad. Sci. Paris* **247,** 24–6 (1958).
38. —— *Les C*-algèbres et leurs représentations.* Gauthier-Villars, Paris (1964).
39. DOMAR, Y. Harmonic analysis based on certain commutative Banach algebras. *Acta Math.* **96,** 1–66 (1956).
40. —— On spectral analysis in the narrow topology. *Math. Scand.* **4,** 328–32 (1956).
41. DUNFORD, N. and SCHWARTZ, J. T. *Linear operators,* Pt. 1, § VI.10. Interscience, New York (1958).
42. EDWARDS, D. A. *J. London Math. Soc.* **36,** 461–2 (1961).
43. EDWARDS, R. E. *Proc. Amer. Math. Soc.* **5,** 71–8 (1954).
44. EYMARD, P. L'algèbre de Fourier d'un groupe localement compact. *Bull. Soc. Math. France* **92,** 181–236 (1964).
45. FØLNER, E. On groups with full Banach mean value. *Math. Scand.* **3,** 243–54 (1955).
46. FREUDENTHAL, H. Review of 1st edn of Pontryagin's *Topological groups. Nieuw Arch. Wisk.* **20,** 311–16 (1940).
47. GATESOUPE, M. *Ann. Inst. Fourier Univ. Grenoble* **17,** 93–107 (1967).
48. GELFAND, I. M. *Mat. Sbornik* **9,** 3–24 (1941).
49. —— and NAIMARK, M. A. *Unitary representations of the classical groups.* In Russian, Steklov Mathematical Institute, Moscow (1950); German translation, Akademie-Verlag, Berlin (1957).

50. Gelfand, I. M. and Raikov, D. *Mat. Sbornik* **13**, 301–16 (1943). *Amer. Math. Soc. Translations*, Series 2, **36**, 1–15 (1964).
51. ——, —— and Shilov, G. *Commutative normed rings*. In Russian, Moscow (1960); English translation, Chelsea, New York (1964).†
52. —— and Shilov, G. E. *Generalized functions*, vol. ii, Chap. I, § 3. In Russian, Moscow (1958); German translation, Akademie-Verlag, Berlin (1962).
53. Glicksberg, I. On convex hulls of translates. *Pacific J. Math.* **13**, 97–113 (1963).
54. Godement, R. Théorèmes tauberiens et théorie spectrale. *Ann. Sci. École Norm. Sup.* **64**, 119–38 (1947).
55. —— Les fonctions de type positif et la théorie des groupes. *Trans. Amer. Math. Soc.* **63**, 1–84 (1948).
56. Greenleaf, F. *Pacific J. Math.* **15**, 1187–1219 (1965).
57. Grosser, S. and Moskowitz, M. *Trans. Amer. Math. Soc.* **127**, 317–40 (1967).
58. Grothendieck, A. *Espaces vectoriels topologiques*, 3rd edn. University of São Paulo (1964).
59. Helgason, S. *Differential geometry and symmetric spaces*. Academic Press, New York (1962).
60. Helson, H. Spectral synthesis of bounded functions. *Ark. Mat.* **1**, 497–502 (1952).
61. —— On the ideal structure of group algebras. *Ark. Mat.* **2**, 83–6 (1952).
62. —— and Kahane, J.-P. *C.R. Acad. Sci. Paris* **247**, 626–8 (1958).
63. Herz, C. S. Spectral synthesis for the Cantor set. *Proc. Nat. Acad. Sci. U.S.A.* **42**, 42–3 (1956).
64. —— Spectral synthesis for the circle. *Ann. of Math.* **68**, 709–12 (1958).
65. —— The spectral theory of bounded functions. *Trans. Amer. Math. Soc.* **94**, 181–232 (1960).
66. —— Review No. 5854. *Math. Rev.* **21**, 1089–91 (1960).
67. —— Fourier transforms related to convex sets. *Ann. of Math.* **75**, 81–92 (1962).
68. —— On the number of lattice points in a convex set. *Amer. J. Math.* **84**, 126–33 (1962).
69. —— *Two problems in the spectral synthesis of unbounded functions*. Mimeographed notes (1964).
70. —— Synthèse harmonique de distribution dans le plan. *C.R. Acad. Sci. Paris* **260**, 4887–90 (1965).
71. —— Remarque sur la note précédente de M. Varopoulos. *C.R. Acad. Sci. Paris* **260**, 6001–4 (1965).
72. —— Review No. 2567. *Math. Rev.* **31**, 462–3 (1966).
73. —— and Leeuw, K. de. *Illinois J. Math.* **9**, 220–9 (1965).
74. Hewitt, E. and Ross, K. A. *Abstract harmonic analysis*, vol. i. Springer, Berlin (1963).
75. Hille, E. and Phillips, R. S. *Functional analysis and semi-groups*, revised edn. American Mathematical Society, Providence (1957).
76. Hobson, E. W. *Theory of functions of a real variable*, vol. i, 3rd edn. Cambridge University Press (1927).

† Cf. also the review in *J. London Math. Soc.* **40**, 573–4 (1965).

77. HOCHSCHILD, G. *The structure of Lie groups.* Holden-Day, San Francisco (1965).

78. HULANICKI, A. Means and Følner condition on locally compact groups. *Studia Math.* **27,** 87–104 (1966).

79. HURWITZ, A. *Math. Werke,* vol. ii, pp. 560–2. Birkhäuser, Basle (1933).

80. IWASAWA, K. On some types of topological groups. *Ann. of Math.* **50,** 507–58 (1949).

81. VAN KAMPEN, E. R. Locally bicompact abelian groups and their character groups. *Ann. of Math.* **36,** 448–63 (1935).

82. KAPLANSKY, I. Primary ideals in group algebras. *Proc. Nat. Acad. Sci. U.S.A.* **35,** 133–6 (1949).

83. —— *An introduction to differential algebra.* Hermann, Paris (1957).

84. KATZNELSON, Y. *C.R. Acad. Sci. Paris* **247,** 404–6 (1958); *Ann. Sci. École Norm. Sup.* **76,** 83–124 (1959).

85. KOOSIS, P. *Pacific J. Math.* **16,** 121–8 (1966).

86. LARSEN, R., LIU, T., and WANG, J. *Michigan Math. J.* **11,** 369–78 (1964).

87. LEPTIN, H. *Math. Ann.* **163,** 111–17 (1966).

88. —— On locally compact groups with invariant means. *Proc. Amer. Math. Soc.* **19,** 489–94 (1968).

89. LÉVY, P. Sur la convergence absolue des séries de Fourier. *Compositio Math.* **1,** 1–14 (1935).

90. LITTMAN, W. *Bull. Amer. Math. Soc.* **69,** 766–70 (1963).

91. LOOMIS, L. H. *An introduction to abstract harmonic analysis.* Van Nostrand, New York (1953).†

92. MACBEATH, A. M. and ŚWIERCZKOWSKI, S. Measures in homogeneous spaces. *Fund. Math.* **49,** 15–24 (1960).

93. MACKEY, G. W. Functions on locally compact groups. *Bull. Amer. Math. Soc.* **56,** 385–412 (1950).

94. —— Induced representations of locally compact groups, I. *Ann. of Math.* **55,** 101–39 (1952).

95. —— *Commutative Banach algebras.* Instituto de Matemática Pura e Applicada, Rio de Janeiro (1959).‡

96. MAHLER, K. *Lectures on diophantine approximations,* Pt. 1. University of Notre Dame (1961).

97. MALLIAVIN, P. Impossibilité de la synthèse spectrale sur les groupes abéliens non compacts. *Inst. Hautes Études Sci. Publ. Math.* No. 2, pp. 61–8 (1959).

98. MANDELBROJT, S. and AGMON, S. Une généralisation du théorème tauberien de Wiener. *Acta Sci. Math. Szeged* **12**B, 167–76 (1950).

99. MIRKIL, H. *Compositio Math.* **14,** 269–73 (1960).

100. MONTGOMERY, D. and ZIPPIN, L. *Topological transformation groups.* Interscience, New York (1955).

101. NAIMARK, M. A. *Normed rings.* In Russian, Moscow, 1956; English translation, 2nd revised edn, Noordhoff, Groningen (1964).

† Corrections and additions are given by D. A. Raikov in the Russian translation, Moscow (1956).

‡ Notes of lectures given at Harvard University in 1951. Obtainable from Livraria Castelo, Ave. Erasmo Braga 227, 2º andar, Rio de Janeiro, Brazil.

102. Namioka, I. *Proc. Amer. Math. Soc.* **17,** 1101–2 (1966).

103. von Neumann, J. *Fund. Math.* **13,** 73–116 (1929).

104. Pitt, H. R. *Tauberian theorems.* Oxford University Press (1958).

105. Pontrjagin, L. The theory of topological commutative groups. *Ann. of Math.* **35,** 361–88 (1934).

106. Pontryagin, L. S. *Topological groups,* 2nd edn. In Russian, Moscow (1954); English translation, Gordon and Breach, New York (1966).

107. Raikov, D. A. *Harmonic analysis on commutative groups with Haar measure and the theory of characters.* In Russian, Steklov Mathematical Institute, Moscow (1945); German translation in *Sowjetische Arbeiten zur Funktionalanalysis,* Berlin (1954).

108. Rajagopalan, M. *Acta Sci. Math. Szeged* **25,** 86–9 (1964).

109. Reiter, H. Investigations in harmonic analysis. *Trans. Amer. Math. Soc.* **73,** 401–27 (1952).

110. —— Über L^1-Räume auf Gruppen, I, II. *Monatsh. Math.* **58,** 73–6, 172–80 (1954).

111. —— Contributions to harmonic analysis, I–VI. *Acta Math.* **96,** 253–63 (1956); *Math. Ann.* **133,** 298–302 (1957); *J. London Math. Soc.* **32,** 477–83 (1957); *Math. Ann.* **135,** 467–76 (1958); *Math. Ann.* **140,** 422–41 (1960); *Ann. of Math.* **77,** 552–62 (1963).

112. —— The convex hull of translates of a function in L^1. *J. London Math. Soc.* **35,** 5–16 (1960).

113. —— Une propriété analytique d'une certaine classe de groupes localement compacts. *C.R. Acad. Sci. Paris* **254,** 3627–9 (1962).

114. —— Sur les groupes de Lie semi-simples connexes. *C.R. Acad. Sci. Paris* **255,** 2883–4 (1962).

115. —— Sur la propriété (P_1) et les fonctions de type positif. *C.R. Acad. Sci. Paris* **258,** 5134–5 (1964).

116. —— On some properties of locally compact groups. *Nederl. Akad. Wetensch. Indag. Math.* **27,** 697–701 (1965).

117. —— Subalgebras of $L^1(G)$. *Nederl. Akad. Wetensch. Indag. Math.* **27,** 691–6 (1965).

118. —— Zwei Anwendungen der Bruhatschen Funktion. *Math. Ann.* **163,** 118–21 (1966).

119. Richards, I. *Bull. Amer. Math. Soc.* **72,** 698–700 (1966); *J. Combinatorial Theory* **2,** 61–70 (1967).

120. Rudin, W. *Fourier analysis on groups.* Interscience, New York (1962).

121. Schwartz, L. Sur une propriété de synthèse spectrale dans les groupes non compacts. *C.R. Acad. Sci. Paris* **227,** 424–6 (1948).

122. Segal, I. E. The span of the translations of a function in a Lebesgue space. *Proc. Nat. Acad. Sci. U.S.A.* **30,** 165–9 (1944).

123. —— The group algebra of a locally compact group. *Trans. Amer. Math. Soc.* **61,** 69–105 (1947).

124. —— The class of functions which are absolutely convergent Fourier transforms. *Acta Sci. Math. Szeged* **12**B, 157–61 (1950).

125. Shilov, G. E. *On regular normed rings* (in Russian). Steklov Mathematical Institute, Moscow (1947).

126. Siegel, C. L. *Ann. of Math.* **46,** 340–7 (especially 340–1) (1945).

127. Stegeman, J. D. On a property concerning locally compact groups. *Nederl. Akad. Wetensch. Indag. Math.* **27**, 702–3 (1965).

128. —— Extension of a theorem of H. Helson. *Int. Congress Math.*, Moscow, Abstracts, Section 5, p. 28 (1966).

129. Stone, M. H. The generalized Weierstrass approximation theorem. *Math. Mag.* **21**, 167–84, 237–54 (1948). Reprinted in *Studies in Mathematics*, vol. i; Prentice-Hall, Englewood Cliffs, N.J., U.S.A. (1962).

130. —— Notes on integration, I–IV. *Proc. Nat. Acad. Sci. U.S.A.* **34**, 336–42, 447–55, 483–90 (1948); **35**, 50–8 (1949).

131. Świerczkowski, S. *Colloq. Math.* **8**, 107–14 (1961).

132. Takenouchi, O. Sur une classe de fonctions continues de type positif sur un groupe localement compact. *Math. J. Okayama Univ.* **4**, 143–73 (1955).

133. Titchmarsh, E. C. *Theory of functions*, 2nd edn. Oxford University Press, (1939).

134. Varopoulos, N. Th. (*a*) *C.R. Acad. Sci. Paris* **260**, 5165–8, 5997–6000 (1965). (*b*) *Proc. Cambridge Phil. Soc.* **62**, 379–87 (1966). (*c*) *Acta Math.* **119**, 51–112 (1967).

135. Warner, C. R. *Trans. Amer. Math. Soc.* **121**, 408–23 (1966).

136. Weil, A. Sur quelques résultats de Siegel. *Summa Brasil. Math.* **1**, 21–39 (1946).

137. —— Sur la théorie du corps de classes. *J. Math. Soc. Japan*, **3**, 1–35 (1951).

138. —— *L'intégration dans les groupes topologiques et ses applications*, 2nd edn, Hermann, Paris (1953).

139. —— Sur certains groupes d'opérateurs unitaires, *Acta Math.* **111**, 143–211 (1964).

140. Weyl, H. Elementary theory of convex polyhedra. *Ann. of Math. Studies* **24**, 3–25 (1950).

141. Whitney, H. *Amer. J. Math.* **70**, 635–58 (1948).†

142. Wiener, N. Tauberian theorems. *Ann. of Math.* **33**, 1–100 (1932).

143. —— *The Fourier integral and certain of its applications*. Cambridge University Press (1933).

144. Zygmund, A. *Trigonometrical series*. Warsaw (1935); 2nd edn, Cambridge University Press (1959).

† For a recent exposition see B. Malgrange, *Ideals of differentiable functions*, Oxford University Press (1966).

SUMMARY OF NOTATIONS

Chapters are indicated by Roman numerals

$|\cdot|$ I, 6.1, VI, 3.4 $|\cdot|_l$ IV, 3.2, Remark 1
$\|\cdot\|_1$ I, 1.1, III, 2.3 (iii) $\|\cdot\|_{1,\alpha}$ I, 6.1, VI, 3.4 $\|\cdot\|_{1,w}$ I, 6.1, III, 7.1
$\|\cdot\|_p$ III, 2.3 (iii) $\|\cdot\|_\infty$ II, 3.1, III, 2.1 (i), III, 2.8 (ii), III, 6.1 $\|\cdot\|_{\infty,w}$ III, 7.3
$\|\cdot\|_S$ VI, 2.1
$\langle\cdot,\cdot\rangle$ I, 1.2, III, 2.8 (iii), III, 6.1, IV, 2.5
$\hat{}$ I, 1.2, IV, 4.1, IV, 4.4 $\check{}$ III, 3.1 (iii) $\tilde{}$ III, 6.1 $\perp$ III, 6.1
$*$ I, 1.2, III, 5.2, III, 5.4 $\star$ I, 1.1, III, 5.1 $\star^G$, $\star^H$ III, 5.1 $\circ$ III, 1.8 (vii)
Ø III, 1.2 **C** III, 1.9

$\int f\,d\mu$ III, 2.1 (ii), III, 2.3 (iv) $\int \mathbf{f}\,d\mu$ III, 2.3 (viii)
$\int^\times f\,d\mu$ III, 2.2 (iii) $\int_A f\,d\mu$ III, 2.3 (vi)
$\int f(x)\,dx$ I, 1.1, III, 3.1 (i) $\int f$ III, 3.1 (i)

β VIII, 1.9
δ_a III, 5.5 Δ III, 3.5 Δ_G, Δ_H III, 3.6
ϑ III, 3.7
λ III, 3.7 λ_y, Λ_y VIII, 1.3
μ III, 2.1 (ii), III, 2.5 $\mu^\times$ III, 2.2 (ii), (iii), (v) μ^* III, 5.4
$\|\mu\|$ III, 2.1 (vi) $|\mu|$ III, 2.1 (viii) μ_h III, 5.4
$\mu\star f$, $\mu_1\star\mu_2$ III, 5.4
π_H III, 1.6
σ V, 1.2 $\sigma(L^1, L^\infty)$, $\sigma(L^\infty, L^1)$ III, 2.8 (iv)
τ V, 1.3
ϕ_A III, 1.9 $\tilde{\phi}$ III, 6.1 $\phi_{x,H}$ VII, 2.1 (i)
χ IV, 2.1

$\mathscr{A}(X)$ II, 1.1 A_y VIII, 2.4, VIII, 6.1 $a\mathbf{Z}$ IV, 1.1

Bdr E II, 4.4

C I, 1.1 $\mathscr{C}(X)$ II, 3.2 $\mathscr{C}^0(X)$ II, 3.1, III, 2.1 (iii) $\mathscr{C}^b(X)$ VIII, 5.1
$C_G(f)$ VIII, 6.1 $C_H(f)$ VIII, 4.1
cosp f II, 1.4, VII, 1.1 cosp I II, 1.4, VII, 1.1

dx I, 1.1, III, 3.1 (i) $d\dot{x}$ III, 3.3 (i) $d\hat{x}$ IV, 4.1 $d\xi^\perp$ V, 5.1
$d_q\dot{x}$ VIII, 1.2 $d_r\dot{x}$ VIII, 1.4 $d_R x$ III, 3.1 (iii) $d_V x$ III, 3.1 (v)

a.e., almost everywhere
F.t., Fourier transform(s)
l.a.e., locally almost everywhere
l.c., locally compact
l.c.a., locally compact abelian
loc. negl., locally negligible
nd., neighbourhood
nds., neighbourhoods
resp., respectively
W, Wiener
WD, Wiener–Ditkin

INDEX

Chapters are indicated by Roman numerals